Atlas and Manual of Plant Pathology

Atlas and Manual of Plant Pathology

Ervin H. Barnes
late of Michigan State University

line drawings by the author

photographs by

Philip G. Coleman

with a new introduction by

Paul G. Pilley
Sir Sandford Fleming College

Plenum Press · New York and London

Library of Congress Cataloging in Publication Data

Barnes, Ervin H

Atlas and manual of plant pathology.

"Except for the correction of a few minor misprints, a substantially unchanged republication of the edition published by Appleton-Century-Crofts in 1968."

I. Includes index.

1. Plant diseases – Atlases. 2. Plant diseases – Laboratory manuals. I. Title.

SB731.B35 632 79-10575

ISBN 0-306-40168-1

The Plenum Press edition of *Atlas and Manual of Plant Pathology* is, except for the correction of a few minor misprints, a substantially unchanged republication of the edition published by Appleton-Century-Crofts in 1968.

A Division of Plenum Publishing Corporation
227 West 17th Street, New York, N.Y. 10011

Printed in the United States of America

Dedicated to my father

WILBUR LOREE BARNES

for he taught me about

"such stuff as dreams are made on"

Introduction

Ideally a textbook should integrate with the lectures and labs in a science course. Selecting such a book can be an onerous (and sometimes impossible) task for the teacher. Students are wary of getting stuck with a "useless" book, i.e., one to which the instructor never refers. The reader probably has some practical appreciation of their concern. I remember an instructor who not only denounced the very text he had chosen, but also informed the class that he wouldn't be using it. This was after I had already purchased a copy!

Being mindful of the foregoing, I decided to try Barnes' *Atlas and Manual of Plant Pathology* in 1973. Six years and 800 students later I have no regrets about my choice. As far as I am concerned it is still the finest book of its kind on this continent.

Barnes' *Atlas* contains an excellent blend of the diagnostic and experimental aspects of plant pathology. His treatment of each disease on an individual basis allows the instructor to omit some pathogens without disturbing the book's continuity. My one-semester course in Forest Pathology is largely descriptive. Strong emphasis is placed on field recognition of symptoms and signs. This is facilitated by Barnes' technique. In a sequence of photographs, the diseased plant or part is first viewed as a whole to show the general symptoms. This is usually followed by a close-up of the signs (i.e., fruiting bodies), which appear as they would under a hand lens or dissecting microscope. Ultimately there is a photomicrograph and diagrammatic representation of the fruiting bodies *in situ*. Labels and line drawings make it easier for the student to discern host tissues and fungus structures. A life cycle diagram concludes the discussion for fifteen of the pathogens. The profusely-illustrated chapter on histological interpretation clearly explains the "bits 'n pieces" and other artifacts that are evident in prepared thin sections, and which are often confusing to students.

I was dismayed to learn that Barnes' *Atlas* was out of print. Fortunately a book of this kind is not as quickly outdated as a general text, and thus reprinting can be justified. It should be noted, however, that the section on viruses includes diseases now attributed to mycoplasmas, e.g., aster yellows. Some of the methods of control have also changed, e.g., pressure injection of solubilized benomyl into roots and stems for control of Dutch elm disease. The absence of such advances does not diminish the pressing need or value of Barnes' work as an atlas and manual. My hope is that it will be around for some time.

PAUL G. PILLEY
Sir Sandford Fleming College

Lindsay, Ontario
May, 1978

Preface

This work is a result of the curiosity of my students to know more about the world in which they live. Their enthusiasm to touch and to see the world about them demands more than I can give in the classroom. Hopefully, this atlas and manual will help them in their search.

This manual is designed to provide experience in both passive observation of prepared materials and active participation in experiments with living plants and causal organisms. Photographs and line drawings are presented to guide the student in the observation of diseased specimens, infected tissues, and structures of causal organisms. Most of this can be accomplished at the student's own rate of speed, and at his own convenience during "free" or unsupervised laboratory time. This not only enables the student to see for himself, but frees fixed laboratory periods for active participation in manipulative experiments where supervision is important in the process of learning skills.

General introductory remarks about specific diseases are presented to provide a fabric into which the student can weave his observations, develop concepts, synthesize principles, and test hypotheses. He is encouraged to seek further details from general texts and references listed after the definitions and at the end of chapters. He is urged especially to browse through the *Phytopathological Classics* published in English by the American Phytopathological Society. They are landmarks in the literature which record the early history of man's understanding of plant diseases.

One way to study symptoms of diseases, abnormal internal structures, and characteristics of causal organisms is by observing prepared mounts of whole specimens and microscope slides of thin sections of diseased tissues. By following the photographs and line drawings which accompany descriptions of specific diseases, a student can help himself to understand that which he observes. Some students will be tempted to scan the illustrations and skip the observations! Such students are myopic for variability from specimen to specimen exists and cannot be captured in a single photograph. Only a study of several views of a number of specimens will help the student to learn how little or how great the variability for each causal organism and disease may be. In addition, laboratory examinations, popularly called practicals or spot tests, can and should be given. The challenge for the students to recognize unlabelled specimens and microscope slides stimulates observations. During their studies, students should use a buddy system and take turns covering the labels of specimens and microscope slides and quizzing each other on their identity.

Each experiment is introduced with a few remarks to present the question or questions of inquiry. Although these remarks may repeat information recorded elsewhere under a specific disease in some measure, it is important in order to crystallize the purpose of the experiment.

Ideally, each student should perform the experiments. They can if classes are small. With large enrollments, however, it is economically prohibitive in supplies and preparation time to provide for each student on an individual basis. Students may be divided then, into groups of two to four students. Groups that are not larger than this still permit reasonably good experience for each member. Demonstrations and field trips should be included to extend the experience of the students. They are supplements to but not substitutes for the involvement of the student in active experimentation.

Preparation of materials for experiments may be handled by the instructor, teaching assistant, or preparator. An advanced student may select some of these experiments as individual projects and prepare the materials himself. Therefore, required materials, plants, cultures of causal organisms, and equipment are listed for each experiment. For convenience of preparation, these are divided into four groups:

Expendable Supplies × Total Number of Students or Groups. These are expendable supplies needed by each student or group of students in the course. It is advisable to include extras to permit selection of uniform plant materials and provide for unforeseen losses. Multiply the items in this list by the number of students or groups of students in the entire enrollment.

Nonexpendable Supplies × Total Number of Students or Groups of Students Which Are Seated in a Laboratory Section. These represent items, such as thermometers, which each student or group of students needs but which can be used again by others in the following laboratory periods. Multiply the items in this list by the number of students or groups of students which are seated in a laboratory section.

General Supplies × Number of Laboratory Sections. These are expendable supplies shared and used by all the students in a single laboratory section. It is contemplated that these will be exhausted and additional materials are needed in each laboratory section. Multiply the items in this list by the number of laboratory sections.

Equipment Available. These are nonexpendable items, such as a balance, which are shared by all. They are needed in the laboratory classroom during the experiments. Multiply the items in this list by one.

For success with experiments, living plant material must be provided on a dependable basis. Seeds should be planted especially for the laboratory exercises and started in anticipation of their need. In each experiment, the average time required for growth of plant material is indicated. Unless otherwise noted, this is based on growing conditions in the greenhouse at temperatures of 72°–78°F. During summer months, the total time from seed planting until class use can be reduced about one week for all plant material because the warmer temperatures increase the rate of growth.

The rate of plant growth cannot be assured nor predicted and the laboratory must be "played by ear." Plant material must be used when ready. This requires a priority of activities in the laboratory: (1) initiation of new experiments; (2) continuation of manipulations or observations of experiments started previously; and, (3) observations of prepared materials which can be made with the self-guiding illustrations another time, either in another laboratory period or during open laboratory hours.

Diagrams accompany some experiments to help the student understand the procedures involved. In addition, data collected by students in previous classes and photographs of experimental results are provided in the event the experiment is omitted or should fail. This is not cheating because these experiments do not constitute original research but exercises for eyes, hands, and minds.

Aberrant experimental data may result from unfavorable environmental conditions, loss of pathogenicity of the causal organism, inadequate replication, or perhaps faulty experimental design! Do not discard the opportunities presented by these occurrences. The real-life stories of investigators include such problems and rewards. The problems arise when distinguishing typical from aberrant data in untried experiments. The rewards are the occasional observations made by the attentive eye and keen mind.

A large number of experiments are designed with the soft rot diseases. These require only supermarket carrots and potatoes, wet paper towels and plastic bags—not greenhouses nor growth chambers. The notions of the Germ Theory of Diseases, the principles of sterile techniques, the concepts of isolation, Koch's Postulates, and insights into the mechanism of pathogenesis can be demonstrated easily with these materials.

The student must realize that no one can see or learn for him. The instructor can only challenge and guide. It is recommended that each student keep a notebook in the laboratory in which to preserve drawings of his observations and supplement the illustrations provided. He should record the details of every experiment as he performs it and not rely on his memory. *What* is done, *how* it is done, *when* it is done, and *why* it is done are the questions to answer in his recorded notes. Experiments overlap and without proper records only confusion can result. The participation of each student in an experiment depends on his own initiative. When there are groups of students performing tasks together, each must contribute and be alert to the contributions of others in order to learn.

A laboratory notebook is the personal property of the student, as are his lecture notes. In general, the notebook should not be required but recommended, and not graded but corrected. It is difficult to ascertain the originality of much of the contents of notebooks and to distinguish the quality of the drawings from the quality of the observations. Achievement on the practical examinations is a better measurement of accomplishment. These examinations also help the student to know what he needs to study more.

Acknowledgments

I am indebted to the many colleagues and friends who provided specimens for the photographs—all of which are originals in order to provide exactly matching color and black-and-white photographs. Dr. John H. Hart and Dr. Nicky A. Smith deserve special recognition for their numerous contributions. Thanks are due to Dr. William B. Allington, Dr. William J. Hooker, and Dr. James E. Kuntz for their constructive suggestions for the manuscript and to Dr. William G. Fields for his suggestions for the chapters dealing with fungi. I am indebted to the gentle ways of my wife during the long hours of writing, typing, and drawing; and, to Dr. William B. Drew, "my boss," who has abetted and aided teaching whenever he could. But for his encouragement and support, I would probably be neither teacher nor author today.

Contents

1

Definitions

Definitions of plant diseases are about as numerous and different as plant pathologists. If diseases were simple and uncomplicated phenomena, we would not expect this. Some definitions are presented to provide clarity and understanding in communication. Definitions for the purpose of memorization serve little real purpose. They should provide tools for helping us to discuss complex issues and develop concepts. A flexible, working concept of a phenomenon is still useful tomorrow when we know more about it and realize that our earlier definition was not wholly adequate.

Disease can be defined partially on the basis of reduction in yield and degradation of quality of our economic crops. Such an economic notion of disease provides no insight into natural relationships of cause and effect. Let us avoid this approach. Let us design our definition as simply as possible: Plant disease is an abnormality. This is one characteristic of all diseases, for how else would we recognize them? The abnormality may be either in structure or function. Thus, the definition becomes: Plant disease is an abnormality in structure or function.

Disease and Injury. We do not consider freshly cut grass to be diseased. Certainly it is abnormal in structure compared to natural, uncut grass. This is analogous to a cut finger which is abnormal but not diseased—it's injured. The notion of injury is useful so let us call the cut grass a plant injury. How shall we distinguish plant disease and plant injury if both are abnormalities?

The abnormalities of the cut finger and the cut grass were brought about swiftly by the sudden appearance and disappearance of the causal agents—the knife and mower blades. Bacteria and fungi which cause abnormalities do not bring about abnormalities with such sudden coming and going. Their effects or symptoms, however, may appear quite suddenly. Incorporation of this idea makes our definition of **plant disease** an abnormality, in structure or function, caused by a continuous irritant. And our definition of **plant injury** becomes an abrupt alteration of structure or function caused by a discontinuous irritant (Figs. 1-1 through 1-7).

Agents of plant injury include lightning, hail, freezing temperatures, lawn mowers, grasshoppers, strolling elephants, herbicides, and dog urine. These can be conveniently classified as mechanical, chemical, biological, or environmental.

Frost cracks are interesting environmental injuries which are brought about on very cold but sunny days. Dark-barked trees absorb the radiant energy of the sun's rays and expand differentially, producing long fissures through the bark and into the wood. Trees which crack often have recurring fissures. Narrow flat bands in the bark mark previous cracks which have healed (Fig. 1-7). Abnormalities caused by environmental factors may be classified as diseases or injuries depending on the duration of the irritant. The death of buds and bark cambium from prolonged low winter temperatures may be diseases, whereas the death of delicate flower parts from sudden drops in temperature in the spring may be frost injuries.

The definitions help us, but injuries and diseases do intergrade. San Jose scale (Fig. 1-1) is not a discontinuous irritant. Injury from herbicides such as 2,4-D (Fig. 11-10) may be caused continuously if a large quantity is spilled on the ground. Is sapsucker injury (Fig. 1-2) biological or mechanical? The causal agent is biological but the injury is mechanical.

Predator. Plant-chewing insects, such as grasshoppers and the cereal leaf beetle (Fig. 1-3), and plant-chewing animals, such as cows, are predators. They are part of a uniform group which is distinct from pathogens, parasites, symbionts, or saprobes. They are agents of plant injury of a special category. Predator is a commonly used term in entomology and zoology but not in plant pathology. The notion is nonetheless useful in helping us to organize our thoughts so that we might better understand how various plant injuries are similar and different.

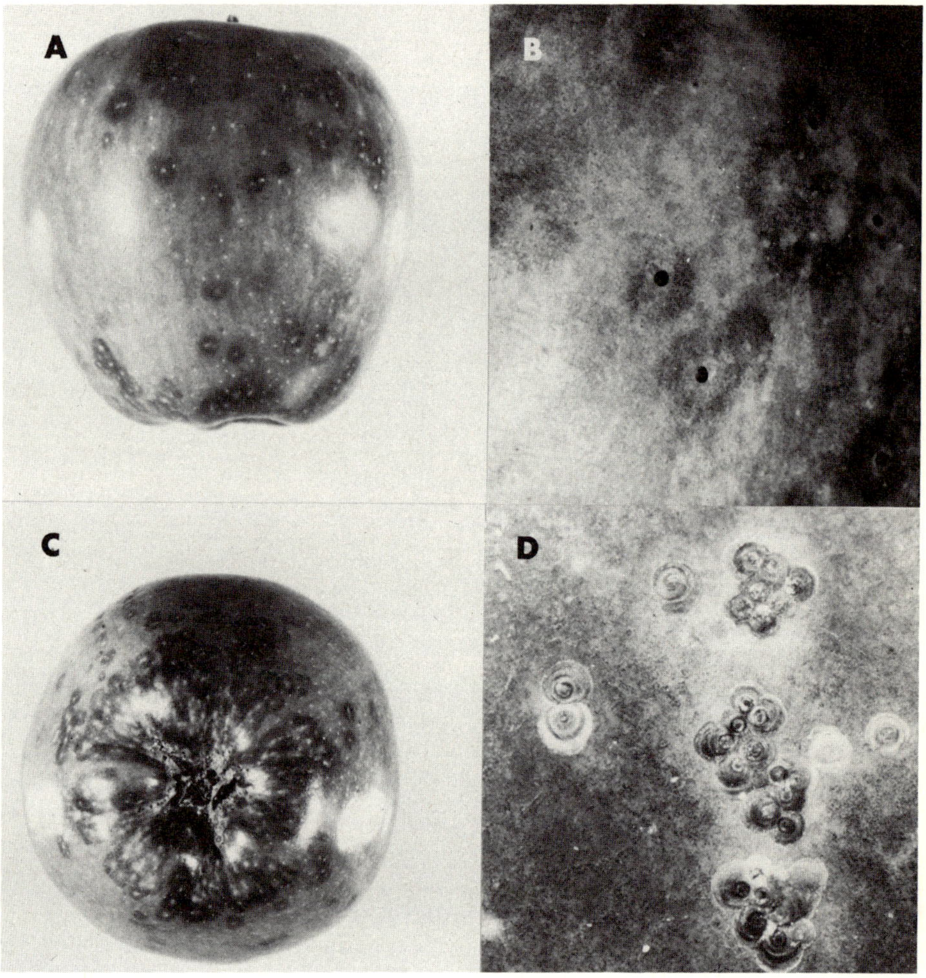

Fig. 1-1. (A) Side and (C) bottom views of an apple showing San Jose scale. (B, D) Higher magnifications show the characteristic clustering and oyster shell pattern.

Fig. 1-2. Sapsucker injury on pine. Note the rows of holes.

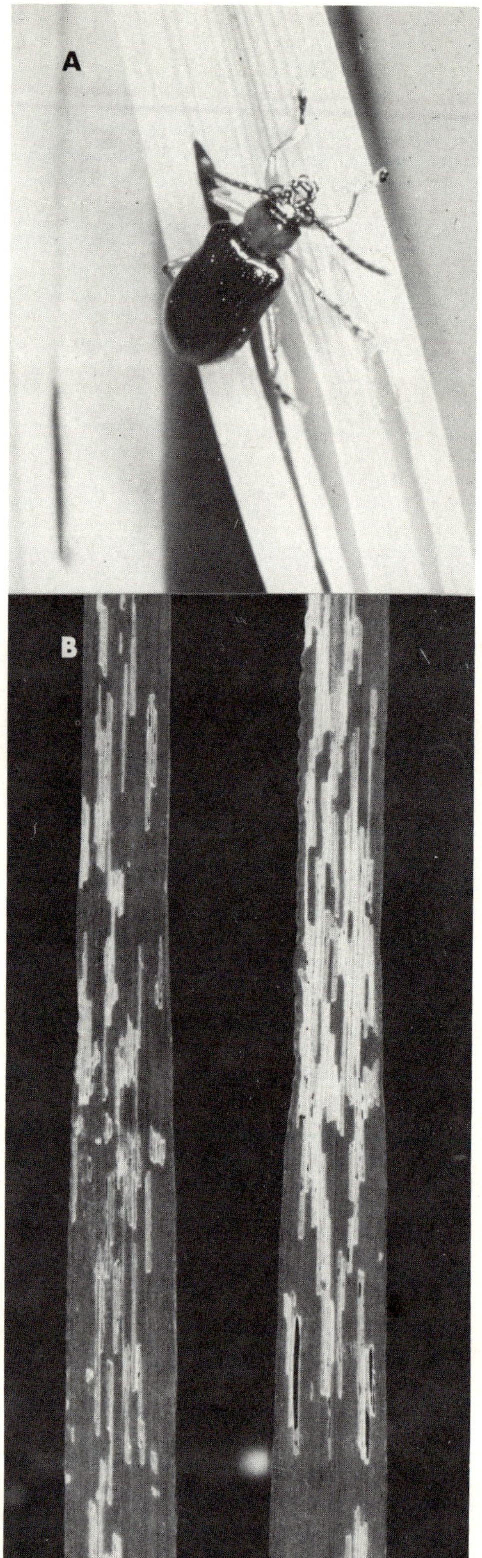

Fig. 1-3 (A) Adult cereal leaf beetle and (B) injury.

Fig. 1-4. (A) Healthy and (B) aphid-infested leaves of sweet cherry.

Fig. 1-5. Leaf miner injury on (A) flowering buttercup and (B) violet.

Fig. 1-6. (A) Mite injury on soybean leaflets and (B) enlarged two-spotted mite (× *230*).

Fig. 1-7. Recent and healed frost cracks on oak.

Nutritional Deficiencies and Diseases. Plants which are deficient in nitrogen are yellow-green and abnormal. By our definition they are diseased because the abnormality is caused by a continuous irritant—the continuous absence of sufficient nitrogen. Symptoms of nitrogen deficiency are often a generalized yellowing. Chlorosis due to a deficiency of iron is also a generalized yellowing (Fig. 11-7). Symptoms of nutritional deficiencies vary with the particular deficiency, the species of plant, and the environmental conditions. Symptoms include color changes (yellow, dark green, purple), necrosis (death of tissues), distortion (especially meristematic regions), and dwarfing. Fruits may be dwarfed, distorted, or seedless. Flowering and fruiting may be absent.

Causal Agent: a general term used to indicate those things, animate or inanimate, which cause injury and disease. A **causal organism** is an animate causal agent. **Pathogen** is often used for causal organism and **pathogenic agent** for inanimate causal agents. **Pathogenicity** is the property or ability of a causal organism to cause disease. A pathogenic organism is **virulent** when it can cause disease and **avirulent** when it cannot cause disease of another cultivar of the same species. Virulence is also used to indicate pathogenicity on any suscept. A **cultivar** is an international term for *culti*vated *var*iety. An organism which cannot cause disease of any suscept is a nonpathogen or nonpathogenic organism. **Nonpathogenic** is also used to describe the inability of a pathogen of one plant to cause disease of another species of plant.

Multiple forms of a pathogen which are virulent and avirulent on a multiple number of cultivars are called **races** or physiological races.

Pathology: the study (*logy*) of suffering (*pathos*) and of disease. A **pathologist** is one who studies diseases. **Pathologic**, or pathological, is a condition of being diseased. **Pathogenesis** is the origin (*genesis*) of disease and therefore the sequential development of disease. The **disease syndrome** is the sum total of all symptoms. In practice, pathology includes the study of plant injuries.

Parasite and Saprobe. Organisms may be classified as saprobes or parasites according to the manner in which they secure their nutrients. A **saprobe** is an organism which lives on nonliving organic matter. A **parasite** is an organism which secures its nutrition, in part or in whole, from another living organism (the host) of a different taxonomic kind. This last characteristic is added to eliminate the foetus and certain plant sporophytes and gametophytes as parasites.

Organisms which are capable of subsisting only as saprobes are **obligate saprobes**, and those which are capable of subsisting only as parasites are **obligate parasites**. Two additional terms are sometimes encountered: **facultative parasite** and **facultative saprobe**. Such organisms are defined to be saprobes or parasites most of the time and the opposite some of the time as circumstances permit. Facultative parasites are usually saprobes and facultative saprobes are usually parasites. There is no clear-cut separation of these two groups.

We use the word parasite, often, when we mean pathogen and pathogen when we mean parasite. This confusion results because the two terms are not exclusive. Most parasites are also pathogens because they cause continuous irritations leading to abnormalities of the hosts from which they secure their nutrients. Pathogens, however, may or may not be parasites. For example, the black walnut tree may be a pathogen of the tomato without being a parasite. Tomatoes do not grow well near black walnut trees. The nearer they are to the trees, the more dwarfed and yellow they become. The roots of the walnut tree release a compound called juglone, after *Juglans*, the generic name of the walnut. Juglone is toxic to the tomatoes. Remember that parasite is defined on the basis of how the organism secures its nutrients and pathogen on the basis of causing abnormality.

Host and Suscept. Host is the term applied to a living organism from which a parasite secures nutrients. To turn around our definition of parasite, a **host** is a living organism, different in taxonomic kind, from which a parasite secures its nutrients, in part or in whole. Suscept is a term used to describe a living organism which may be diseased. Thus, there are two kinds of interactions which concern us: host-parasite interactions and suscept-pathogen interactions. Both host and suscept are used to indicate plants which may be parasitized or diseased as well as plants which are parasitized and diseased.

A suscept is **resistant** when it cannot be afflicted with disease and **susceptible** when it can. Pathogens may penetrate resistant plants but do not become established. Susceptibility and resistance depend on the pathogen and the environment as well as the suscept. A plant may be susceptible to one race of a pathogen and resistant to another. It may be resistant at one temperature and susceptible at another.

Symbiosis and Disease. Symbiosis—life (*bios*), together (*sym*)—is usually considered to be an intimate, mutually beneficial association of two living organisms of different taxonomic kind. The participants are called **symbionts.** A classical example of symbiosis is the nitrogen-fixing nodules on the roots of legumes caused by bacteria in the genus *Rhizobium*. The bacteria, while in the nodules, secure their nutrients wholly from the plant. The plants are hosts and the bacteria are parasites. The host secures usable nitrogen from the activities of the bacteria and, technically, are parasites too. Thus, symbiosis in this case is mutual parasitism!

Although the occurrence of the nitrogen-fixing nodules is common on legumes, they are not normal, inherited structures of the plants. They are abnormal structures caused by continuous irritants—the bacteria. These nodules are diseases, as well as symbiotic associations of bacteria and plants. This is reaffirmed by two observations: (1) Some strains of nodule-forming bacteria are ineffective in fixing nitrogen in some cultivars of the host even though nodules are formed. There is no benefit to the host and no symbiosis—the plant is diseased. (2) Some nodules are effective when the plants are grown in full sunlight but are ineffective in full shade.

Symbiosis and disease, by definition and concept, are not exclusive. Both are intimate associations of living organisms and the result of their interactions. We tend to think of symbiosis as something good and disease as something bad. Subjective value judgments of good and bad give no indication of the natural relationships of the organisms to each other. The common denominator of disease and symbiosis is captured with the definition that disease is **antagonistic symbiosis.**

Signs and Symptoms. **Symptoms** are visible responses of the suscept to the effects of the pathogen. They include changes in color, size, and shape of the parts, organs, and entire structures of plants as well as the time of emergence of seedlings, rate of growth, and date of maturity. A **sign** is a manifestation of the causal agent, its presence, some structural part of it, residues from it, or remains of it.

Environment and Disease. The occurrence of disease is not solely dependent on the presence of the suscept and the pathogen. The environmental factors, especially moisture and temperature, must be favorable to the causal organism and its interactions with the suscept. Disease, in reality, is a product of the interaction of suscept and pathogen with each other and with the environment. The environmental factors can increase or decrease the susceptibility of the suscept and the pathogenicity of the pathogen. In addition, the environment can affect the interaction between suscept and pathogen.

The effects of the environment, acting prior to infection and which alter the susceptibility of the suscept are called **preconditioning.** The effects which increase the susceptibility of the suscept, or in some way make it more likely to be diseased, are called **predisposition.** There is no special term for the effects which decrease the susceptibility.

Those diseases which are caused by persistent, unfavorable environmental conditions are sometimes called **environmental diseases.** Persistent unfavorable temperatures and persistent air pollution commonly cause continuous irritations leading to the development of disease.

Classification of Disease. Diseases may be classified by *suscept* (apple and vegetable diseases), *age of suscept* (seedling and pole diseases of trees), *plant part* (root and fruit diseases), type of *disease symptom* (dry rot, soft rot, wilt), *location of occurrence* (transit, storage), *geography* (tropical, Michigan, and Asian diseases) or by *causal agent* (viral, nutritional, bacterial).

Diseases may be divided into two major groups: infectious and noninfectious or parasitic and nonparasitic. Infectious diseases are considered, often, to be those caused by microorganisms, including viruses, whether the causal organism is parasitic or not. Parasitic diseases are those caused by pathogens which are also parasites.

Diseases are organized herein according to causal agent for a pragmatic reason. This manual is intended for the novice who has had no training in plant pathology and a limited exposure to supporting sciences. I assume, for example, that the student knows no mycology; thus, I provide brief, introductory remarks on each kind of causal organism. For this reason, there is an efficiency to studying the causal organisms and the diseases they cause, together. This classification may be old-fashioned but that does not lessen its utility. Where appropriate, diseases are discussed by type, such as damping-off, wood rots, and vascular wilts.

Life Cycle and Disease Cycle. A **life cycle** is the sequence of stages through which an organism grows, reproduces, and perpetuates itself. All of the parts, fragments, and structures of an organism which can grow and reproduce the organism are called **propagules** because they propagate the organism. (Refer to the Chapter on Fungi for a detailed account of a life cycle.) The **disease cycle** is the chain of events by which a disease develops and is perpetuated. Because perpetuation of the disease depends on the pathogen, the suscept, and in many cases biological vectors, the life cycles of these organisms as well as the environmental factors are involved in the disease cycle.

The propagules which survive adverse periods are called **overwintering** or **oversummering forms.** In temperate zones the winters are cold and adverse, whereas in the subtropical and tropical zones, the unfavorable periods are usually the dry, hot summers.

The infectious propagules which initiate the first infections at the starting of a growing season and which are derived from overwintering (oversummering) sources are **primary inoculum.** The overwintering propagules may be the primary inoculum if they are infectious. Often, however, the overwintering propagules are not infectious but produce other propagules which are infectious and constitute the primary inoculum. The infections which result from primary inoculum are **primary infections.** The infectious propagules produced as a result of primary infections constitute the **secondary inoculum.** There may be a series of **secondary infections** which produce more propagules which cause more infections, and so on. This series is called the **repeating cycle.**

Agents of Dispersal. Any object, a passing cat's tail, a turning tractor wheel, or a scampering chipmunk, can carry infectious propagules in a passive, incidental way. These are the **carriers** of pathogens. The **vectors,** on the other hand, are active agents of transmission and often exhibit specificity of transmission. Some authors consider vectors to be living organisms whereas others include inanimate agents of dispersal, such as wind and rain. There is merit to restricting the use of the term vector to living organisms.

Transmit, Disseminate, and Spread. Technically, the causal organism and not the disease is spread, disseminated, or transmitted. Laymen and popular reports commonly use disease spread even though spread of the causal organism is what is really meant. Spread and disseminate mean to distribute, sow, or disperse. There is a tendency to use dissemination for long distance dispersion and spread for local distribution. Transmission implies active transfer of the causal organism from one suscept to another, such as transmission by insects or grafting.

Infect, Infest, and Contaminate. When nonliving things contain or carry pathogenic propagules they are contaminated or infested. **Contamination** is the loss of purity whereas **infestation** implies pollution by large numbers of the pollutant. Nonliving things cannot be infected.

Though living things may be infested or contaminated, they may be infected as well. **Infection** is the establishment of a causal organism in a suscept following penetration. **Penetration** is the initial invasion of the suscept whereas **establishment** is successful procurement of nutrients. The interval of time between establishment and the appearance of symptoms is the **incubation period.** In practice, incubation period is often used to indicate the period of time from inoculation to appearance of symptoms. **Inoculation** is the process of applying inoculum to a suscept. **Inoculum** is material placed on or in the tissues of a suscept, with the intent of producing disease. It is also the naturally occurring propagules of causal organisms in nature. The inoculum ordinarily consists of propagules which may or may not be infectious. Disease may or may not result. Suscepts may be inoculated to produce disease, test susceptibility of the suscept, test pathogenicity of the pathogen, or evaluate the effects of environmental conditions on disease development.

Epidemic, Epiphytotic, and Epizootic. An **epidemic** is the occurrence of an unarrested increase and spread of disease in a significant portion of a community of suscepts. Ten per cent may be considered a significant portion. The rate of increase is not a factor in defining epidemic. Many epidemics develop suddenly and rapidly, but some, such as swollen shoot of cacao, may take years to develop. Epidemic means *among the people* but is commonly used with any disease of any organism—plant, animal, or man. Perhaps it is well to leave it as such and use epidemic in the broad sense, and the terms **epiphytotic** and **epizootic** to indicate epidemics of plants and animals when specificity is desired. An epidemic disease, as a type of disease, is acute, exhibits characteristics of an epidemic, and generally occurs at interrupted and infrequent intervals. Sometimes introduced diseases, such as Dutch elm disease and chestnut blight, cause epidemics.

Endemic diseases are indigenous, or native, to an area or a population of suscepts; they are not introduced. Some introduced diseases, however, become endemic in nature and for all practical purposes are endemic. They may increase and decrease slowly or rapidly. They are chronic and tend to be with us year in and year out. They may affect only a few members of a community each year or most of them. Endemic diseases occasionally become epidemic when only a few members are initially affected and unarrested increase and spread of disease occurs.

GENERAL REFERENCES

Anderson, H. W., *Diseases of Fruit Crops.* New York, McGraw-Hill, 1956.

Boyce, J. S., *Forest Pathology,* 3d ed. New York, McGraw-Hill, 1961.

Butler, E. J., and Jones, J. G. *Plant Pathology.* London, MacMillan, 1949.

Carter, W., *Insects in Relation to Plant Disease.* New York, Interscience, 1962.

Chester, K. S., *The Nature and Prevention of Plant Diseases,* 2d ed. Philadelphia, Pa., Blakiston, 1947.

Chupp, C. and Sherf, A. F., *Vegetable Diseases and their Control.* New York, Ronald, 1960.

Dickson, J. G., *Diseases of Field Crops.* New York, McGraw-Hill, 1956.

Garrett, S. D., *Root Disease Fungi.* Waltham, Massachusetts, Chronica Botanica, 1944.

Gäumann, E., *Principles of Plant Infection,* transl. by W. B. Brierly. London, Lockwood, 1950.

Heald, F. D., *Introduction to Plant Pathology.* New York, McGraw Hill, 1937.

Hunger Signs in Crops, H. B. Sprague, ed., 3d ed. New York, David McKay Co., 1964.

Index of Plant Diseases in the United States. Agricultural Handbook No. 165, U. S. Dept. of Agriculture, Washington, D. C., 1960.

Kamat, M. N., *Handbook of Tropical Crop Diseases.* Poona, India, Prakash Publ. House, 1958.

Leach, J. G., *Insect Transmission of Plant Disease.* New York, McGraw-Hill, 1940.

Lucas, G. B., *Diseases of Tobacco.* New York, Scarecrow Press, 1958.

Pirone, P. P., Dodge, B. O., and Rickett, H. W., *Diseases and Pests of Ornamental Plants,* 3d ed. New York, Ronald, 1960.

Plant Diseases: The Yearbook of Agriculture. U. S. Dept. of Agriculture, Washington, D. C., 1953.

Plant Pathology: An Advanced Treatise, J. G. Horsfall, and A. E. Dimond, eds., Vols. I, II, III. New York, Academic Press, 1959-1960.

Plant Pathology: Problems and Progress, 1908–1958, C. S. Holton, and others, eds., Madison, Wisconsin, The University of Wisconsin Press, 1959.

Pyenson, L. L., *Elements of Plant Protection.* New York, Interscience, 1951.
Sharvelle, E. G., *The Nature and Uses of Modern Fungicides.* Minneapolis, Burgess, 1961.
Shurtleff, M. C., *How to Control Plant Diseases in Home and Garden.* Ames, University of Iowa, 1962.
Smith, J. D., *Fungal Diseases of Turf Grasses.* Bingley, Yorkshire, England, The Sports Turf Research Institute, 1959.
Sprague, R., *Diseases of Cereals and Grasses in North America.* New York, Ronald, 1950.
Stakman, E. C., and Harrar, J. G., *Principles of Plant Pathology.* New York, Ronald, 1957.
Stevens, N. E., and Stevens, R. B., *Disease in Plants: An Introduction to Agricultural Phytopathology.* Waltham, Massachusetts, Chronica Botanica, 1952.
van der Plank, J. E., *Plant Diseases: Epidemics and Control.* New York, Academic Press, 1963.
Walker, J. C., *Plant Pathology*, 2d ed. New York, McGraw-Hill, 1951.
—— *Diseases of Vegetable Crops.* New York, McGraw-Hill, 1952.
Wescott, Cynthia, *Plant Disease Handbook.* Princeton, New Jersey, Van Nostrand, 1950.
—— *Plant Doctoring is Fun.* Princeton, New Jersey, Van Nostrand, 1957.
—— *Are You Your Garden's Worst Pest?* Garden City, New York, Doubleday, 1961.

2

Interpretation of Microscope Observations

Artifacts. Some of the things we see in the microscope are not a part of the original structure of the living specimen but are artifacts. An **artifact** is a product of man's workmanship. Workmanship may bring to mind the craftsmanship of the Eskimo's polar bear carved out of walrus tusk or the graceful gazelle of African ebony. Man's workmanship also includes the embedding, sectioning, and staining of tissues for microscope observations. The effects of these manipulations often produce distortions and color changes which are not characteristic of the tissues when alive. In order to interpret what we see and to understand the structure of living tissues, we need to know what artifacts are and what makes them.

Dehydration and Shrinkage. Some specimens dry out naturally and shrink out of proportion (Figs. 2-1, 2-2). Sections of these materials are distorted and do not represent the living specimen. Such effects are not artifacts but are natural distortions. Distortions are also produced by the process of embedding material in paraffin for sectioning. The infiltration of paraffin, or other materials, is necessary to give support to the tissues when sliced into thin sections. Paraffin does not mix with water but with gasoline-related solvents. Therefore, the material must be dehydrated by rinsing in mixtures of alcohol and water of increasing alcohol concentration. Distortion from this dehydration process may be more subtle than natural desiccation of whole specimens. It may cause shrinkage of the protoplasts of cells rather than entire organs or tissues.

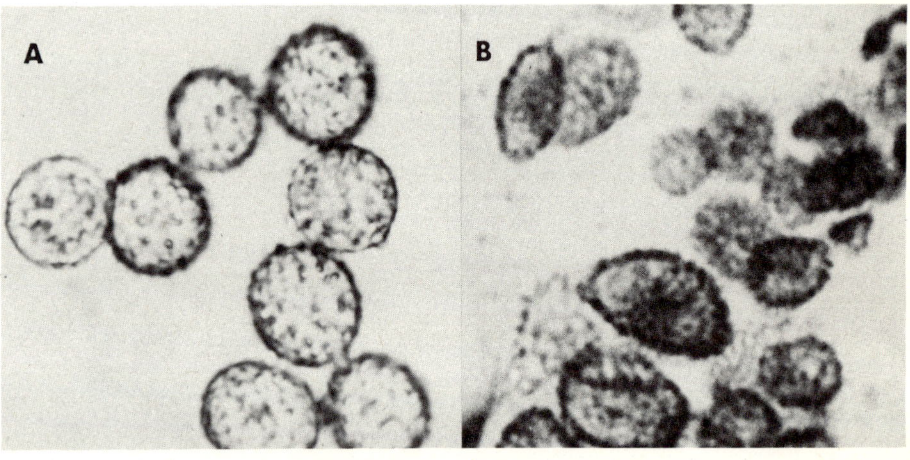

Fig. 2-1. (A) Freshly prepared and (B) dried teliospores of *Ustilago maydis* (× *1000*).

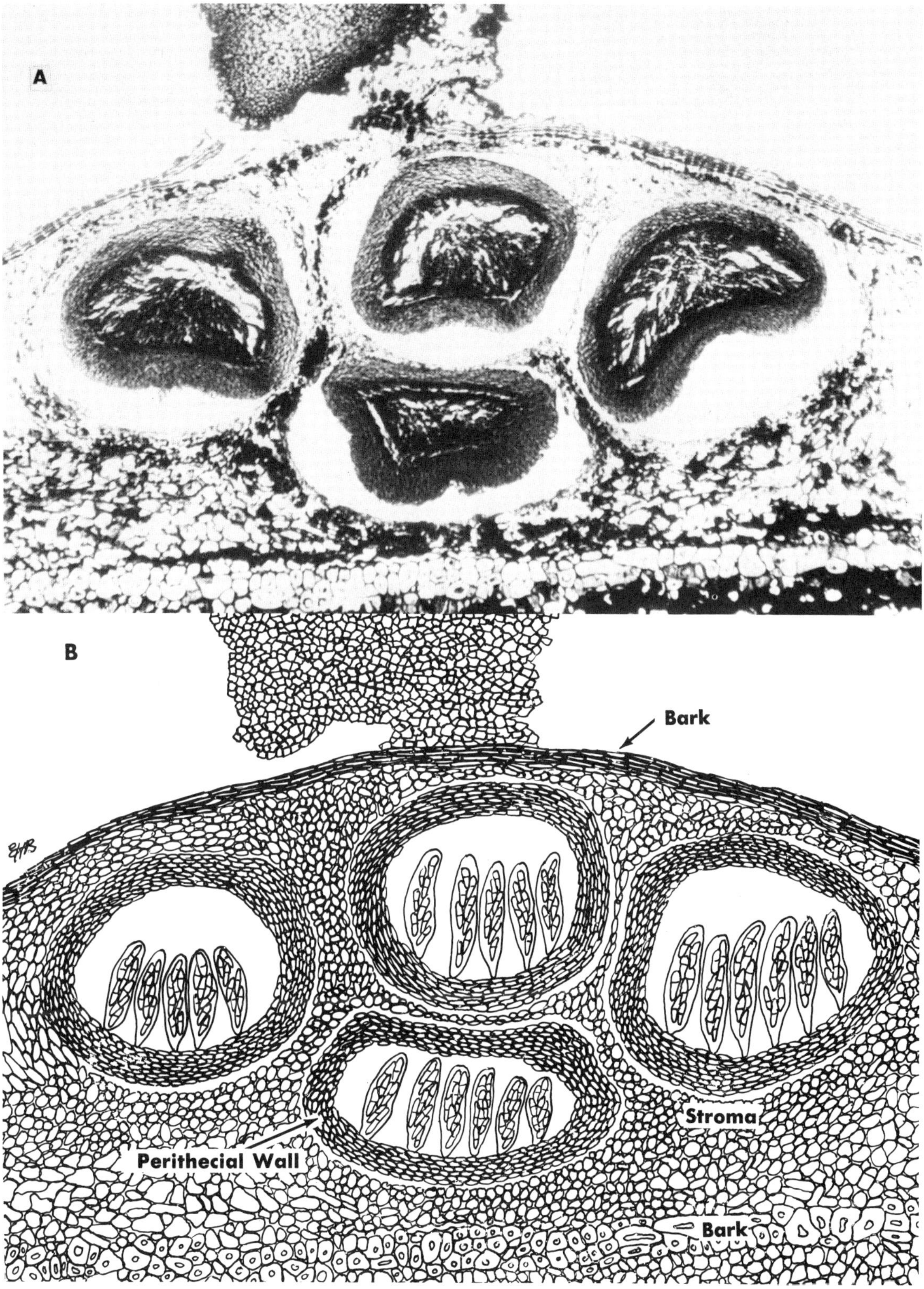

Fig. 2-2. (A) Dried and shriveled perithecia of *Nectria* and (B) reconstruction (× *125*).

Uneven Knife Blade and Streaks. Dark or light lines, which are not part of the structure, may appear in specimens (Fig. 2-3). If they are straight and not in relation to other structural features, they may be the result of an uneven knife blade used to make the sections. If there is a nick in the blade, there is a line of greater density because the nick permits a thicker slice to be cut. If there is a burr on the knife edge, the section is gouged and has a line of lesser density.

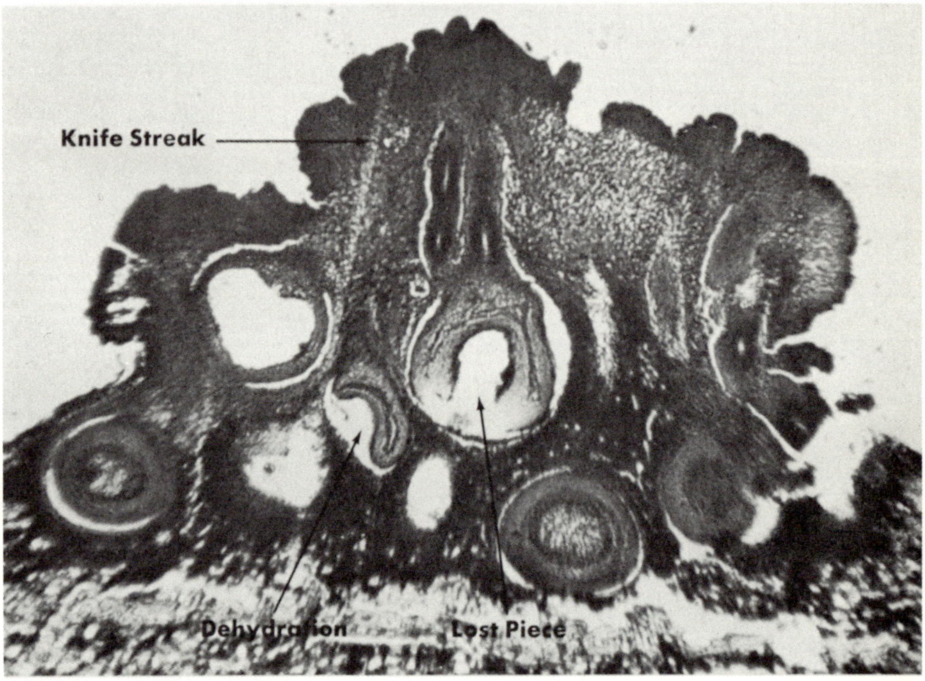

Fig. 2-3. Sections through a stroma of *Endothica parasitica* showing a knife streak. Shrinkage from dehydration and lost pieces are evident. The lack of sharp definition across the bottom of the photograph is due to the use of an objective lens which was not optically flat (× *64*).

Inadequate Tissue Support and Displacement. Thin sections of tissue may have little support even with infiltration of paraffin or other embedding materials. Displacement of pieces of the section may occur during manipulation of the section on the microscope slide (Figs. 2-4, 2-5, 8-1). Sometimes larger pieces are displaced or lost during manipulation (Figs.2-6, 2-7)—e.g., the cuticle may drift free of the epidermis. (Fig. 2-7).

Staining and Color Changes. Often different structures in tissues have little contrast when viewed in the microscope. Stains, which color only certain tissues and not others, are used to distinguish various tissues. These are called differential stains. For example, certain stains are used to color xylem elements red. They are not red in living tissues. Other tissues may be stained another color, and thus the tissues are differentially stained. On the other hand, some structures have natural colors which are removed or altered during the dehydration process. Preparation of sections may result, therefore, in the loss or the gain of color.

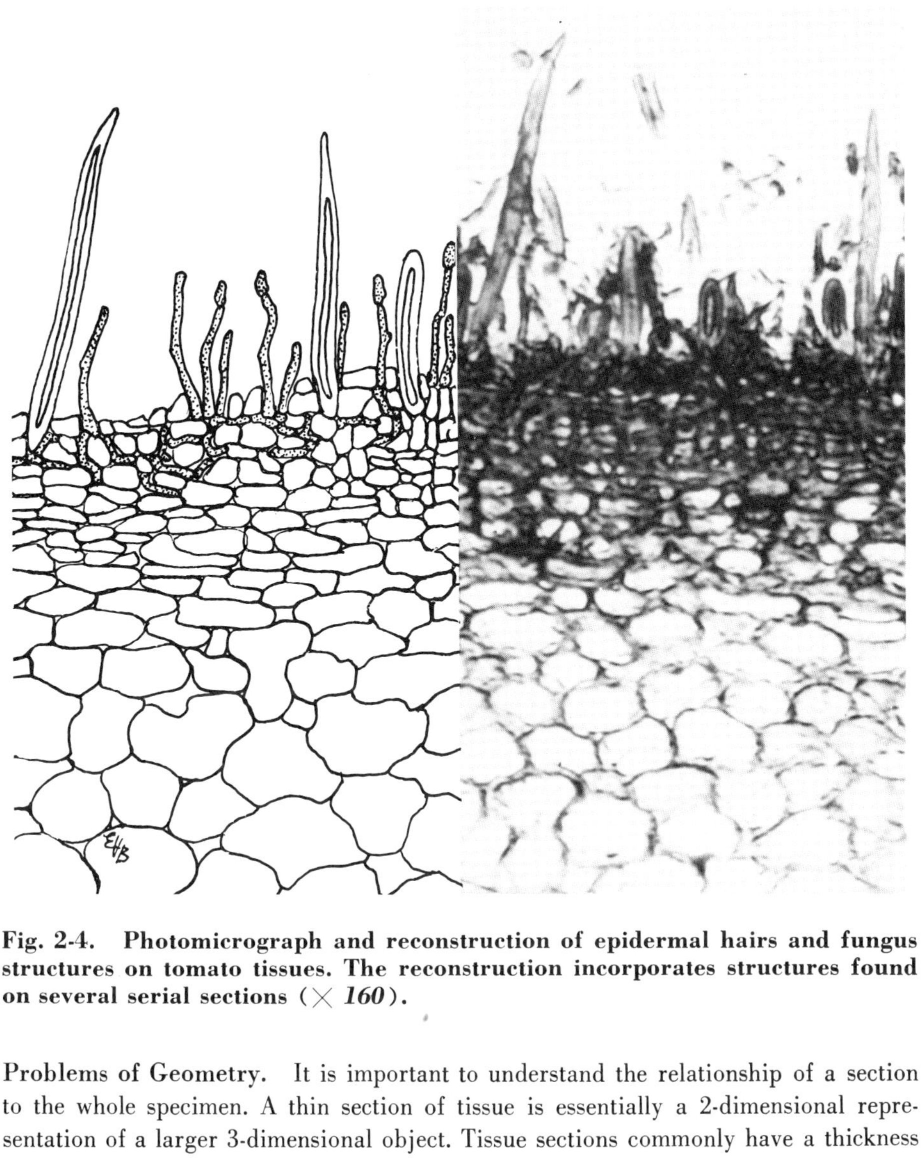

Fig. 2-4. Photomicrograph and reconstruction of epidermal hairs and fungus structures on tomato tissues. The reconstruction incorporates structures found on several serial sections (× *160*).

Problems of Geometry. It is important to understand the relationship of a section to the whole specimen. A thin section of tissue is essentially a 2-dimensional representation of a larger 3-dimensional object. Tissue sections commonly have a thickness of 10 to 100 microns. By focusing the microscope up and down we can see the third dimension of the section. This is not the third dimension of the whole specimen.

Serial Sections. A series of sections, taken one after the other through a flask-shaped fruiting structure of a fungus, gives different representations of the same object (Fig. 2-8, 2-9). Only by studying a series of sections can the whole object be understood. There are some other fungus structures which are round and without a neck like a basketball. A section through the flask-shaped body at some point away from the neck would appear to be a section through a neckless structure.

Cross Section, Longitudinal Section, and Tangential Section. Since objects are 3-dimensional there are a number of different angles by which a specimen can be sliced. Comparison of cross sections and longitudinal sections helps us to visualize the 3-dimensional structures (Fig. 2-10). Tangential sections (Fig. 2-11B) often give a distorted view of longitudinal sections (Fig. 2-11A).

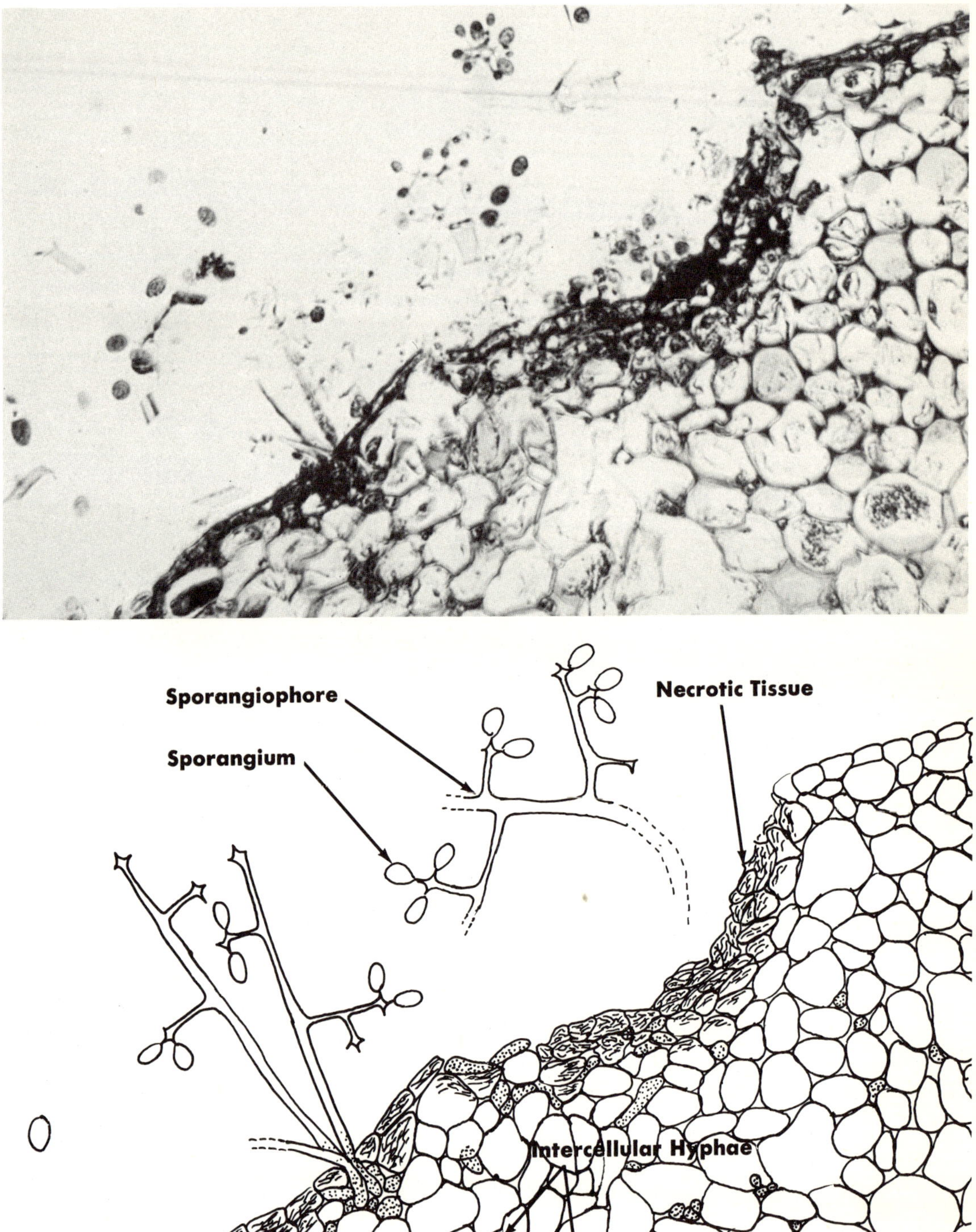

Fig. 2-5. Photomicograph and reconstruction of sporangiophores and sporangia of *Plasmopara viticola*. The reconstruction incorporates structures found on several serial sections (× *250*).

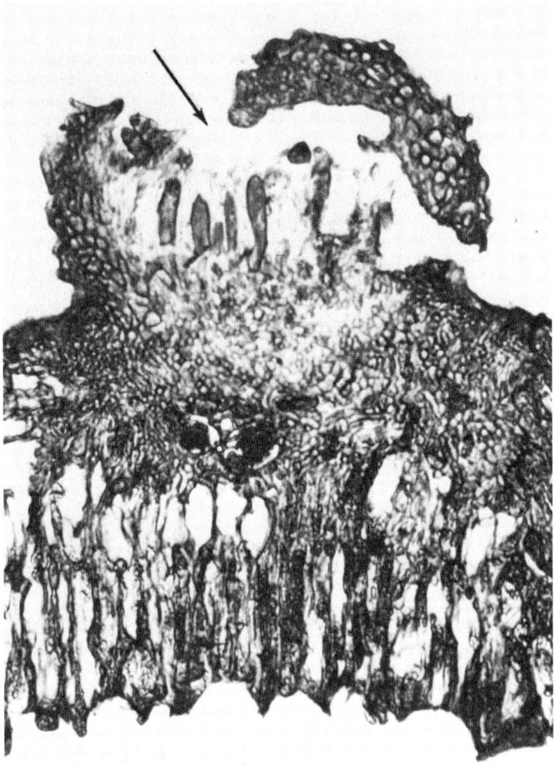

Fig. 2-6. Photomicrograph of a section through a young perithecium of *Dibotryon morbosum* showing loss of fairly large sections. Arrow designates missing piece (× *250*).

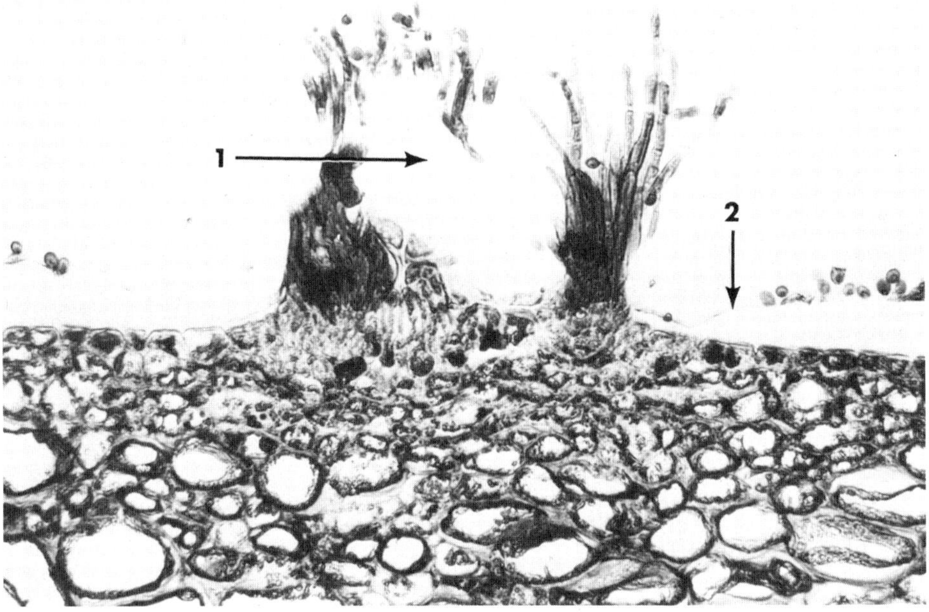

Fig. 2-7. Photomicrograph of a section through an infected peach showing conidiophores and conidia of *Monilinia fructicola* with a large section lost (arrow 1). Displacement of the cuticle (arrow 2) with adhering conidia is evident (× *208*).

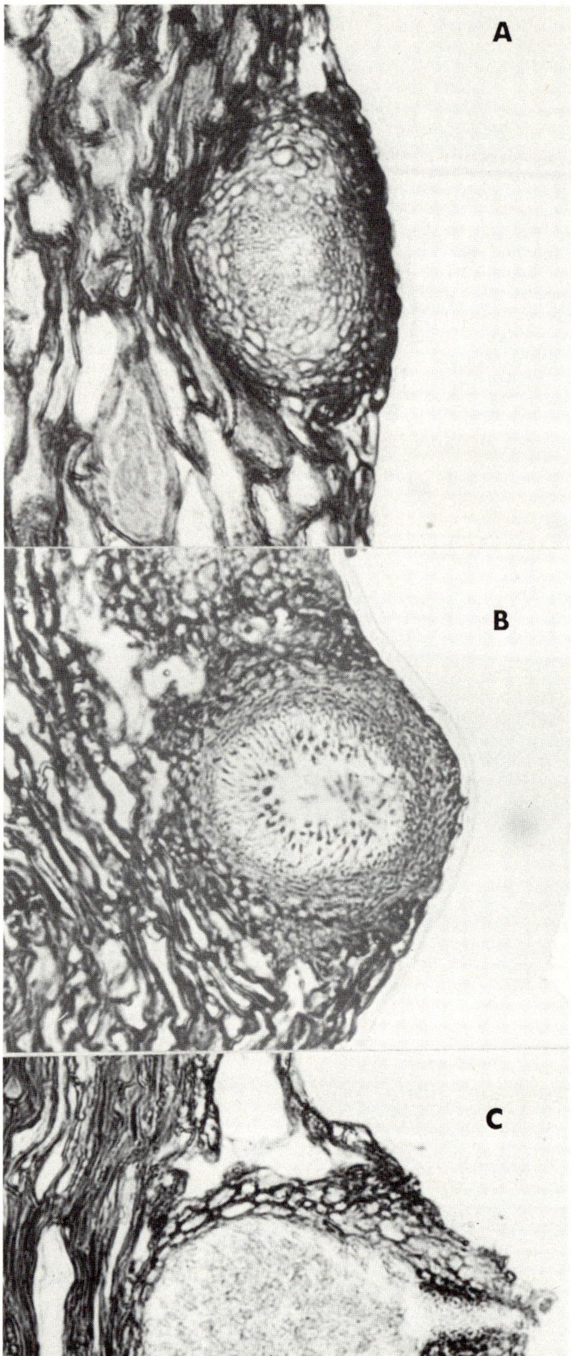

Fig. 2-8. Serial sections through a pycnidium of *Guignardia bidwellii* (× *250*).

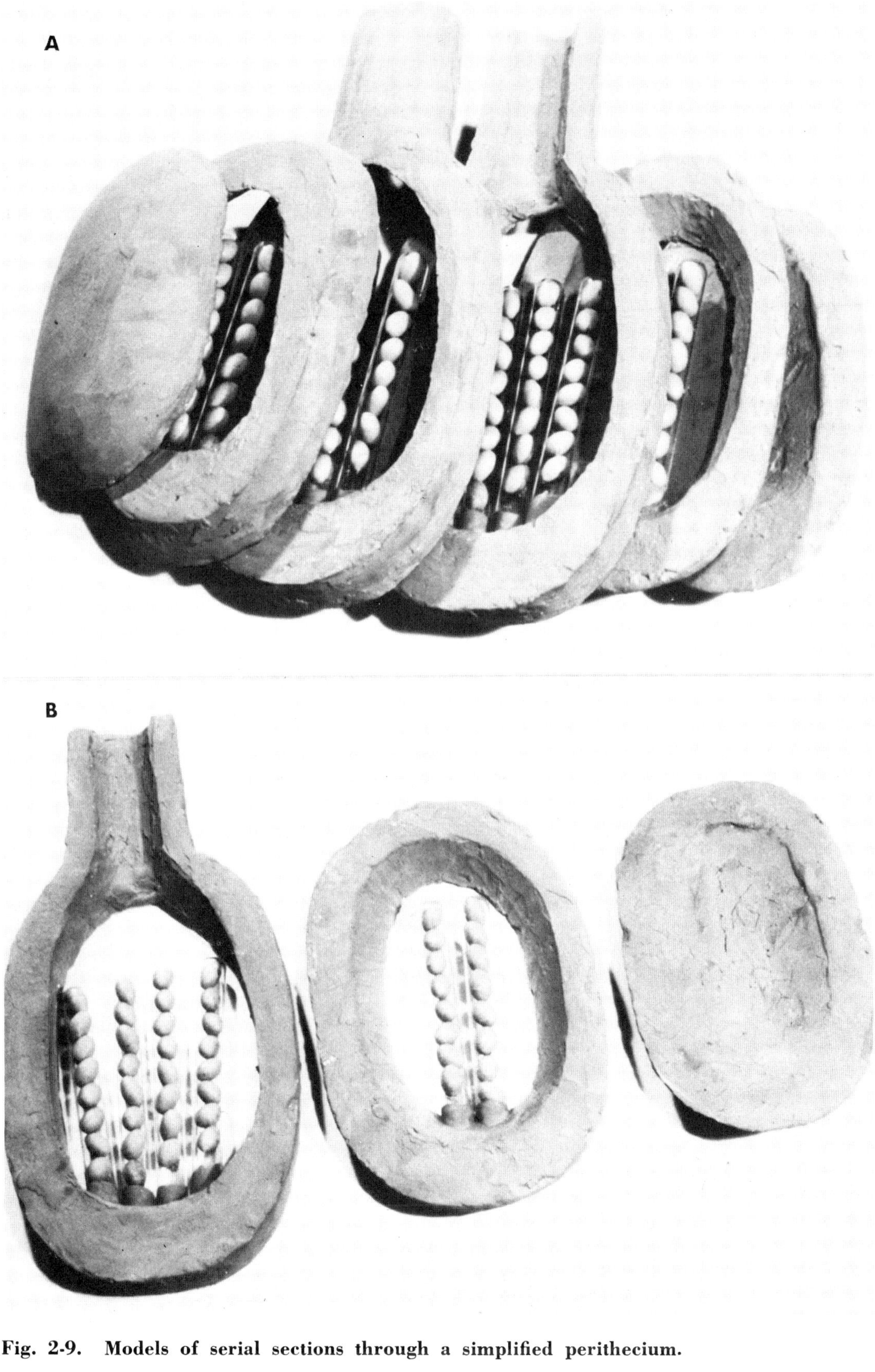

Fig. 2-9. Models of serial sections through a simplified perithecium.

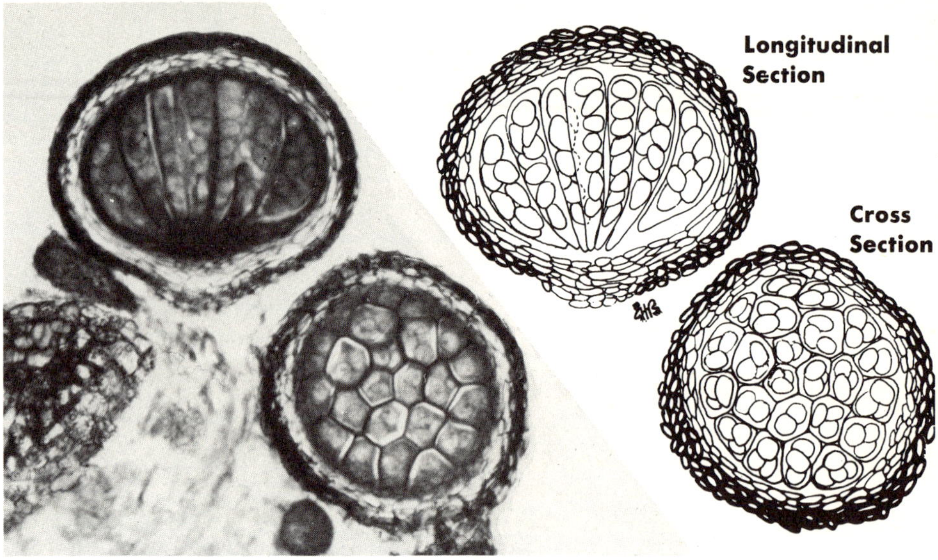

Fig. 2-10. Photomicrograph and representation of sections through cleistothecia of *Erysiphe aggregata* (× *200*).

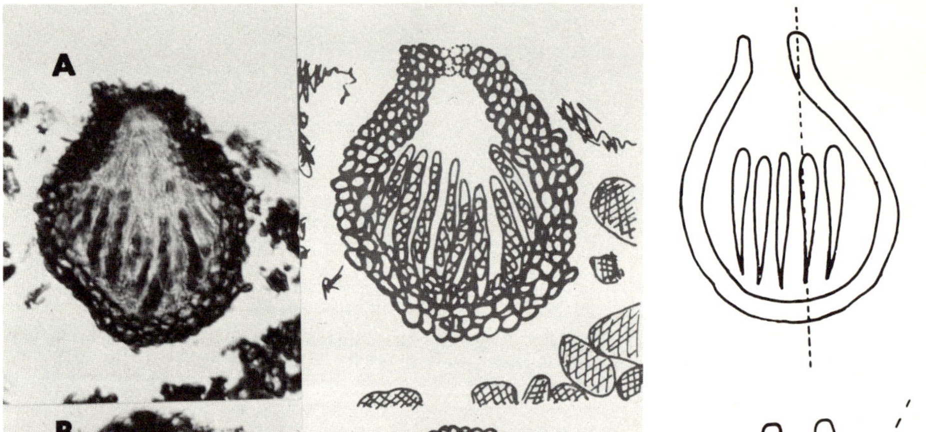

Fig. 2-11. (A) Longitudinal and (B) tangential sections through perithecia of *Venturia inaequalis.*

Debris and Miscellaneous Pieces. Debris adhering to a specimen becomes embedded and sectioned along with the specimen. Sometimes it is difficult to distinguish debris from miscellaneous or isolated pieces of the specimen. Long, slender structures often are twisted, curled, or bent and sections through them give rise to apparently unrelated bits (Fig. 2-12). A study of serial sections helps to reconstruct the original structure.

Fig. 2-12. Model of sections through a twisted epidermal hair.

Aberrations and Illusions. Techniques of microscopy and photography can create aberrations and illusions. The optical system can cast halos around objects under certain conditions of focus (Fig. 5-3C). Very small particles may cast larger, clear-looking configurations that might be mistaken for structural parts of some organism (Fig. 5-3C).

The relative contrast of the background and subject may affect the appearance of the subject. Colorless epidermal hairs, for example, disappear in a photograph of a bean stem with a light background but are quite distinct when photographed against a dark background (Fig. 2-13). Some specimens are photographed on glass supported above the background, with lights placed so that shadows are not cast directly below the specimen. Reflections in the glass can occur, however (Fig. 5-1, 6-1). In black-and-white photographs, when the background is close to the subject, shadows are cast which could be confused with the structures of the subject (Fig. 6-1).

Then there are the classics—the eyelash and the air bubbles. . . .

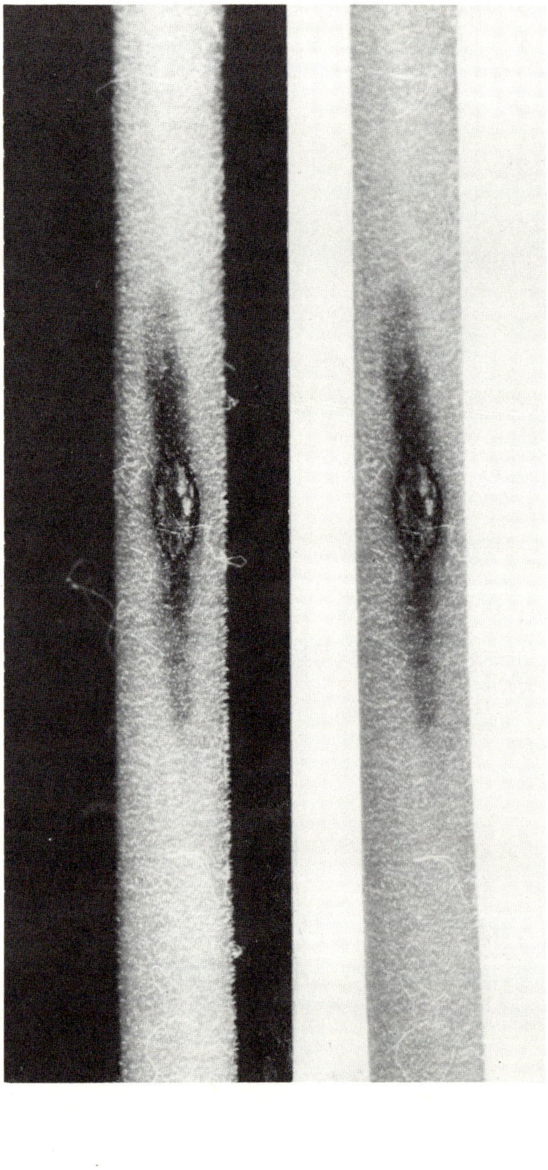

Fig. 2-13. Visibility of colorless epidermal hairs photographed against a dark and light background.

3

Principles of Sterile Technique

Bacteria and fungi are members of the microbiological world and are especially significant because of their universality. They are found everywhere: in the air, soil, hair, clothes, water, crevices in bark, paint, glue, crank-case oil, snow, and pickle juice. These little creatures also provide interesting paradoxes: (1) There are places where they are not normally found: the blood of healthy animals, the internal tissues of healthy plants and animals, and most conspicuous of all, sterilized objects. (2) Some are so undemanding in nutritional requirements as to survive and grow in stale distilled water containing traces of dust; others are demanding obligate parasites which require a certain cultivar of a particular species as host. (3) Some, like intestinal flora, are necessary whereas others, such as the tubercle bacillus and the Dutch elm fungus, cause severe and sometimes fatal diseases. (4) Some are red; others are yellow or white. (5) Some are uncommon while others are everywhere.

Early observations in the study of microorganisms led to the development of the Theory of Spontaneous Generation. The application of principles of sterile technique permitted a Frenchman, Louis Pasteur, to demonstrate conclusively that spontaneous generation was not responsible for the apparent spontaneity of the occurrence of microorganisms. Their apparent spontaneity was due to their ubiquitous presence.

The Five Principles of Sterile Technique. Understanding the principles of sterile technique requires, first, the recognition of the *ubiquity of microorganisms*—the first principle of sterile technique. The second principle is the *creation of sterility* by its natural occurrence or the destruction or removal of microorganisms. The third principle is the *maintenance of sterility* by isolation from unsterile objects and contaminants. The fourth principle is the *prevention of contamination* of sterile objects and pure cultures during manipulations. Contaminants are microorganisms which are present where they are not wanted; just as weeds are plants where they are not wanted. The fifth principle is the *confinement of microorganisms.* This is not usually an issue when working with saprobes and plant pathogens, but can be a critical problem when working with human pathogens. The problem of keeping microorganisms from getting out of a container can be quite different from keeping others from getting in.

Sterilization: Application of the First Three Principles. Creation of sterility by the destruction of microorganisms can be achieved by chemical means—e.g., mercuric chloride, sodium hypochlorite, alcohol, ethylene bromide; and by physical means—e.g., ionizing radiations, heat. Microorganisms are destroyed by these chemical and physical treatments because these treatments irreversibly disrupt protoplasmic structure. The appropriate application of these treatments destroys all microorganisms because they are all arrangements of protoplasm. Media and other materials are sterilized routinely in an autoclave for 20 minutes at 15 psi (pounds per square inch). This provides a temperature of 120°C. Once materials are sterile, they do not remain so unless isolated from the ever-present microorganisms. Thus, a medium in a test tube, plugged with cotton and sterilized, remains sterile.

If the plug or other closure is removed, contaminants may gain entrance. The longer the tube or other vessel is open, and the more frequently it is opened, the higher the probability of contamination. Even tubes which have been sterilized and never opened can be contaminated by small vermin such as mites and ants. Their legs are contaminated with bacteria and the agar surface is seeded with them with every little footstep! Their tracks are clearly visible later, as trails of small colonies of bacteria (Fig. 3-1).

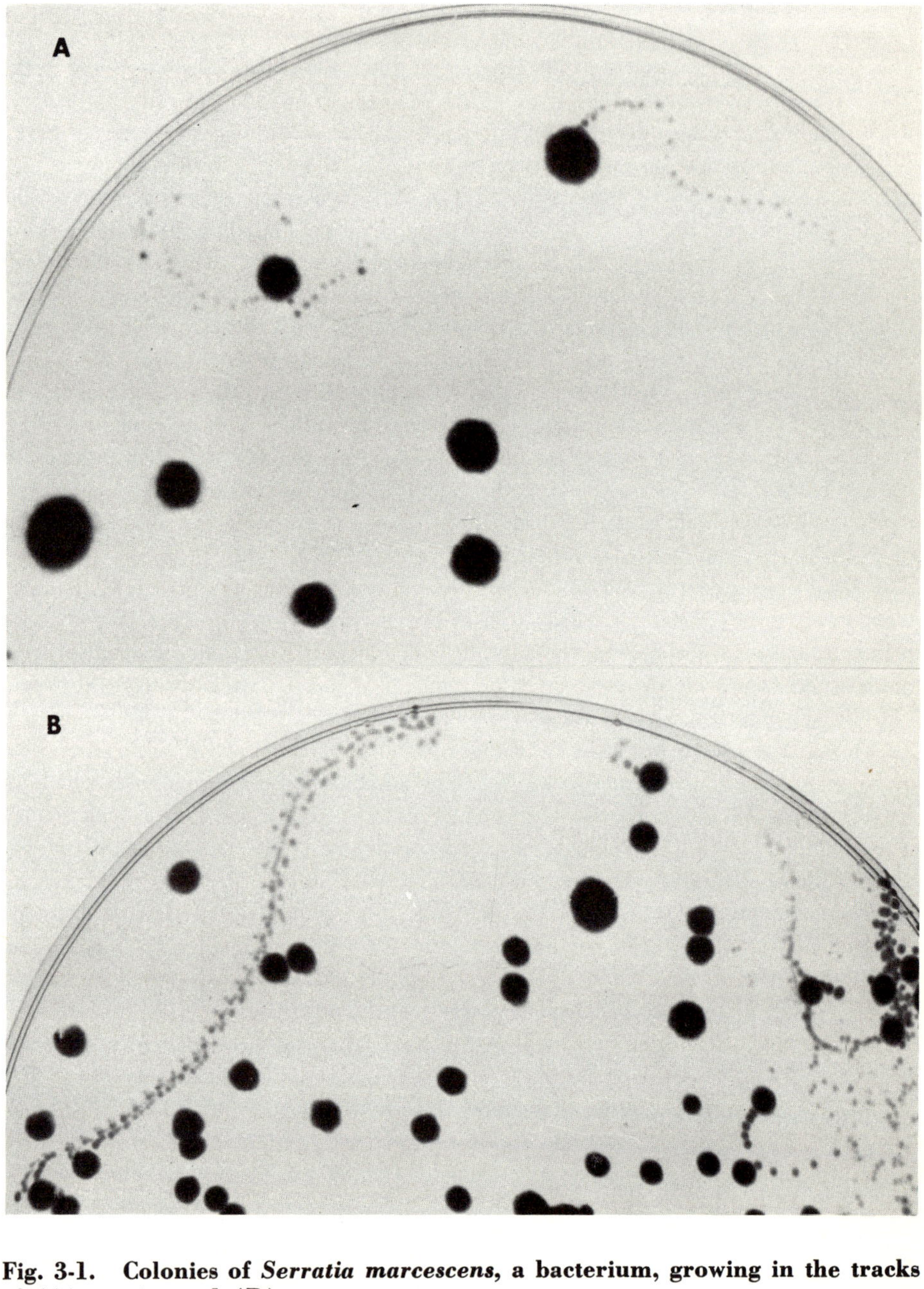

Fig. 3-1. Colonies of *Serratia marcescens*, a bacterium, growing in the tracks of (A) a mite and (B) an ant.

Transfer of a Culture: Application of the Last Two Principles. A description of a standard method of transferring a culture of bacteria, a basic technique in the microbiological laboratory, provides insight into the last two principles of sterile technique.

A few simple preparations are necessary for successful transfer without contamination. Windows and doors should be closed to avoid air currents which carry dust and contamination. The working surface should be washed with soap to remove dust and dirt. A moist piece of cheesecloth (18 x 24 inches) laid on the working surface fixes falling contaminants and keeps them and dust particles from swirling about during manipulations. Sudden, rapid, and unnecessary movements should be avoided. Microbes are carried on dust particles.

The tube with the culture is grasped in one hand and the transfer loop in the other (Fig. 3-2A). The needle should be held near the balance point of the handle. The loop is sterilized by heating in the Bunsen burner until it is red hot and then allowed to cool in the air or plunged into the sterile agar after the plug is removed. The cotton plug is removed by grasping it between the third and fourth fingers of the hand holding the loop (Fig. 3-2B). It may be necessary to twist the plug to loosen before removing it from the tube. The tube-end of the plug and the loop must touch neither your hand, nor the table, nor any unsterile object. The loop is then scraped across the bacterial colony and removed without striking the sides of the tube (Fig. 3-2C). The mouth of the tube is flamed and the plug replaced. The tube of sterile medium is opened, streaked, and closed in a similar manner. The tubes should be kept level or tilted no more than 30° from a horizontal position while open to prevent contaminants from falling into the tube. Those which fall on the lip or just inside are destroyed by flaming before replacement of the plug.

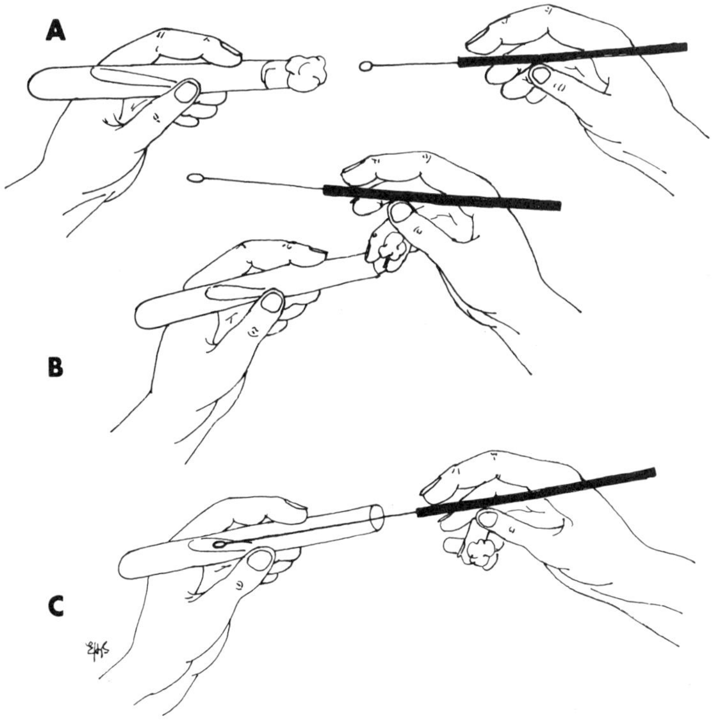

Fig. 3-2. Techniques of handling tubes and transfer loops during transfer of bacteria from one tube to another: (A) position of the tube and transfer loop in the hands; (B) removal of the cotton plug; and, (C) insertion of the transfer loop.

Experiment I

Application of Sterile Techniques

EXPERIMENT

1. Transfer bacteria from a slant with bacterial growth to a slant of sterile medium. (A slant is agar which is permitted to solidify in a slanted position to afford a working surface.)
2. Repeat the motions of transferring bacteria with another slant but use a sterile transfer needle.
3. Repeat the motions of transferring bacteria with another slant but use a contaminated transfer needle.
4. Preserve a slant without disturbance as a control.
5. Transfer bacteria to a petri dish of agar (called a plate) by the streak method of isolation described in Experiment IV.
6. Sprinkle soil on a plate.
7. Expose a plate to the air for 1 hour without a lid.
8. Preserve a plate without disturbance as a control.
9. Label all slants and plates with name and date. Put away all equipment. Cleanup the laboratory bench. Turn off all appliances. Record complete notes.

Serratia marcescens is a good bacterium to use in this experiment because it is bright red in color and easily noted by beginning students. The red pigmentation is not produced on certain media such as potato-dextrose agar but is very marked on commercial nutrient agar.

Expendable Supplies × Total Number of Students or Groups

4 slants of nutrient agar
4 plates of nutrient agar
1 culture of *Serratia marcescens*

Nonexpendable Supplies × Number of Students or Groups Seated in a Laboratory Section

1 test-tube rack
1 transfer loop
1 wax pencil
1 piece of cheesecloth (18 × 24 inches)

DISCUSSION QUESTIONS

1. If one tube of medium out of 100 which had been prepared, plugged, sterilized, and left undisturbed became contaminated after a few days, how would you prove that this was not spontaneous generation?
2. If all 100 tubes of medium prepared as above became contaminated what would you conclude? How would you verify your conclusions?
3. Why are the fourth and fifth principles of sterile technique different?

Experiment II

Media Preparation

The minimal nutritional requirements for the growth of many bacteria and fungi are carbohydrates, fixed nitrogen, and minerals. Because these nutrients are supplied in culture media from synthetic and natural sources, four arbitrary types of media can be designated: (1) synthetic, (2) semi-synthetic, (3) fortified nonsynthetic, and (4) nonsynthetic.

Synthetic media contain only known compounds in known quantities and can be precisely reproduced for careful experiments. When this is not necessary, the economy and ease of preparation of natural media are advantages. Some organisms grow adequately only on media with natural ingredients. Juices or extracts of plant and animal tissues are used for natural ingredients. Examples of the four kinds of media:

Synthetic: Czapek's Medium

Saccharose	30.00 g.
Sodium Nitrate	3.00 g.
Dipotassium Phosphate	1.00 g.
Magnesium Sulfate	0.50 g.
Potassium Chloride	0.50 g.
Ferrous Sulfate	0.01 g.
Distilled water	1.00 liter

Semi-Synthetic: Czapek's Medium Plus 1% Yeast Extract

Fortified Nonsynthetic: Potato-Dextrose Agar (PDA)

Potatoes, Infusion from	200 g.
Dextrose	20 g.
Agar	15 g.
Distilled water to total	1 liter

Nonsynthetic: 2% Malt Extract

Media may be used in liquid culture with the ingredients as given above without agar, or as solid media by the addition of 1.5% agar, although 0.75% to 2% agar may be used. Organisms in liquid culture may be grown in still culture (stationary) or in shake-culture in which the culture is continuously agitated by a shaking device or bubbling, sterile air. Agitation provides for aeration and uniformity of growth.

Media are prepared by dissolution of the ingredients in distilled water, dissolution of agar by heat in a steamer, distribution of media to culture containers such as tubes or flasks, plugging or capping, and sterilizing in an autoclave for 20 minutes at 15 psi. With large volumes in a single container, longer periods of sterilization are necessary so that the central portions are adequately heated.

Many synthetic and fortified nonsynthetic media are available commercially as dehydrated preparations. These are reconstituted by dissolution in water.

Students should learn how to prepare media. Frequently, however, classes are too large to permit the participation of each student. If this is so, conduct the experiment as a demonstration. Students should be aware of the general procedures used to prepare media, the time required for their preparation, and their uses.

Prepare Agar Culture Media. Weigh-out dehydrated potato-dextrose agar from a commercial bottle and add it to distilled water according to the proportions indicated on the label. Use a flask which is twice the capacity of the volume of medium being prepared. Plug the flask with cotton, and sterilize in an autoclave at 15 psi for an amount of time in accordance with the volume of the medium.

milliliters	*minutes*
15	15
75	20
250	25
500	30
1000	35
2000	45

The agar is dissolved by heating during sterilization but remains settled on the bottom of the flask. After sterilization, allow the flask to cool and then gently swirl the contents to mix the agar. Do not shake or swirl immediately after removal from the autoclave. The contents are at the boiling point and a sudden movement may make the contents boil. Boiling media may push the plug out of the flask, run over the top, and burn your hand as well as become contaminated and not useful.

Prepare Slants. Prepare medium as described above but do not autoclave it. Melt the agar in a steamer (30 to 40 min. for 500 to 1,000 ml.). Pour the dissolved agar into a funnel apparatus (Fig. 3-3) for distribution to tubes. Do not allow the delivery tip to touch the sides of the test tube. Agar deposited on the sides causes the cotton plug to stick, making removal and replacement of the plug during manipulations difficult. Also, it allows the contaminants to grow down into the agar along the side and defile the contents. Test tubes are commonly filled with 6 to 8 ml. of medium. If the tubes are to be retained for prolonged periods, 10 to 12 ml. of medium may be used to delay dehydration. Practice filling tubes with distilled water.

After the medium is dispersed, plug the tubes with cotton and sterilize for 20 minutes in an autoclave at 15 psi. Following sterilization, place the tubes in a slanted position to harden. The slanted position provides a working surface.

Prepare Plates. Prepare 250 ml. of PDA in a 500-ml. flask. Plug with cotton, sterilize, and allow to cool. Pour plates at a temperature below boiling but above body temperature. (Some workers pour plates at boiling temperatures to reduce contamination. Boiling agar is dangerous for the beginner to handle.) To pour the plates, remove the plug by grasping it between the 4th and 5th fingers (or the 3d and 4th fingers). Grip the plug near the knuckles with the palm up. This places the plug above your hand during manipulations and out of the way of the table top and other un-

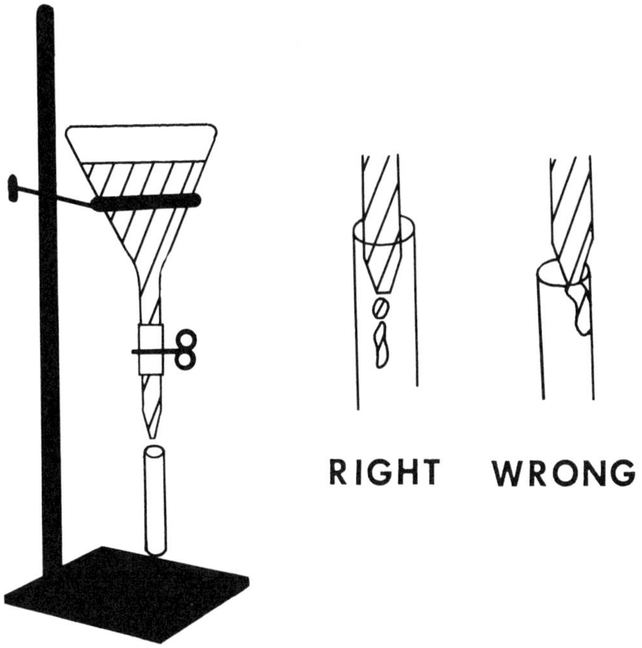

Fig. 3-3. Apparatus and procedure for pouring slants.

Fig. 3-4. Procedure for pouring plates.

sterile objects. Tilt the lid of the petri dish only enough to insert the mouth of the flask (Fig. 3-4) after flaming the mouth. Do not wave the petri dish lid in the air or slop agar from an excessive height. Practice with water.

Equipment

- 1 ringstand
- 1 ring clamp
- 1 pinch clamp
- 1 500-ml. flask with water and cotton plug
- 1 basket of clean test tubes
- 1 funnel with tubing
- a few prepared slants
- a few prepared plates

SUGGESTED READING

Difco Manual of Dehydrated Culture Media and Reagents. Detroit, Mich., Difco Laboratories, Inc., 1964.

Riker, A. J., and Riker, Regina S., *Introduction to Research on Plant Diseases.* Chicago, John S. Swift Co., Inc., 1936.

4

Bacterial Diseases

The leadership of the United States in the development of an understanding of bacteria as plant pathogens was launched in 1878 by Burrill of Illinois with his studies on fireblight of apple and pear. This was only two years after Koch's demonstration of the role of the anthrax bacilli in animal diseases. Each created a first.

The period of 1890 to 1900 was like a great political campaign—the years of the great debates between Erwin F. Smith of the United States Department of Agriculture and Alfred Fisher, a German botanist at the Königlichen Gymnasium at Leipzig. Smith presented evidence that bacteria were plant pathogens and Fisher, in 1897, challenged the reliability of the claim. Smith replied in 1899 that the eminent German botanist lacked knowledge on the subject, to which Fisher replied twice more in 1899. By 1900, however, the world accepted bacteria as plant pathogens.

The bacteria which cause plant diseases are, without exception, all nonspore-forming rods. In addition, all are gram negative, except for species in the genus *Corynebacterium*. The plant pathogenic bacteria may or may not be motile and the attachment of flagella is variable. Of approximately 1700 species of bacteria, about 160 are plant pathogens. That's nearly 10 per cent! (Refer to the Chapter on Fungi for a review of the binomial system of nomenclature.)

There are six kinds of symptoms that bacterial plant pathogens induce: soft rots, vascular wilts, galls, root nodules, leaf spots, and blights. Table 4-1 summarizes the characteristics of the bacteria that cause various kinds of plant diseases.

TABLE 4-1. General characteristics of the genera of bacteria that cause various kinds of plant diseases.

Genus	Disease	Color	Gram Stain	Motility	Number of Species
Agrobacterium	galls	white	negative	+	4
Corynebacterium	spots, blights, wilts	white	positive	—	10
Erwinia	spots, blights	white	negative	±	6
Pectobacterium	soft rots	white	negative	±	1
Pseudomonas	wilts, spots, blights	white, fluorescent	negative	+	85
Rhizobium	root nodules	white	negative	±	6
Xanthomonas	wilts, spots blights	yellow	negative	+	47

Plant pathogenic bacteria overwinter in hibernating insects, perennial cankers, infected plant debris, infested materials, and in the soil as colonizers. They are spread by rain, insects, planting stock, infected or contaminated seed, contaminated implements, and infested soil clinging to the roots of transplants and nursery stock. Wind is relatively unimportant in the spread of plant pathogenic bacteria compared to these other means and compared to the effectiveness of wind in the dissemination of fungus spores.

Bacterial plant diseases are controlled by sanitation, insect control, resistant crop cultivars, nursery plant inspection, crop rotation, and seed certification.

GENERAL REFERENCES

Breed, R. S., Murray, E. G. D., and Smith, N. R., *Bergey's Manual of Determinative Bacteriology*, 7th ed. Baltimore, Williams and Wilkins, 1957.

Dowson, W. J., *Plant Diseases Due to Bacteria.* London, Cambridge, 1957.

Elliott, Charlotte, *Manual of Bacterial Plant Pathogens,* 2d ed. Waltham, Massachusetts, Chronica Botanica Co., 1951.

Stapp, C., *Bacterial Plant Pathogens,* trans. by A. Schoenfeld. London, Oxford, 1961.

Thimann, K. V., *The Life of Bacteria.* New York, Macmillan, 1955.

5

Bacterial Soft Rot

Bacterial soft rots are best known as diseases of fleshy plant structures such as potato tubers, fruits, and flower bulbs. They are also diseases of stems and flowers and other above-ground parts. Perhaps no other group of related diseases causes more damage to vegetables. Soft rots attack a large number of plants. They strike in the fields, in storehouses, in the marketplaces, and in homes throughout the world.

The most striking symptoms of soft rot are mushy texture (Fig. 5-1B) and the dramatic olfactory stimulus—it stinks! When the bacteria cause decay of potato seed pieces, the shoots become streaked with black and die (Fig. 5-1A). This syndrome, or set of symptoms, is called blackleg.

Several closely related forms of bacteria cause soft rot. The first described were *Bacillus carotovorus* from carrot in 1901 by L. R. Jones of Vermont, *Bacillus atrosepticus* on potato by van Hall in 1902 in Holland, and *Bacillus aroideae* on potato by Appel in 1902 in Germany. Still other forms have been described and all transferred to the genus *Erwina* where some authors recognize them today. Speciation is in a great state of confusion. Graham has reviewed this problem and logically grouped the various forms into subspecies of *Pectobacterium carotovorus* on the basis of eight biochemical reactions and pathogenicity to potato stems at high and low temperatures. These can be collectively referred to as *Pectobacterium carotovorus.* For our purposes, all can be so named, but recognizing that variations occur in the biochemical reactions and the pathogenicity.

Disease Cycle. Bacteria which cause soft rot usually enter the plant body through wounds and multiply, initially, in the intercellular spaces (Fig. 5-2A). Freezing injuries, harvest bruises, and insect punctures are frequently the open doors of entry. The bacteria can never penetrate the intact epidermis nor, ordinarily, the natural openings. However, in moist storage chambers or in frequently wet soils, uninjured roots and storage organs may be infected through lenticels. In any event, very high humidity (probably greater than 90%) is required for successful and continued progress in the early stages of the disease.

Once bacteria have gained entrance and temperatures are warm, moisture is high, multiplication in the intercellular spaces begins (Fig. 5-2B). (Note that bacteria divide and double their numbers as rapidly as every 30 minutes!) The bacteria are never able to penetrate the living cells but produce pectolytic enzymes which diffuse in advance of the bacteria and dissolve the middle lamella. This substance, composed of pectates, cements the cells together. Destruction of this cementing substance brings about a separation of cells (Fig. 5-2C, 5-3D). Once deprived of continuity, and possibly poisoned by the pectolytic enzymes and other metabolic products of the bacteria and the dying plant cells, the cells die and disintegrate. The dead and disintegrating plant cells provide nourishment for the ever-growing hordes of bacteria. A teaspoonful of rotten potato may contain millions of bacteria. Although highly pathogenic and capable of causing death and destruction, these bacteria are not very good parasites, but are saprobes living off the refuse of the dead cells.

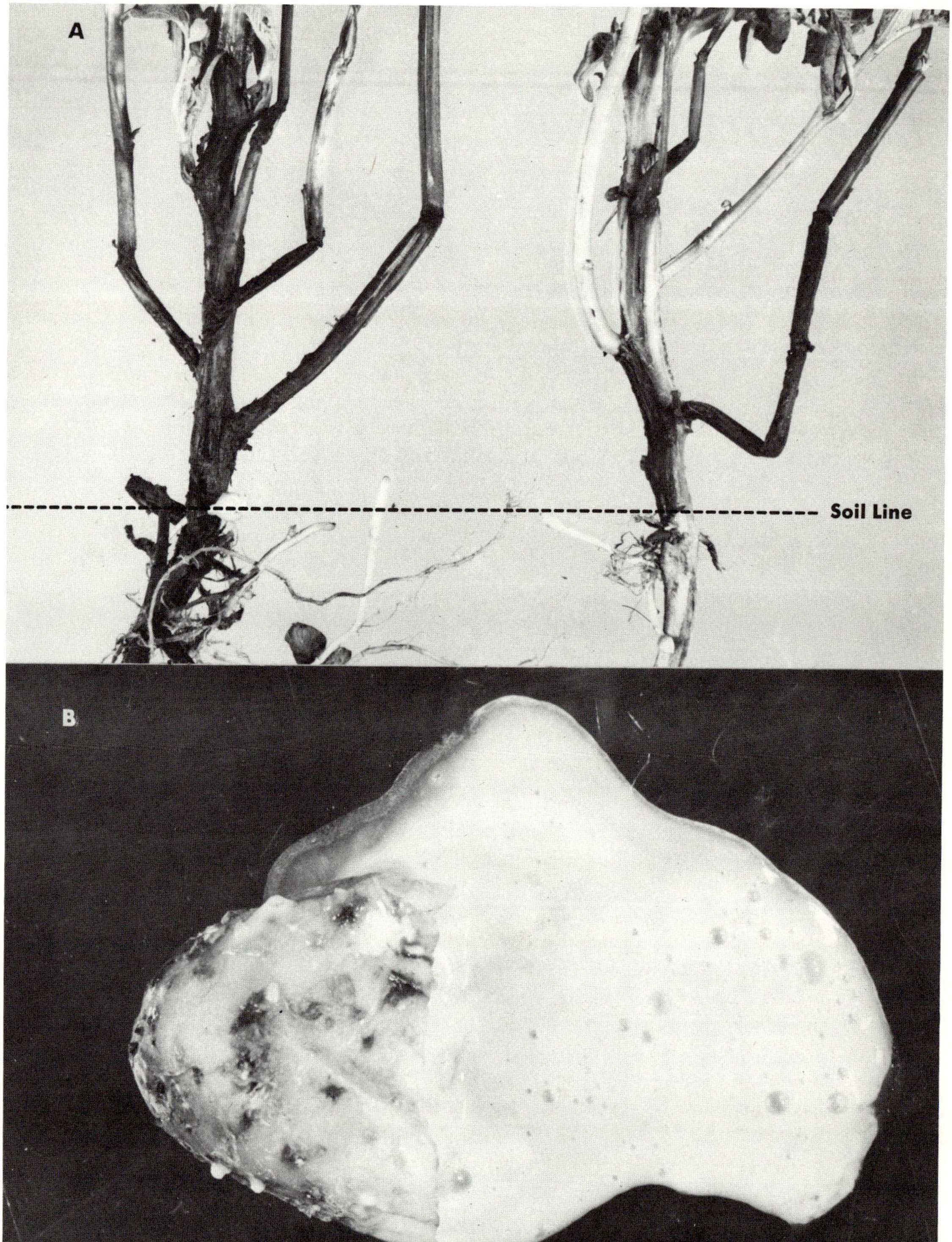

Fig. 5-1. (A) Symptoms of blackleg of potato. Note the severe black streaking of the infected tissues. (B) Symptoms of soft rot of a potato tuber. Note the bubbly, watery mass that has flowed from the cut tuber onto the glass surface supporting it. (Viewed from the top.)

Several species of maggot fly have special relationships to the soft rot diseases. The cabbage maggot spreads the causal bacteria among members of the cabbage family and the seed-corn maggot among potatoes, celery, and other crops. Adult flies are attracted to the rotten, smelly plant debris and soon their legs and mouthparts are smeared from the orgy on this preferred food. The bacteria are ingested and can be found in their alimentary system. Here they survive in the midintestine and the cast-off linings of the foregut and hindgut. The eggs are contaminated during oviposition and, subsequently, the bacteria can be found in the internal portions of the insect during larval and pupal stages. Various saprobic bacteria are essential for normal development of the maggot flies. Although there have been claims that the soft rot bacteria are essential for normal development of the maggot flies, pure cultures of the soft rot bacteria have not permitted full development of insects free of all other microbes.

Soft rot bacteria are spread also by other insects, tools, contact with contaminated storage containers, hands, rain, and running water. They overwinter in decayed refuse in the soil. No information is available as to the longevity of bacteria in soil which is free of decaying plant parts.

Control: avoid wounds during cultivation, harvesting, and handling. Vegetables that are washed, such as potatoes, should be rinsed with clean water and dried rapidly. Susceptible materials should be stored under properly refrigerated conditions.

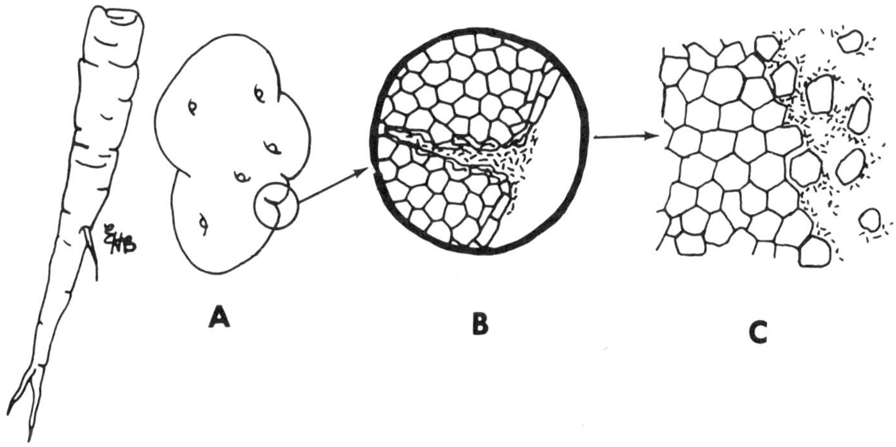

Fig. 5-2. Sequence of events in the soft rot of vegetables. (A) Penetration of wounds; (B) multiplication of bacteria in the wound; and, (C) separation of cells by enzymatic action of the bacteria.

DISCUSSION QUESTIONS

1. How do soft rot symptoms differ from those of blackleg of potato?
2. How conclusive are the signs and symptoms of the soft rot disease in the determination of the cause of the disease?

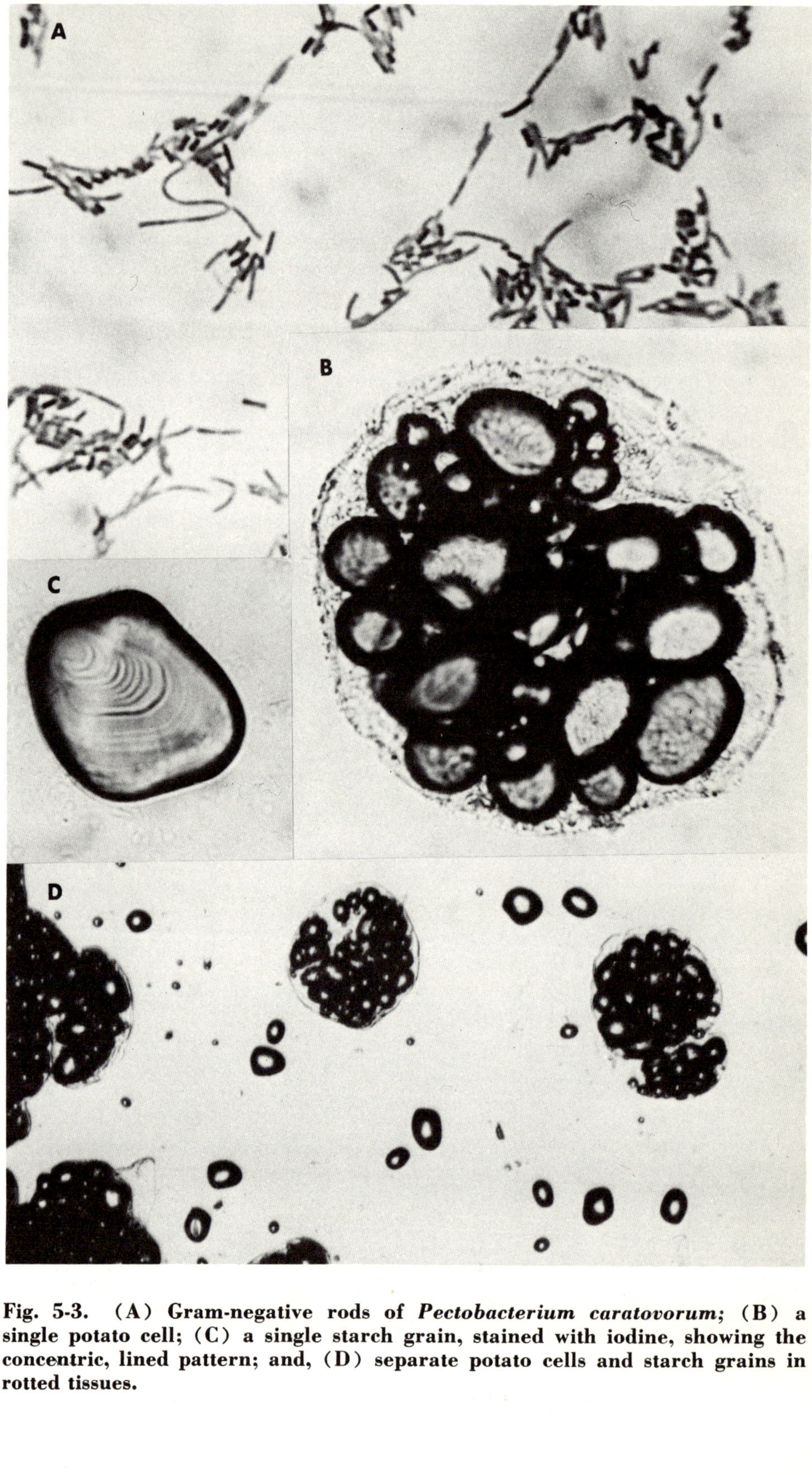

Fig. 5-3. (A) Gram-negative rods of *Pectobacterium caratovorum;* (B) a single potato cell; (C) a single starch grain, stained with iodine, showing the concentric, lined pattern; and, (D) separate potato cells and starch grains in rotted tissues.

3. Why is a knowledge of the method of penetration and conditions necessary for infection important in understanding control measures for soft rot diseases?

4. What is the Gram stain? What is its significance?

Experiment III

The Germ Theory of Disease

One of the important precepts in biology is the Germ Theory of Disease which developed from a recognition of the infectious nature of disease. This concept is important in understanding the cause, dissemination, and control of diseases.

EXPERIMENT

Prove to yourself that disease is contagious: that rotten things make similar, neighboring objects rotten. Inoculate healthy potatoes and carrots by inserting mushy, diseased tissues from rotten potatoes, carrots, or lettuce into holes made by removing plugs about ¼ inch in diameter and ½ inch long. Add 3 or 4 drops of sterile water to each inoculated hole. Cut off ⅛ inch of the tip of the plug, replace it, and seal the edges of the plug with vaseline. Place the inoculated materials and a wet paper towel in a plastic bag and secure firmly with a rubber band or a wire Twistem.

Expendable Supplies × Total Number of Students or Groups

1 healthy potato
1 healthy carrot
1 plastic bag
1 rubber band or Twistem

Nonexpendable Supplies × Number of Students or Groups Seated in a Laboratory Section

1 dropper bottle of sterile water

General Supplies Available

jar of vaseline
soft rot inoculum (for inoculum it is convenient to place 1 cup of rotten vegetables in a finger bowl with ½ cup water. Place in a warm location for a few days before needed in class—use a cover!)

DISCUSSION QUESTIONS

1. What is the Germ Theory of Disease? Who were the important contributors to its development and demonstration? When?

2. What is the significance of this theory in the diagnosis of diseases? In controlling diseases? In the spread of diseases?

3. The plastic bags, if tightly sealed, expand like balloons as the vegetables rot inside. Why?

SUGGESTED READING

Dubos, R., "Second Thoughts on the Germ Theory." *Scientific American*, Vol. 192, (May, 1955), pp. 31–35.

Graham, D. C., "Taxonomy of the Soft Rot Coliform Bacteria." Vol. 2, *Annual Review of Plant Pathology*, 1964.

Singer, C., *A History of Biology*. New York, Henry Schuman, Inc., 1951.

Experiment IV

Examination of Diseased Tissues

Two important phases in the diagnosis of diseases are (1) examination for signs and symptoms and (2) demonstration of causality. Symptoms of diseases are responses which the host exhibits. These include abnormal color, size, and shape. Plants or plant parts may be malformed or bear overgrowths. Signs of diseases are manifestations of the causal agent—structures (bacterial cells, fungus fruiting structures and spores), deposits (insect frass) or direct effects (insect galleries, insect punctures, and boreholes in cell walls from wood decay fungi).

EXPERIMENT

1. Determine the symptoms of soft rot. Use your senses—look, touch, sniff, and taste if you like!

2. Observe bacteria, a sign, in diseased tissues by staining.

3. Compare stained bacteria from diseased tissues with those from a pure culture.

Procedure for Gram's Stain

(a) Clean a microscope slide thoroughly, flame to remove traces of grease, and allow to cool.

(b) Macerate a very small amount of diseased tissue or scrapings from a culture of bacteria in a drop of distilled water on the flamed side of the microscope slide. A quantity the size of ¼ the center of this "o" is still 5 times too much material to use!

(c) Heat gently to fix the bacteria to the slide. From time to time touch the slide with the back of your hand. Heat the slide no more than your hand can tolerate.
(d) Flood with equal parts of solutions of gentian violet and sodium bicarbonate for 10 seconds by adding approximately equal amounts, one after the other. Drain.
(e) Flood with a solution of iodine for 10 seconds. Rinse with distilled water. Watch closely for purple streaks to run off when the decolorizer is added, one drop at a time. At the very moment that the streaks stop coming, rinse with distilled water. This is the most critical step in the Gram stain procedure.
(f) Decolor with a mixture of acetone and alcohol. Hold the slide at a slant and watch closely for purple streaks to run off when the decolorizer is added, one drop at a time. At the very moment that the streaks stop coming, rinse with distilled water. This is the most critical step in the Gram stain procedure.
(g) Flood with a solution of basic fuchsin for not longer than 2 to 3 seconds. Rinse with distilled water.
(h) Allow to air dry—add a drop of immersion oil directly to the slide for microscope observation. Be sure that the bacteria are on the upper surface of the slide or you will ram the objective lens against the slide in an attempt to focus on the underside—which is out of focus for the oil immersion lens.

4. *Observe* the effects of the disease by microscope observations of free-hand sections through an area where firm tissue and disintegrating tissues meet. Separate, but apparently healthy cells, as well as disintegrating cells and bacteria can be seen (Fig. 5-3D). Potatoes are excellent material for this. Be sure to distinguish starch grains (Fig. 5-3C) from potato cells (Fig. 5-3B). Stain with iodine and the starch grains, whether alone or within cells, turn purple to black and have concentric lined patterns. The plants cells have a distinguishable cell wall encompassing the starch grains and the protoplasm.

Expendable Supplies × Total Number of Students or Groups

soft rotted carrots and potatoes from Experiment III

Nonexpendable Supplies × Number of Students or Groups Seated in a Laboratory Section

1 staining pan (can be made with a bread pan and ¼ inch hardware cloth for slide support)
1 250-ml. flask of distilled water for rinsing
1 dropper bottle of each Gram stain reagent
 gentian violet (2.0 g. in 1000 ml. distilled water)
 sodium bicarbonate (12.5 g. in 1000 ml. distilled water)
 iodine (20.0 g. in 900 ml. distilled water plus 10 ml. 1 *M* sodium hydroxide)
 decolorizer (150 ml. acetone and 750 ml. ethyl alcohol)
 basic fuchsin (100 ml. saturated alcoholic solution plus 900 ml. distilled water)

Experiment V

Isolation of Bacteria

The isolation of causal organisms is necessary in order to study them. Sometimes we speak of isolating diseases when we actually mean the causal organisms. Isolation of a causal organism is the placement of the organism in isolation from all other living organisms, including the suscept. This is called a pure culture.

Two reliable methods of isolating bacteria are the dilution plate (Fig. 5-4) and the streak plate (Fig. 5-5) techniques. Successful isolation of pure cultures depends especially on the capacity of the technique to separate the bacterial cells from each other and from propagules of other microorganisms so that the colony which each bacterial cell forms is separate and apart from any other. The proficiency of the operator is important.

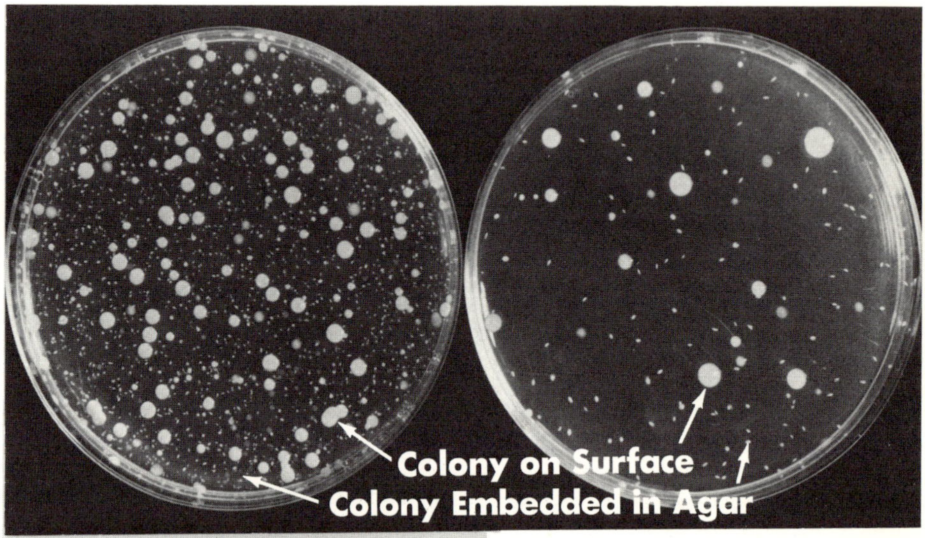

Fig. 5-4. Dilution plates of *Xanthomonas phaseoli* var. *sojensis*.

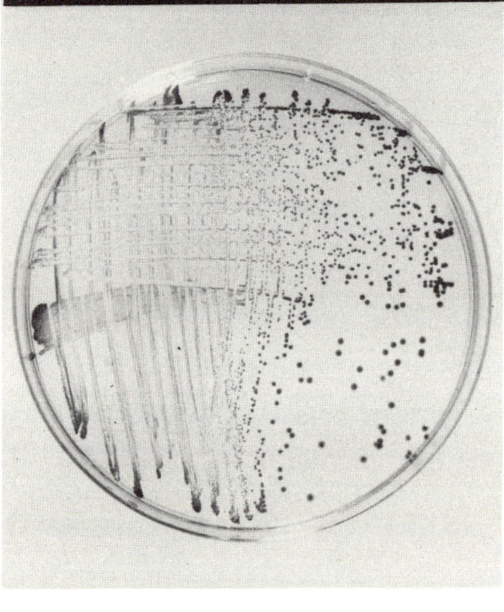

Fig. 5-5. Streak plate of *Serratia marcescens*.

EXPERIMENT

1. **The Dilution Plate Method.** (This portion of the experiment can be combined with Experiment VII if desired.)

(a) Place a small amount of crushed diseased tissue in sterile 1% sodium chloride. The sample may be weighed if one wishes to calculate the number of bacteria per unit of weight. (See Experiment VII.)

(b) Make ten-fold dilutions through 10^{-10}. This is accomplished easily by using a series of test tubes with 9 ml. of sterile 1% sodium chloride solution in each. Sodium chloride solutions are used because some bacteria are rapidly killed in distilled water. Utilizing a sterile pipette, transfer one ml. of the mixture of crushed diseased tissue and sodium chloride to a test tube containing 9 ml. of sterile 1% sodium chloride solution and mix. One ml. from this tube is transferred with another sterile pipette to the next tube of 9 ml., and so forth. (Use the index finger and not the thumb to control flow.)

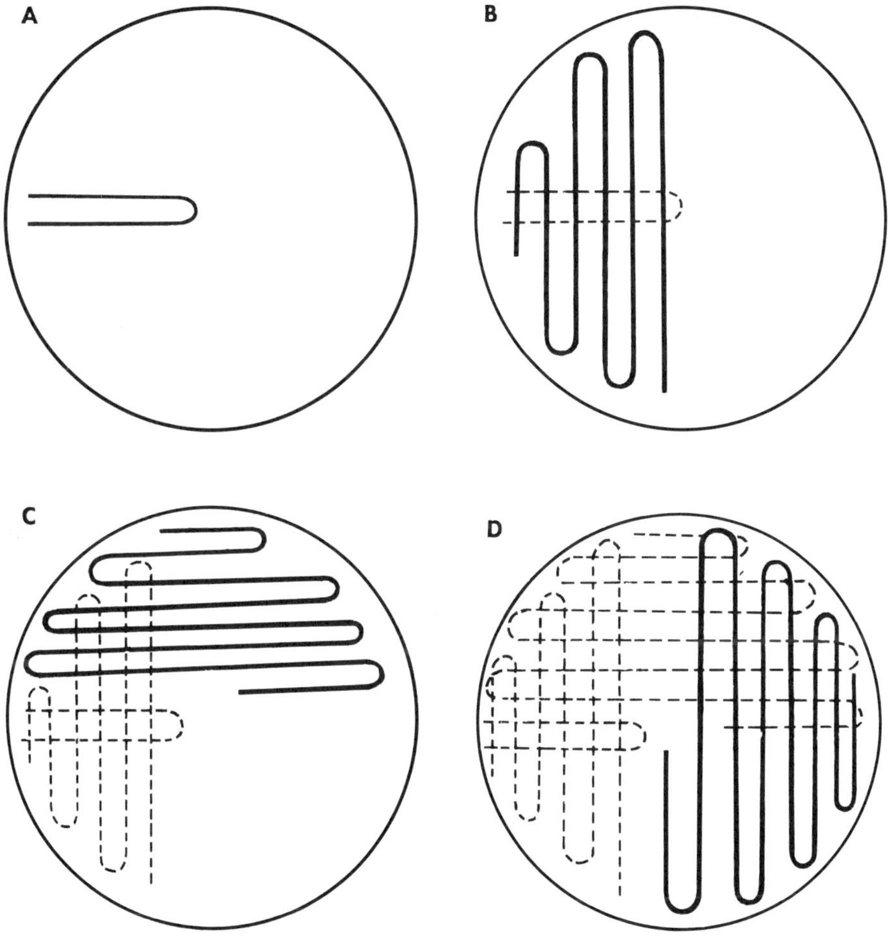

Fig. 5-6. Execution of a streak plate. The solid lines indicate the new streaks to be made with a sterile loop at each step. The broken lines indicate the location of streaks previously made.

(c) Deposit 1 ml. of each of the 3 highest dilutions in separate sterile petri dishes. Add sufficient nutrient agar (about 20 ml.), which has been cooled to a temperature which feels slightly warm to the back of the hand (42–45° C.), to cover the bottom of the petri dish. A proper amount is roughly measured if pouring is stopped when ¾ of the bottom of the petri dish is covered. With the dishes flat on the table, swirl them *gently* to mix the bacteria and agar. Do not spill the agar over the edge of the petri dish since contaminants in the air can grow up the nutrient agar and into the dish. Each bacterial cell, or clump of cells, will produce a colony (Fig. 5-4). The most accurate estimates of the concentration of bacteria in a sample are obtained with dilutions which give rise to 50 to 200 colonies per plate.

2. **The Streak Plate Method.** Insert a sterile loop into diseased tissue and streak lightly across a potato-dextrose agar plate containing 2½% agar (Fig. 5-6A). The extra agar makes the plates firm and easy to streak without gouging into the agar with the loop. This first streak should be less than halfway across the plate. Sterilize the loop and streak back and forth at right angles to the first streak. Be sure the loop is allowed to cool after sterilizing. Half of the plate should now be covered with streaks (Fig. 5-6B). Do no streak repeatedly in one area but make a zig–zag line. Turn the plate 90° and repeat with a sterilized loop (Fig. 5-6C). Be sure not to touch the first streak, only the second zig–zag streaks. Turn the plate another 90° and streak again with a sterilized loop, touching only the last streaks made (Fig. 5-6D). The bacteria are diluted by the streaking (Fig. 5-5).

DISCUSSION QUESTIONS

1. Of what value are these techniques? What can be accomplished with them?
2. Why must the agar be cooled before use in the dilution plate method?
3. Why is 1% sodium chloride and not distilled water used for making the dilutions?
4. Can you think of other ways of isolating pure cultures of bacteria from diseased tissues? (There are some!)
5. Which techniques are most suitable for isolation of the causal organisms of soft rot, crown gall, cucumber wilt, and bacterial root nodules?
6. Why are the colonies of bacteria which are embedded in the agar in dilution plates lens-shaped?

Experiment VI

Koch's Postulates

Typically, if bacteria are cultured from rotted tissue, several kinds are found. Which is the cause of the disease and which are secondary invaders subsisting on the refuse of the diseased tissues? If isolations were made from several rotted specimens, only those bacteria which occurred each time, or nearly each time, would be most suspected of being the causal organism. In other words, the organism which causes the disease should be associated with it. This is the *first* rule of Koch's Postulates. The other three follow logically. *Second:* Isolate the suspected causal organism in pure culture. *Third:* Reproduce the same symptoms on inoculation of the suscept with the pure culture. *Fourth:* Re-isolate the same organism from the experimental inoculations.

EXPERIMENT

Follow Koch's Postulates using soft rot of potato and carrot. (Crown gall is another excellent disease for this exercise.) This entails 6 phases of activities spread over as many or more laboratory periods. Keep up-to-date and accurate notes as you proceed or you will become confused. The instructor may choose to make the activities of Experiment III, IV, and V part of this comprehensive exercise in the fulfillment of Koch's Postulates, or he may choose to repeat them for practice. The progress of activity is outlined in phases with Koch's Postulates indicated where appropriate.

First Koch Postulate: Associate organism with disease symptoms.

Phase I. Examination of Diseased Tissues (Experiment IV).

Second Koch Postulate: Isolate the suspected causal organism in pure culture.

Phase II. Isolation of Bacteria (Experiment V).

Phase III. Transfer isolated bacteria on streak plates to PDA slants to establish pure cultures. Transfer only half of each different colony and make Gram stains of the remaining half of each colony. Examine for cocci and spore-forming rods. We know that these should not induce soft rot and they provide a good check on techniques if we inoculate potatoes with them. As an additional control, include an uninoculated potato.

Third Koch Postulate: Reproduce the same symptoms as first observed upon inoculation with pure culture. Slices of carrots and potatoes can be used in plastic petri dishes with wet paper towels rather than whole vegetables.

Phase IV. Inoculate potatoes and carrots (Experiment III).

Fourth Koch Postulate: Re-isolate the same organism in pure culture.

Phase V. Isolate bacteria (Experiment V).

Phase VI. Transfer bacteria to PDA slants for pure cultures. Gram stain and determine if they have the same characteristics in reaction to the Gram stain and shape as the first ones isolated and used for inoculum.

DISCUSSION QUESTIONS

1. Why are all four steps in Koch's Postulates necessary?
2. How can proof of causality be demonstrated with pathogens which are obligate parasites? With nutritional deficiencies? With viruses?

3. Why does the first rule of Koch's Postulates refer to bacteria which occur each or *nearly* each time as the suspected pathogens? Should only those that occur each time without fail be considered and not the others?

4. If the uninoculated potato developed soft rot during the experiment, what would you conclude? How would you verify your conclusions?

SUGGESTED READING

Keitt, G. W., "History of Plant Pathology" in J. G. Horsfall and A. E. Dimond, eds., *Plant Pathology*, Vol. I (New York, 1959), pp. 67–70.

Experiment VII

Inoculum Potential: An Epidemiological Factor

Bacteria multiply by fission and do so rapidly. They often double their numbers every 30 minutes. When rotted potatoes disintegrate, hordes of bacteria are released and may cause many new infections. Clearly, the number of new infections is related to the number of units of inoculum available. The concentration of infectious propagules in a given sample is a simple way of expressing the inoculum potential for soft rot.

Disease potential is a product of inoculum potential, suscept potential, vector potential, and environment. The numbers of suscepts, their distribution, and degree of susceptibility is the suscept potential. The vector potential is the number, efficiency, and movement of the vectors. Disease potential is the probability that infection will occur and could be expressed as the number of suscepts that become infected in a unit of area in a unit of time, under given environmental conditions.

EXPERIMENT

Determine the number of viable bacteria in a gram of rotted potato tissue by the dilution plate technique. This can be a separate exercise or a part of Experiment V.

1. One student in each group should weigh-out 1 g. of healthy potato tissue in a sterile aluminum weighing dish or beaker. Another student should weigh-out 1 g. of diseased tissue. The healthy tissues are not likely to be contaminated by handling if different students handle the diseased materials.

2. The two student preparators then macerate their specimens separately in a sterilized blender cup containing 99 ml. of 1% solution sodium chloride.

3. Transfer 1 ml. of the suspension of healthy tissues in each of 3 sterile petri dishes. Use a sterile pipette and sterile techniques. Each ml. has 1/100 of the original material from the healthy tissue. This is a dilution of 10^{-2}. Label the petri dishes during the experiment.

4. Make 8 ten-fold dilutions of the suspension of rotted tissues. The eighth dilu-

tion will be a total dilution of 10^{-10} because the suspension originally was diluted 10^{-2} in the 99 ml. of solution of 1% sodium chloride in the blender. Place 1 ml. of the 10^{-10} dilution in each of 3 sterile petri dishes. Label the petri dishes as they are used.

5. Place the dirty pipettes, soon after their use, with the delivery tip up, in a pipette soaking jar provided. If permitted to dry they are more difficult to clean.

6. Pour about 20 ml. of sterile, melted agar cooled to a temperature that feels slightly warm to the hand.

Note: The total number of viable bacteria varies with the age of rotted tissue. Highest counts occur about the time that full maceration develops and becomes progressively lower thereafter. Why?

Expendable Supplies × Total Number of Students or Groups

rotten potatoes from previous experiments
8 tubes of sterile solution of sodium chloride, 9 ml.
6 sterile petri dishes (plastic ones are throw-aways)
1 flask of 180 ml. of sterile PDA
9 sterile 1-ml. pipettes

Nonexpendable Supplies × Number of Students or Groups Seated in a Laboratory Section

1 test-tube rack

Equipment and Supplies Available

healthy potatoes
2 sterile blender cups (steam between classes)
aluminum weighing dishes or sterile beakers
1 blender
1 pipette soaking jar
1 balance

DISCUSSION QUESTIONS

1. The number of colonies in three plates with one milliliter each from the same suspension will differ. Why?

2. Why does the number of viable bacteria per gram of rotted potato vary with age of the diseased specimen?

3. Why is an understanding of inoculum potential important in understanding plant disease control?

4. If the number of bacteria in one gram of rotted potato could be spread three feet in radius and to a depth of six inches uniformly, how many bacteria would be in one cubic inch of soil? (Obviously this is an artificial distribution.) What should be a normal distribution of bacteria from a single rotted potato in the field?

5. If the bacteria in one gram of rotted potato were spread out so that there was one bacterial cell per square inch, how large an area would be covered?

SUGGESTED READING

Garrett, S. D., "Inoculum potential," in J. G. Horsfall and A. E. Dimond, eds., *Plant Pathology,* Vol. III (New York, 1960), pp. 23–53; especially pp. 28–37.

Experiment VIII

Pathogenesis: Tissue Maceration by Pectic Enzymes

The most striking symptoms of soft rot are, as we have mentioned before, the soft mushy texture and the dramatic olfactory stimulus. A step by step account of the processes leading to these morbid conditions is pathogenesis, the sequential development of disease. Pathogenesis is the origin (*genesis*) of suffering (*pathos*).

Pectobacterium carotovorus, which causes soft rot, enters primarily through wounds and multiplies in the intercellular spaces. The bacteria elaborate various kinds of pectolytic enzymes which diffuse in advance of the bacteria and dissolve the calcium pectates of the middle lamella which cement the cells together. The cells separate and die because of the loss of tissue integrity and perhaps because of toxicity of other metabolic products of the bacteria and the dying host cells. Although soft rot bacteria produce unpleasant gases when they grow, the stench of rotten vegetables is due often to the sulfurous odors produced by secondary tramp bacteria.

EXPERIMENT

Investigate the action of pectolytic enzymes by measuring the change in viscosity of a solution containing pectin and pectolytic enzymes. The viscosity of such a mixture decreases with time because the enzymes reduce the long polymerized molecules of pectin to shorter polymers of polypectates and still shorter and smaller molecules of pectic acid. The reduction in viscosity can be measured by determining the rate of flow of the mixture through a small tube.

1. Connect a piece of capillary tubing (0.5 mm. bore and 4 in. long) to the delivery tip of a 10-ml. pipette. Suspend the pipette, right side up and vertically, from a ring stand with a ring clamp. Place a slit piece of heavy-walled rubber tubing around the pipette where the clamp secures it. Connect a sterile piece of rubber tubing (10 in. long) to the top of the pipette.

2. Mix 1 ml. of 0.5% pectolytic enzyme solution with 10 ml. of 1% pectin solution.

3. Quickly draw the mixed solutions up the pipette to the 3 or 4 ml. mark of the pipette by suction on the end of the tubing at the top. You can use either a suction device or the mouth. Permit the solution to flow freely through the apparatus and drip into a 50-ml. beaker.

4. Record the rate of flow of the solution through the capillary tubing in ml./min. After securing a reading, force the solution through the pipette by applying pressure on the tubing at the top.

5. Repeat measurements of the rate of flow at 5, 10, 15, 20, and 40 minutes after mixing the pectin and pectolytic enzyme solutions. Be sure to take readings of rate

of flow each time from the same starting point in the 10-ml. pipette. Why is this necessary?

6. Plot your data on the graph provided in the manual. Plot the following typical data: 1 min. after mixing, 0.7 ml./min.; 5 min. after mixing, 1.25 ml./min.; 10 min. after mixing, 1.5 ml./min.; 15 min. after mixing, 1.6 ml./min.; and, 28 min. after mixing, 1.7 ml./min.

7. Rinse the beaker, add 1 ml. of pectolytic enzyme solution (0.5%), and heat until nearly boiling. Repeat the above experiment using the heated pectolytic enzyme solution after it has been filtered.

8. The instructor may elect to have the experiment repeated using culture filtrates of *Pectobacterium carotovorum* as a souce of pectolytic enzymes.

Expendable Supplies × Total Number of Students or Groups

2 ml. of 0.5% pectolytic enzyme solution, filtered
20 ml. of 1% pectin solution, filtered

Nonexpendable Supplies × Number of Students or Groups Seated in a Laboratory Section

1 ringstand
1 ring clamp
1 10 ml. pipette
1 piece of rubber tubing (1 in. long)
1 piece of heavy-walled rubber tubing, slit (1 in. long)
1 piece of capillary tubing, 0.5mm. bore, 4 in. long
1 50-ml. beaker

DISCUSSION QUESTIONS

1. Are secondary parasites pathogens?
2. From where does all the water come that makes the soft rotted potatoes so slimy?
3. Would the rate of flow through the apparatus increase or decrease if the capillary tube were longer? Larger in diameter?
4. Would the rate of flow through the apparatus increase or decrease if the concentration of pectolytic enzyme were 0.25%?
5. Would the rate of flow through the apparatus increase or decrease if the temperature were 15° C. rather than room temperature?

SUGGESTED READING

Brown, W., "Toxins and Cell-wall Dissolving Enzymes in Relation to Plant Disease." *Annual Review of Phytopathology*, New York, 1965, pp. 1–18.

Husain, A., and Kelman, A., "Tissue is Disintegrated," in Horsfall, J. G., and A. E. Dimond, eds., *Plant Pathology*, Vol. I (New York, 1959), pp. 149–152; 159–164.

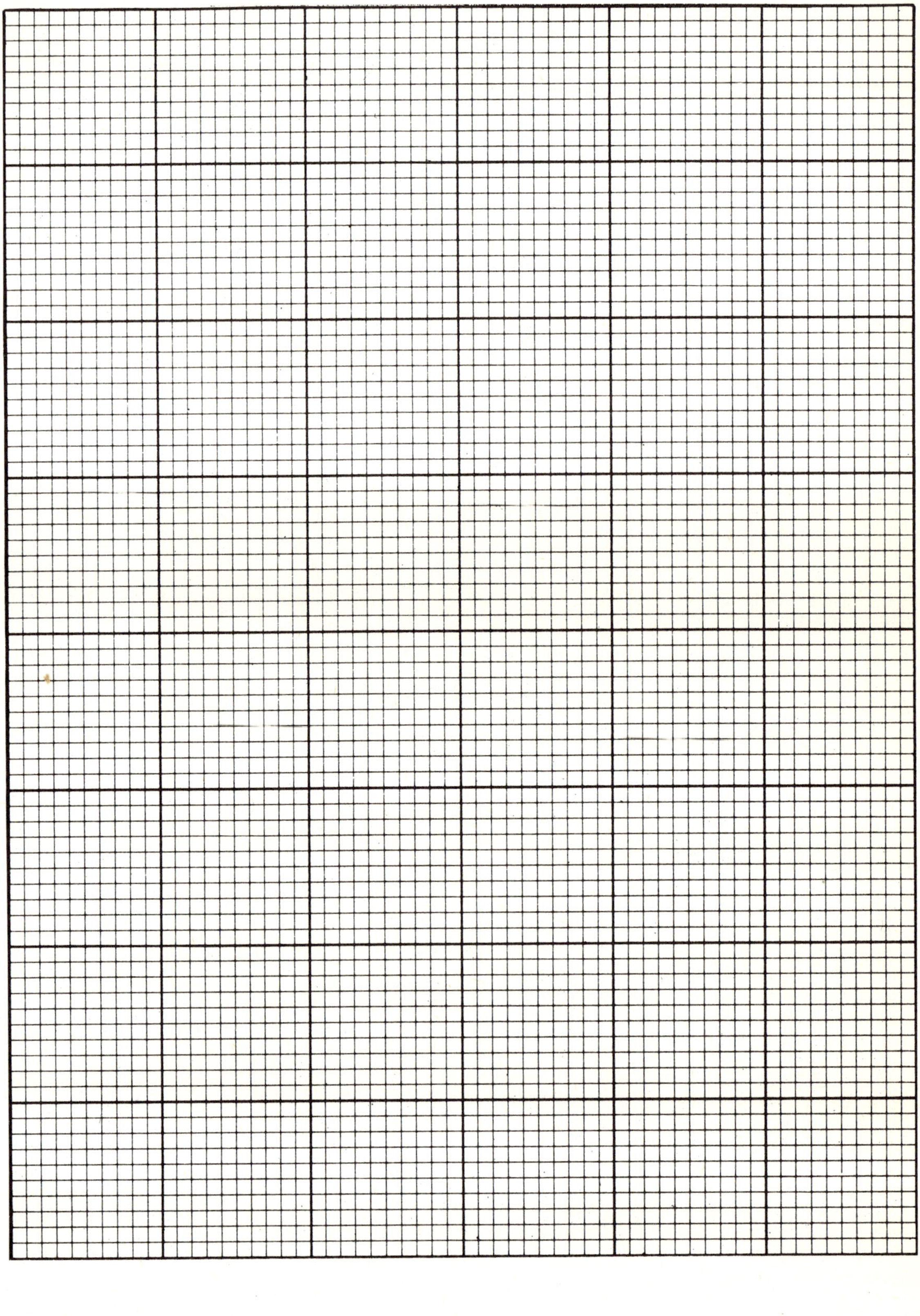

Experiment IX

Dispersal by Rain

Rain is important in the epidemiological development of plant diseases because it provides moisture for successful infection and for production, liberation, and dispersal of propagules. It is one of the most important agents of dissemination.

Rain spreads small objects by carrying them, in dripping and running water, from high surfaces to lower ones. The effectiveness of splashing, however, is usually unrecognized. To understand the signifiance of splashing it is necessary to examine what happens when a drop strikes the surface of a body of water (Fig. 5-7).

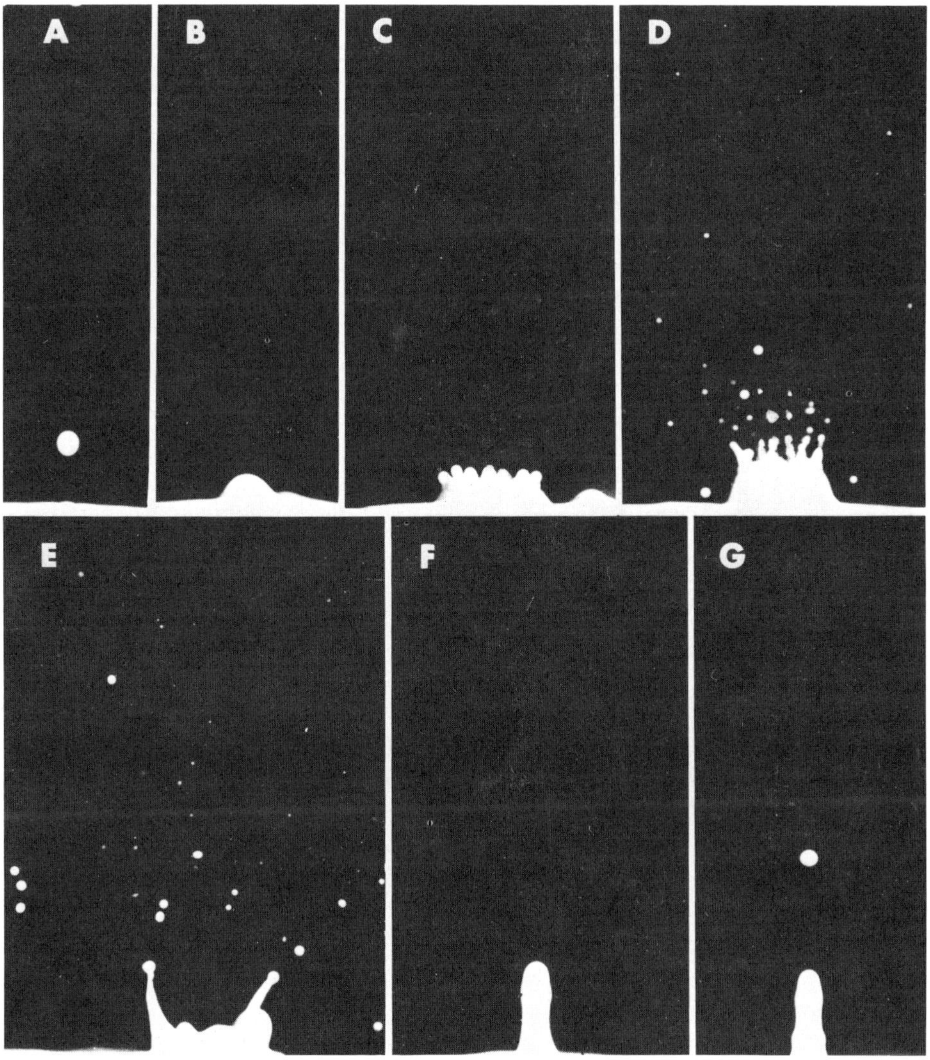

Fig. 5-7. Sequence of events when a falling drop of milk strikes a body of milk. (A) Double exposure of a falling drop of milk; (B) surface of the milk rising after the drop has struck; (C) Formation of a crown; (D) Small droplets flying off the crown after formation by action of surface tension; (E) Collapse of the crown; (F) Rise of a central column in equalization of the pressure from the collapsing crown and indirectly from the pressure of the atmosphere on the surface of the milk; and, (G) Small drop flying off the top of the central column from action of surface tension.

When a drop (A) strikes the surface of a liquid, a crater is formed. The sides of the crater rise above the surface of the liquid and create a crown (B, C) 0.003 seconds after impact. Small jets, shot out from the crown, break up into numerous spray droplets which fly off into the air (D) 0.006 seconds after impact. The spray droplets may number 50 or more. The upward surge of the crown diminishes and the crown collapses (E) 0.014 seconds after impact. A center column (F), called the Rayleigh jet, rises in equalization to the pressure from the falling crown and the upward-directed pressure applied indirectly by the pressure of the atmosphere on the surface of the liquid (0.04 seconds after impact). A few large drops may be squeezed off the top of the column (0.07 seconds after impact), and they continue upward for some distance in their momentum (G).

The formation of the drops from the terminal end of the Rayleigh jet is due to amplification of an unstable wave on the surface of the jet. The spray droplets that are thrown from the crown are similarly formed by the amplification of unstable waves propagated in 2 directions—vertically and along the circumference of the crown. Duration of the crown may be too short for a break-away of the spray droplets if the force of the falling drop is too little. On the other hand, if the force of the falling drop is too great, the mouth of the crater closes to form a bubble and the splashing process is modified.

Watch striking drops closely. What happens can be seen with the unaided eye if observations are repeated a number of times with concentration on the early and late stages of the crown and column formations.

SUGGESTED READING

Hobbs, P. V., and Kezweeny, A. J., "Splashing of a Water Drop." *Science*, Vol. 155 (March, 1967), pp. 1112–1114.

EXPERIMENT

Suspend a separatory funnel by the neck 10 or 12 feet above the floor. Tie 2 additional strings to the stem of the funnel and secure them in 2 different directions with enough tension to slant the separatory funnel from the perpendicular. This gives the funnel 3-point suspension which prohibits swinging. This is important so that the falling drops strike repeatedly in one spot. Sometimes the drops from funnels drip off at angles from the tip and do not strike in one spot. If this happens, add a short length of rubber tubing with a drawn piece of glass tubing. Place a screw clamp on the rubber tubing and use it to control the flow.

1. Place a large sheet of white paper, or even brown wrapping paper, on the floor beneath the separatory funnel. Set a large container below it to catch the drips. Adjust the stopcock (or clamp) so that 1 drop falls about every 2 or 3 seconds. The 2 to 3 second delay permits the surface of the water which the drops strike to come to rest between drops. Remove the container and allow one drop to strike the paper. This marks the bull's eye. Place a watch glass on the paper over the bull's eye and fill nearly full with a dark blue or red dye. Allow several drops to fall into the dye and then replace the catching container to keep the drops from striking the dye. Observe the pattern of splashing: (1) There are both large and small drops. (2) Only very small drops are at the greatest distances. (3) Most drops fall quite closely to the puddle of dye.

2. Repeat the above experiment with the substitution of a suspension of *Serratia marcescens*, a bright red bacterium, in place of the dye. Place petri dishes of nutrient agar at 4½, 9, 18, and 36 in. on a straight line from the watch glass. Place another series at 90° to the left and a third series at 90° to the right. The surface of the suspension is put in motion when the drops begin to strike the surface. If the surface is not horizontal when the drops strike, the splashing drops are greater in one direction than another. Placement of the petri dishes along 3 radiating lines, 90° from each other, provides a method of averaging the variability of the distribution of the splashing drops.

After the drops are started, uncover the petri dishes and allow 2 drops to fall before covering the dishes at 4½ in., 5 drops for those at 9 in., 20 drops for those at 18 in., and 200 drops for those at 36 in. Incubate the plates and count colonies when they appear. Average the counts for plates located at the same distance and adjust the data to counts per 100 drops. For example, if the average number of colonies on plates at 9 in., which receive splashes from 5 drops, was 12, then the adjusted number would be $12 \times \frac{100}{5}$, or 240.

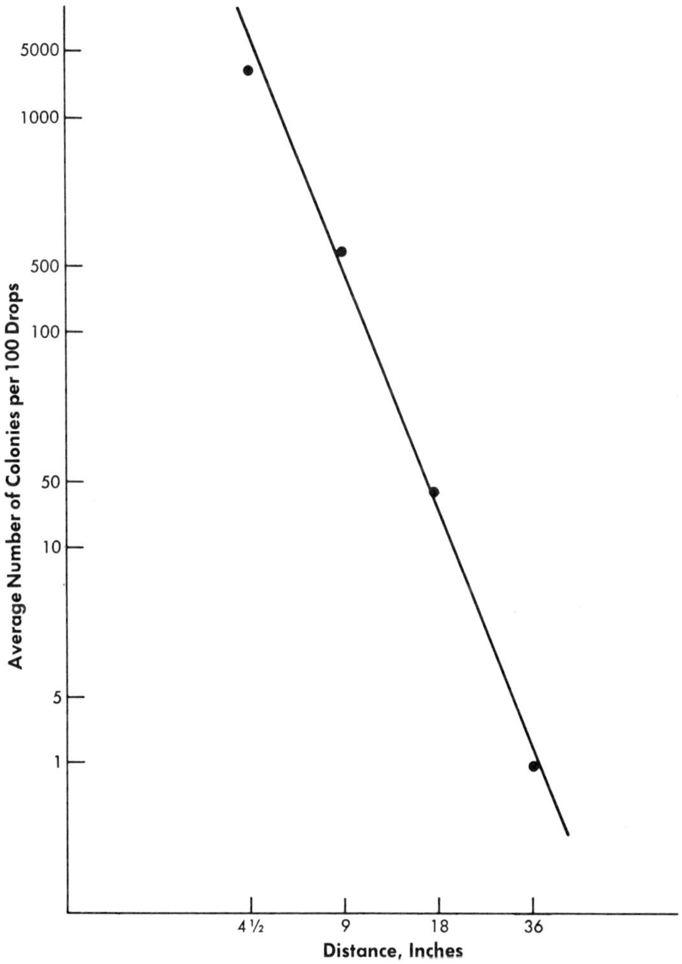

Fig. 5-8. Distribution of splashing drops as a function of distance from source.

Expendable Supplies × Number of Experiments. Instructor should decide if each student or student groups will perform this experiment.

12 plates of nutrient agar
1 large sheet of paper, preferably white
2 large watch glasses
blue or red dye (enough to fill the watch glass)
2 slants of 24-hour growth of *Serratia marcescens*
1 large separatory funnel
string

DISCUSSION QUESTIONS

1. What are two reasons for using a bright red bacterium as a simulant plant pathogen?

2. What is the relationship of the distribution of splashing drops to distance with the inverse-square law? (The inverse-square law is used in physics in relation to foot-candles with distance.)

SUGGESTED READING

Savile, D. B. O., "Spore Discharge in Basidiomycetes: A Unified Theory." *Science*, Vol. 147 (January, 1965), pp. 165–166.

Worthington, A. M., "The Splash of a Drop and Allied Phenomena." *Annual Report of the Smithsonian Institution,* 1894.

K+E LOGARITHMIC 3 X 5 CYCLES

KEUFFEL & ESSER CO.

46 7520

MADE IN U.S.A.

6

Fireblight of Apple and Pear

The history of fireblight is a significant story. This was the first plant disease for which the causal organism was proved to be a bacterium—*Edwinia amylovora.* Beginning in 1878, Burrill, of Illinois, published a series of papers reporting his studies which demonstrated the relationship of bacteria to the disease.

Ornamentals such as *Sorbus, Pyracantha, Cotoneaster,* and *Malus* are affected, as well as commercial cutivars of apple and pear. A rather large number of species in the Rosaceae are susceptible.

Infected blossoms turn brown and then develop black streaks which extend down the pedicel to the spur (Fig. 6-1B). The entire cluster may shrivel and blacken. The fruit also becomes discolored and shriveled (Fig. 6-1A). The edges or the entire area of the leaves may be affected. Such lesions are good examples of *necrosis*—death and collapse of cells and tissues. Close examination reveals disorganization and the collapse of cells (Fig. 6-2).

Dieback—a condition in woody plants in which peripheral parts are killed,—extends down new shoots and typically involves all the current season's growth. The dead leaves are often oriented to one side and the tip of the shoot curls and forms the characteristic *shepherd's crook* (Fig. 6-3A, 6-3B). When twig blight extends from fruiting spurs and small branches to limbs and larger branches, cankers develop which become perennial (Fig. 6-3A). These serve as a persistent means of overwintering and a perennial source of primary inoculum.

Disease Cycle. Primary infections of blossoms and foliage are caused by bacteria disseminated primarily by rain from the perennial cankers (Fig. 6-3C). Bacteria enter through natural openings such as the nectaries in blossoms and stomates, glandular trichomes, and hydathodes in leaves. Nectaries usually have a direct phloem connection. Glandular trichomes and hydathodes are associated with tracheal tissue and the ends of veinlets. Thus, the bacteria have ready access to vascular tissues and invade the plant systemically. Symptoms appear in six days to two weeks.

Bacteria proceed into the veins of the petiole and multiply in the phloem, causing complete destruction of the cambium. The enucleate sieve tubes of the phloem are the major pathway of spread. The bacteria spread as rapidly as two centimeters per hour in young shoots and about half that speed in old shoots. The xylem parenchyma is invaded but the thick-walled xylem vessels are rarely penetrated. Large cavities may be dissolved in the cortex.

Honeybees are especially effective in dissemination of the bacteria from blossom to blossom. Although survival of the bacteria in the hive through the winter is an interesting idea, evidence indicates that this is not so. The pear psylla, an insect, is associated with fireblight infections in leaf axils. This is a location where first generation nymphs congregate.

Control. Sanitation reduces disease incidence. Holdover cankers must be removed, especially those which are high in the trees because rain scatters the bacteria through-

Fig. 6-1. Symptoms of fireblight. (A) Comparison of healthy green fruit and shriveled infected fruit and leaves of crabapple. (B) Blackening of infected tissues of pear.

out the tree from such heights. Incidence of the disease in apple is much less than in pear, but is greatly increased by the prevalence of inoculum from nearby pear trees. For this reason, apples should not be grown in proximity to pears. Modest success with antibiotic sprays has been reported; however, the efficacy of general control measures is not as good as desired.

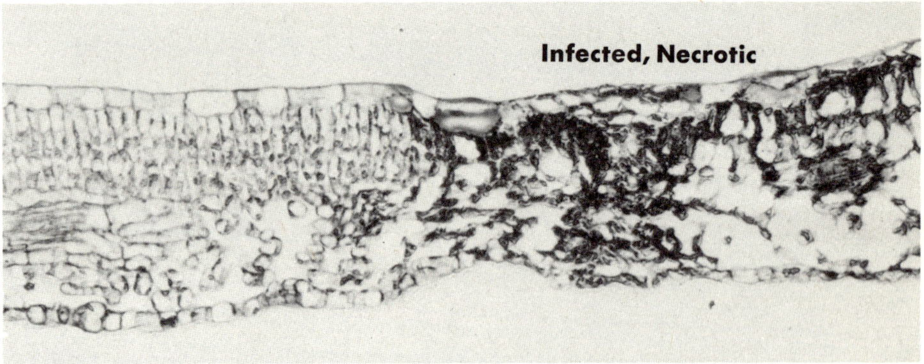

Fig. 6-2. Cross section of a leaf infected with *Erwinia amylovora* showing necrotic collapsed tissues (× *208*).

Fig. 6-3. (A) Shepherd's crook and overwintered canker of fireblight; (B) Shepherd's crook; and, (C) Canker oozing gummy drops containing bacteria.

SUGGESTED READING

Lewis, Sally, and Goodman, R. N., "Mode of Penetration and Movement of Fireblight in Apple and Leaf Stem Tissue." *Phytopathology*, Vol. 55 (July, 1965), pp. 719–723.

Fig. 6-4. Disease cycle of fireblight: (A) overwintering canker; (B) droplets of ooze containing bacteria (springtime); (C) blackening of infected tissues; (D-I) multiplication of bacteria by fission to produce secondary inoculum; and (J) Shepherd's crook.

7

Common Bacterial Blight of Bean

Bacterial blight of bean is a typical leaf spot and leaf blight disease. At first, the lesions on leaves are small, translucent, water-soaked spots which later develop dry brown centers and narrow yellow halos. Lesions coalesce into irregularly-shaped areas which may include the whole leaflet. Lesions on stems and pods tend to be more restricted and sunken (Fig. 7-1A). Those on the pods frequently turn from red to brown with age. With continued development, the vascular system may also turn brown and surface cankers form on the stem.

The seeds in lightly infected pods develop normally and may show no signs of the disease or only small, polished, yellow spots. The yellow discolorations may be very small spots, follow the veins of the seed coat, or engulf the entire seed (Fig. 7-1B). Seeds in severely infected pods are undeveloped and shriveled. Infected seeds provide the primary inoculum which causes the charactristic spots on stems, leaves, and cotyledons of seedlings.

Bacteria infect the seed either by passing through the vascular system of the pedicel of the pod, the vascular tissues of the pod, and then the funiculus, or by growing from an external infection of the pod (Fig. 7-2) through the parenchyma, and into the conducting tissues. Infections of the seed in either case are usually not deep, but primarily in the surface to subsurface regions of the cotyledons.

Disease Cycle. The bacteria overwinter in the seed and infected plant refuse which has not been destroyed by decay. Bacteria are carried on and within the infected seed and contaminate the surface of the expanding cotyledons when the seed germinates. The bacteria penetrate rifts in the cuticle of the cotyledons and are splashed onto open wounds and stomates causing primary infections. They progress through the parenchyma and into the vascular tissues, migrate through the large xylem elements, break out into parenchyma from place to place, and give rise to leaf lesions and stem cankers.

Continued spread of secondary inoculum from primary lesions is brought about by driving rain, irrigation water, implements, and chewing insects. Moisture is essential for infection and is important in dissemination. Warm temperatures are favorable for disease development.

Control. The bacterial diseases have been controlled by the use of disease-free seed grown in nearly disease-free areas of California and Idaho. Usually there is insufficient rainfall and humidity in these areas to permit disease development except where overhead sprinkler systems are used for irrigation. Climatic changes have brought about increased amounts of moisture in Idaho and hence there has been more disease and more seeds carry more infections. Remember that the major source of primary inoculum is carried on the seed. Consequently, the bacterial diseases of bean are not being controlled and the diseases are of major importance in the production of beans.

Crop rotation of three years to avoid overwintering bacteria can be helpful where

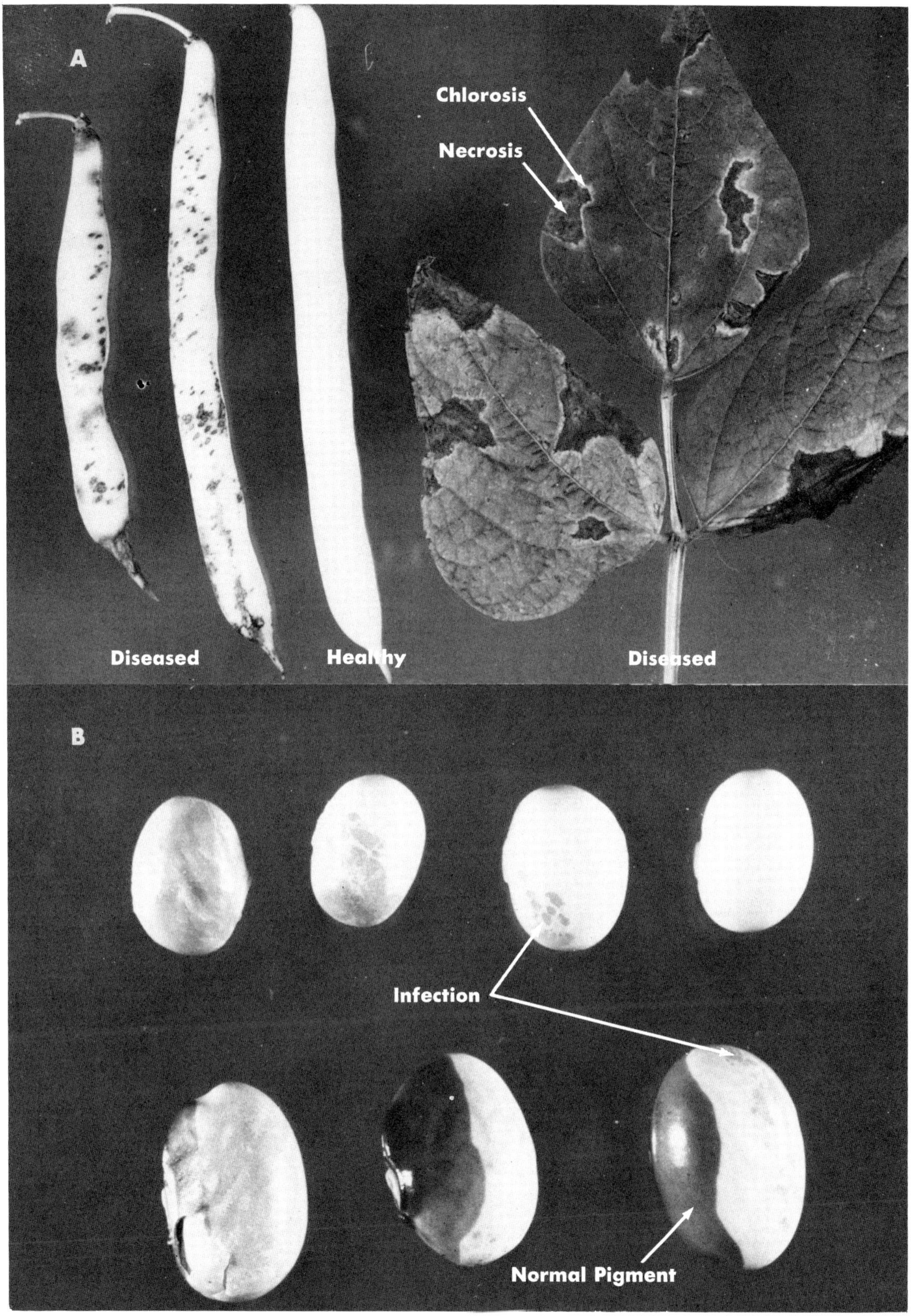

Fig. 7-1. (A) Symptoms of common bacterial blight of bean on pods and leaflets. (B) Symptoms on seeds of navy and cranberry bean.

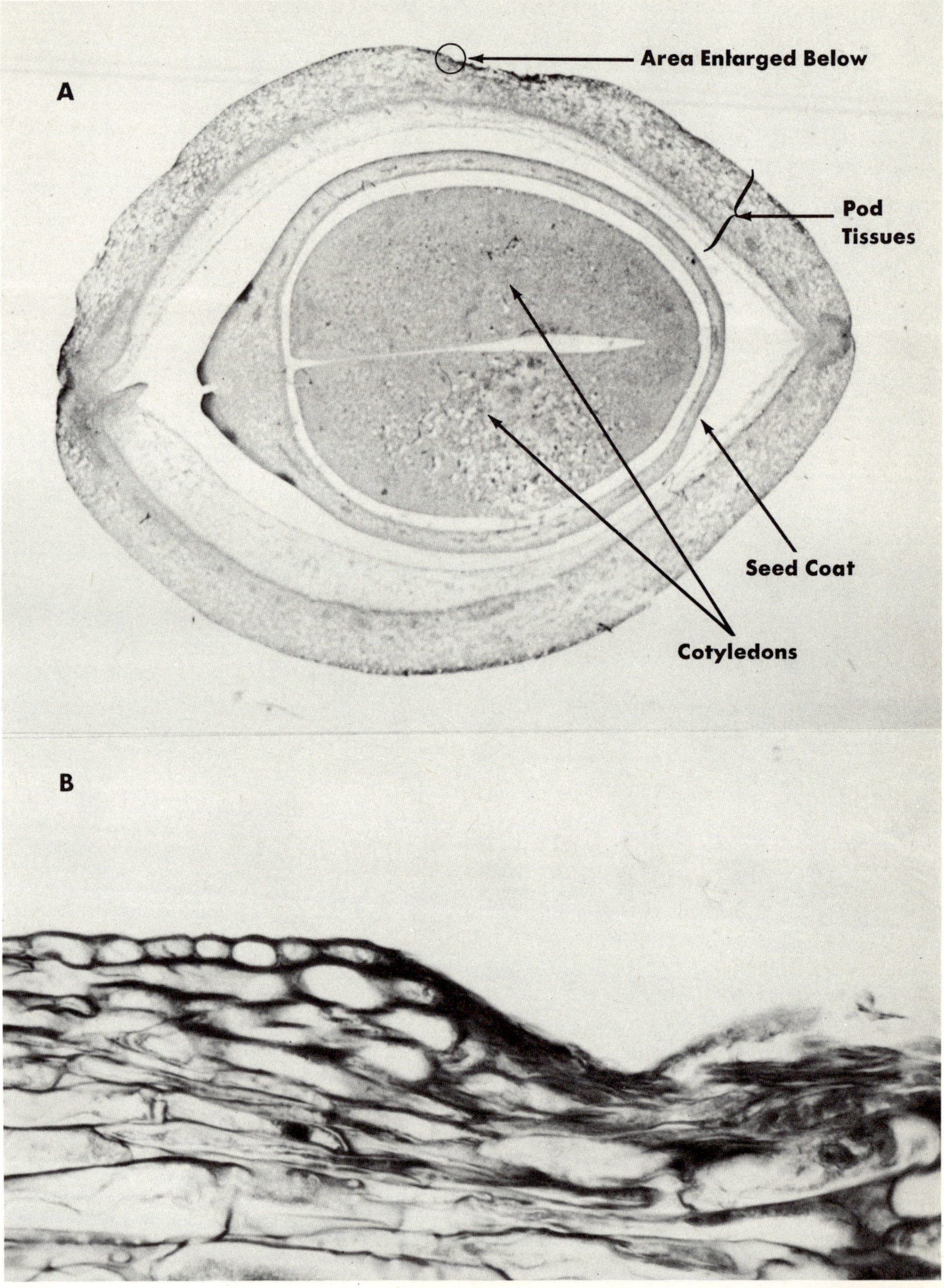

Fig. 7-2. **(A) Cross section of a bean pod showing the site of a lesion on the outer surface of the pod (× *10.4*). This lesion would correspond to one small spot on the diseased pods in Fig. 7-1A. (B) Enlargement of the infected area showing indentation made by dead and collapsed cells (× *260*).**

the disease has been a serious problem. Rigid control of the dispersal of bean straw is a sanitary measure that should be enforced. Although fully resistant cultivars are not available, the preferred use of tolerant ones is justified.

Diagnosis. Common bacterial blight of bean cannot be diagnosed, with certainty, in the field. This disease must be distinguished from four other bacterial diseases of bean which have overlapping symptoms. This problem of differentiating different diseases from overlapping symptoms is commonly encountered in the positive diagnosis of plant diseases.

Halo blight, caused by *Pseudomonas phaseolicola*, characteristically exhibits a halo. The disease, however, cannot be distinguished from common blight on the basis of its large halo for the halo is not produced if there have been periods of very warm weather. Although the wilt disease caused by *Corynebacterium flaccumfaciens* does not cause leaf spots, the leaf spot diseases caused by four different bacteria can also include wilting symptoms. *Corynebacterium* species, however, are the only plant pathogenic bacteria which are gram-positive. Fuscous blight can be distinguished only by the brown pigment that the causal organism, *Xanthomonas phaseoli* var. *fuscans*, produces on certain media such as PDA. Table 7-1 indicates characteristics of the five species of bacteria which are necessary to distinguish one from the other.

TABLE 7.1. Differential characteristics of five species of bacteria that cause different diseases of bean.

Bacterium	Disease	Gram Stain	Color in Culture
Corynebacterium flaccumfaciens	bacterial wilt	positive	yellow
Pseudomonas phaseolicola	halo blight	negative	white, green florescent
Pseudomonas syringae	bacterial brown spot	negative	white
Xanthomonas phaseoli	common bacterial blight	negative	yellow
Xanthomonas phaseoli var. *fuscans*	fuscous blight	negative	yellow, brown pigment on PDA

SUGGESTED READING

Grogan, R. G. and Kimble, K. A., "The Role of Seed Contamination in the Transmission of *Pseudomonas phaseolicola* in *Phaseolus vulgaris*." *Phytopathology*, Vol. 57 (January, 1967), pp. 28–31.

Zaumeyer, W. J., and Thomas, H. R., *A Monographic Study of Bean Diseases and Methods for their Control.* U. S. Dept. Agr. Tech. Bull., 1957.

8

Bacterial Wilt of Cucumber

Wilting is the outward expression of bacterial wilt of cucumber and related plants in the Cucurbitaceae, the cucumber family. This includes squash, pumpkins, gourds, muskmelons, and wild cucumbers. Watermelon, another relative, is relatively resistant. The causal organism is *Erwinia tracheiphila.*

Dwarfing, yellowing, and wilting are symptoms of another bacterial wilt disease —bacterial wilt of alfalfa caused by *Corynebacterium insidiosum* (Fig. 8-3).

An early symptom of many wilt diseases of herbaceous plants, including bacterial wilt of cucumber, is a recurring, temporary wilt on hot dry days. The severity of wilt and the portion of the plant affected increases until the wilt is permanent. (See the Chapter on FUSARIUM wilts for more details on mechanisms of wilting.) Less susceptible plants, like squashes, may show dwarfing of growth before the wilt pattern develops. In more advanced stages, especially with cucumbers, milky drops ooze from cut stems when they are squeezed. These drops tend to string out when touched and slowly drawn out. None of the symptoms are applicable solely to this disease. The positive diagnosis is made only by demonstrating the presence of the causal bacteria by microscopic and cultural methods. Without such demonstrations, diagnoses must be considered provisional.

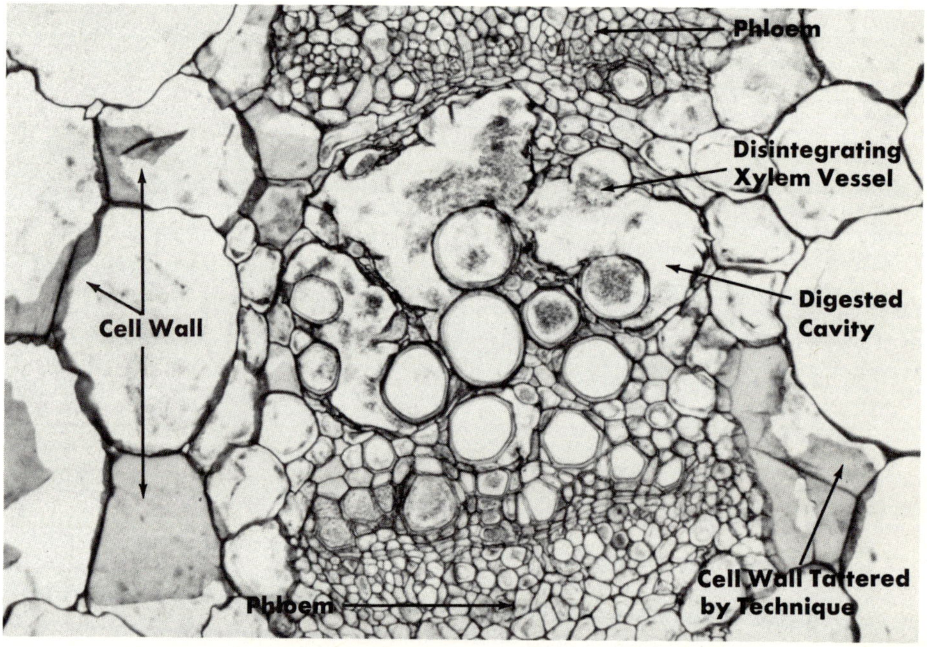

Fig. 8-1. Cross section through the stem of a cucumber plant with bacterial wilt. The xylem but not the phloem is plugged with bacteria, ooze, and debris. Severe dissolution of the xylem elements is evident (× *200*).

Disease Cycle. The bacteria can enter the plant only through deep wounds and only when a sufficient film of moisture is present. They enter the ends of xylem tubes (Fig. 8-1). The first infections in the spring and subsequent secondary infections are associated with feeding of the striped and 12-spotted cucumber beetles (Fig. 8-2). The beetles contaminate their mouthparts with bacteria when feeding on infected plants. Subsequently, other plants become infected by the beetles when they feed.

The dependence of the bacteria on these beetles for dissemination and spread is marked by the disappearance of the disease when the beetles are eliminated. Experiments with caged cucumber plants show that wilt develops when beetles are also in the cage. Proof that the beetles are an overwintering site is lacking because of the inability to maintain the beetles experimentally through a winter period. Successful transmission has been made with beetles which were held at low temperatures for 6 weeks only. J. G. Leach recently has raised some interesting questions concerning the beetles and the overwintering mechanism of the bacteria: (1) The beetles ordinarily feed on blossoms of early blooming plants, chiefly in the rose family, for 40 to 50 days in the early spring before any cucumber plants, cultivated or wild, are available. (2) The beetles do not regurgitate and do not have salivary glands; therefore, how do the bacteria get into the wounds of the plants? (3) Bacteria are found only in the gut of the beetle. There is no evidence for multiplication of the bacteria in the beetle. Whether or not the bacteria overwinter in the beetles, in the wild plants that they feed on in early spring, or in both places is unanswered. However, the role of the insects in bringing about primary infections of cucumber and subsequent secondary infections is unquestioned.

Control. The bacteria are not seed- or soil-borne and control depends on control of the beetles. One to four generations of the beetle may occur and frequent applications of insecticides may be necessary.

Fig. 8-2. Twelve-spotted cucumber beetle.

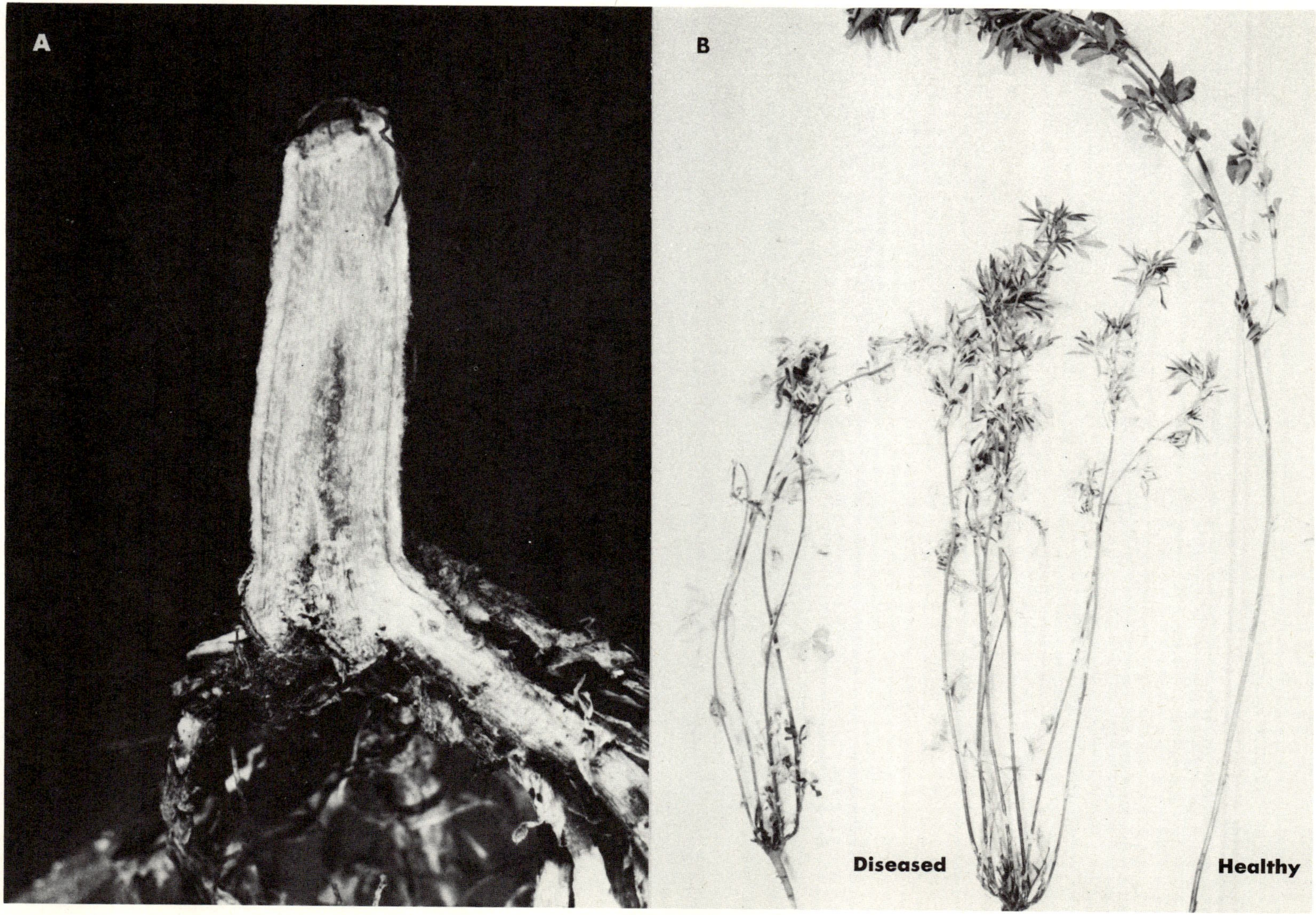

Fig. 8-3. **Bacterial wilt of alfalfa. (A) Discoloration of crown and lower stem. (B) Yellowing and dwarfing of infected plants.**

SUGGESTED READING

Burkholder, W. H., "Some Observations on *Erwinia tracheiphila,* the Causal Agent of the Cucurbit Wilt." *Phytopathology,* Vol. 50 (February, 1960), pp. 179–180.

Leach, J. G., "Observations on Cucumber Beetles as Vectors of Cucurbit Wilt." *Phytopathology,* Vol. 54 (May, 1964), pp. 606–607.

Rand, F. V., and Cash, L. C., "Some Insect Relations of *Bacillus tracheiphilus. Erw. Sm.*" *Phytopathology,* Vol. 10 (March, 1926), pp. 133–140.

9

Crown Gall

One of the most fascinating abnormalities exhibited by protoplasm is the development of autonomous or self-governing tumors which do not limit themselves in size (Fig. 9-1). They contrast with the neat, regular, highly architectured galls caused by mites and wasps. These galls require the continued presence of the causal agent and are self-limiting in size and shape (Figs. 9-2, 9-3).

Crown gall tissue is composed of disorganized, randomly proliferating cells (Fig. 9-4) that divide to give rise to similar cells without the continued presence of the causal organism. The growth of the gall is controlled by the host plant only indirectly—the host is the sole source of nutrients for the gall tissue. This is in contrast to normal cells which develop into definite, ordered, and limited patterns of tissues and organs because of strict obedience to morphogenetic laws of differentiation.

Crown gall is caused by *Agrobacterium tumefaciens* and occurs on plants in more than 140 plant families. The diseases caused by this bacterium are most important on woody nursery plants such as apple, peach, cane fruits, and roses. Although many vegetables are attacked, the diseases are rarely of economic importance on such crops. Other species of *Agrobacterium* cause hairy root of apple and cane gall of brambles.

The galls consist of round and sometimes elongate overgrowths which occur most frequently on the roots, crown, and stems near the ground. They may be pea-size or may reach inches in girth and weigh several pounds. Galls on plants such as tomato and Jimson weed are relatively soft, whereas those on woody plants are quite woody and hard and are covered with a cork-like layer. The galls disintegrate readily and are invaded commonly by secondary organisms—frequently the soft rot bacteria.

On occasion, sterile galls free of bacteria are found above the galls caused by the bacteria. It is not clear if these are brought about by: (1) the passage of a very few bacteria through the vascular system, (2) connecting strands of gall tissues, or (3) compounds produced by the bacteria or the host plant. Perhaps they are brought about by some manner yet to be found and described by some student now discovering this gap in our knowledge!

The bacteria can enter only through wounds and must invade a wound before the fifth day following its occurrence to be successful in the induction of a gall (Fig. 9-5A). The bacterial cells multiply in the intercellular spaces in the vicinity of the wound. In a wound without the presence of these bacteria, meristematic activity is initiated with the production of a cork cambium (**phellogen**) which produces cork cells (**phelloderm**) which become impregnated with suberin (Fig. 9-6). Meristematic activity ceases after this scar tissue is laid down. In the presence of the bacteria, rapid and continuous cell division of the plant tissues persists (Fig. 9-5B). Once initiated, this meristematic activity can continue without the presence of the bacteria. Large masses of cells are produced in overabundance (**hyperplasia**: an increase in cell numbers) (Figs. 9-5C, 9-5D). Many of these cells enlarge abnormally (**hypertrophy**: an increase in cell size).

Differentiation in gall tissue is limited generally to discontinuous and randomly oriented vessels (Figs. 9-4B, 9-5C). The bacteria occur in fairly low numbers in contrast to the hordes of *Pectobacterium carotovorus* cells in a soft-rotted potato.

Fig. 9-1. **Crown gall on *Euonymous japonicus*.**

Fig. 9-2. Insect galls on oak leaves.

Fig. 9-3. Insect galls on oak leaves (A). Arrows designate areas enlarged (B, C).

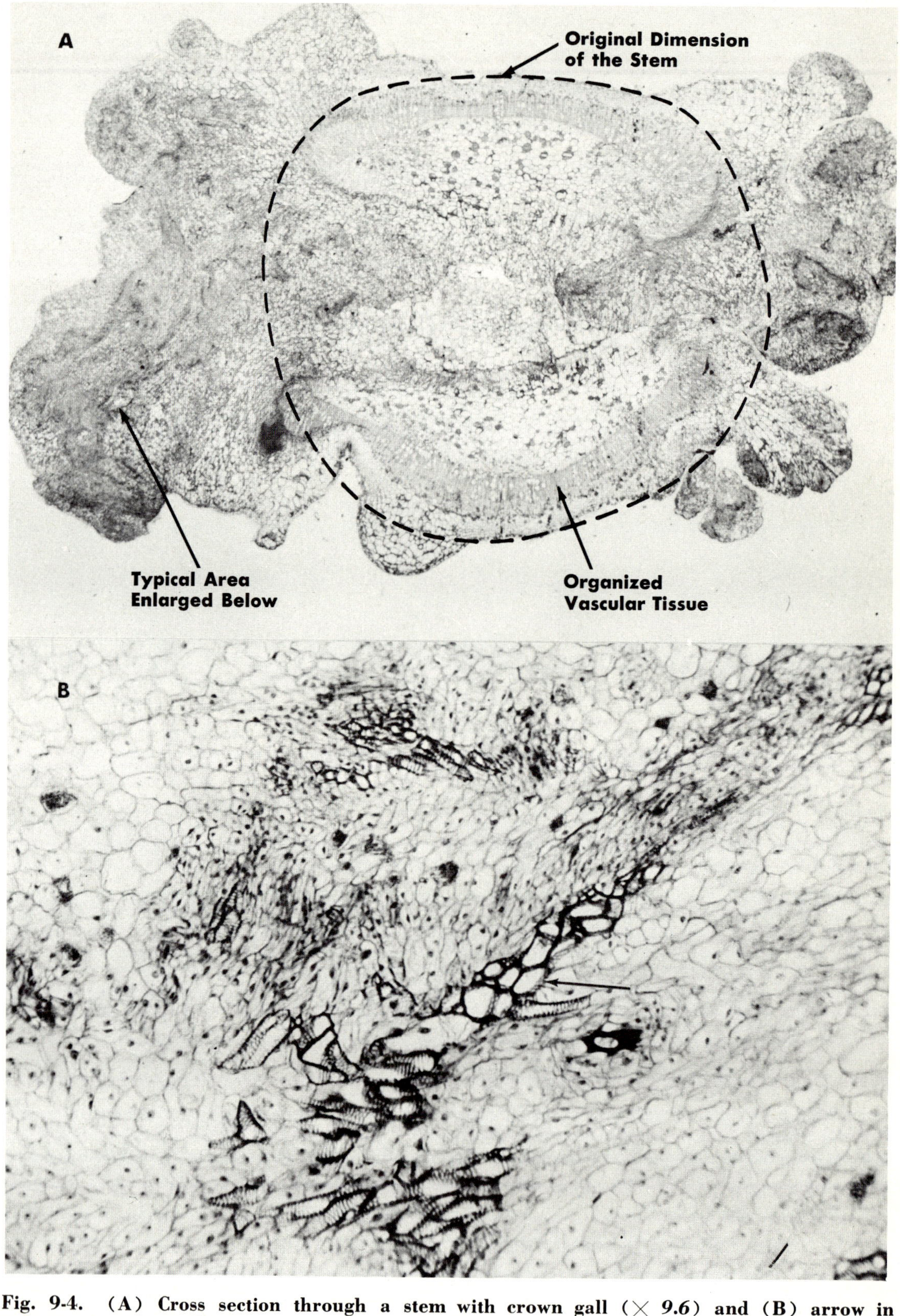

Fig. 9-4. (A) Cross section through a stem with crown gall (× *9.6*) and (B) arrow in enlargement shows disorganized vascular elements (× *120*).

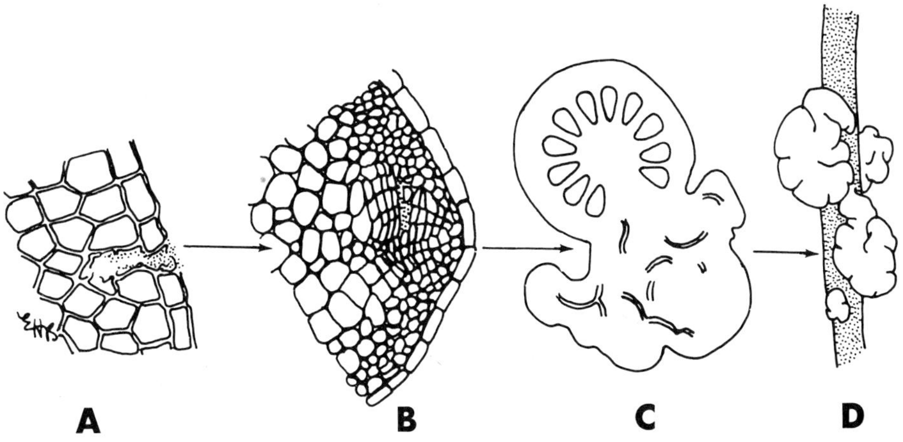

Fig. 9-5. Sequence of events in the development of a gall. (A) Multiplication of bacteria in a wound; (B) division of the cells of the suscept; and, (C) proliferation of the gall tissues.

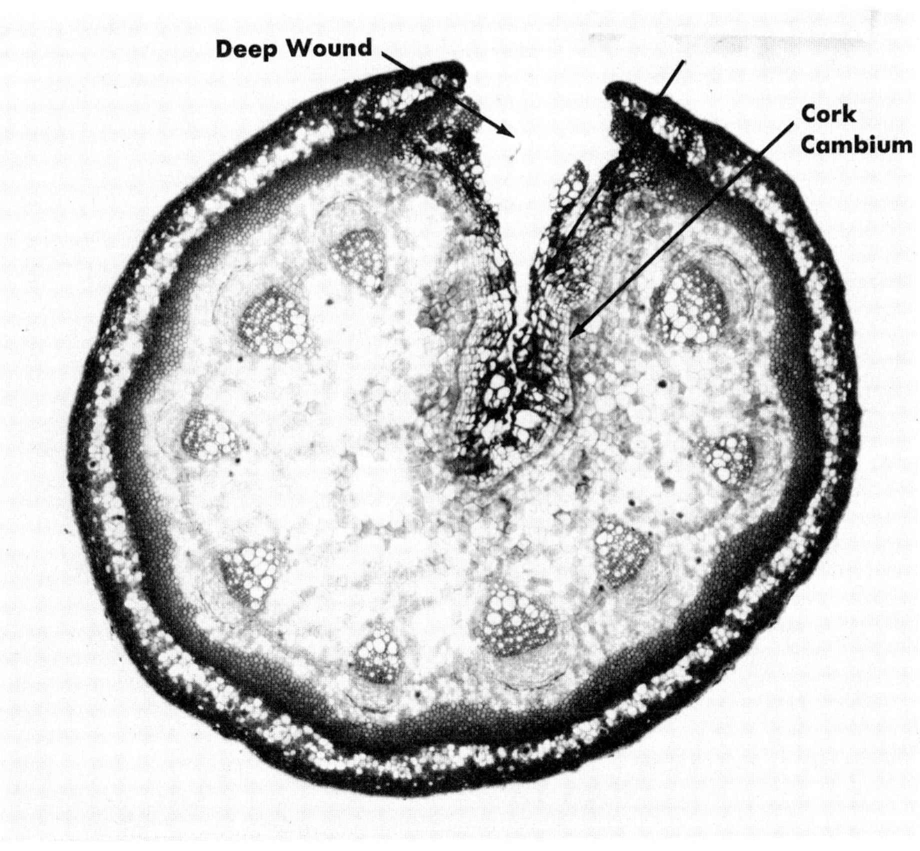

Fig. 9-6. Cork cambium and cork tissue in a wounded stem of *Aristolochia* without the presence of *Agrobacterium tumefaciens* (× 9).

TABLE 9-1. A comparison of different kinds of galls.

Disease or Injury	Host	Shape of Gall	Plant Part Affected	Distinguishing Characteristics
Crown Gall	Rosaceae mainly	irregular	stems, roots	bacteria, irregular xylem, autonomous
Club Root	Cruciferae only	irregular	stems, roots	wilting, fungus structures, Krankheitsherde
Nodules	Leguminosae mainly	uniform, small	roots	uniform cells, structured gall
2,4-D Injury	Dicotyledonous plants mainly	irregular	stems, leaves, flowers, roots	mottling, no organism, noninfectious, leaf distortion

Disease Cycle. Primary infections result from the planting of infected nursery stock. The bacteria are spread from established infections by means of transplants, machinery, and chewing insects. Further spread during the growing season gives rise to secondary infections. The organism grows and multiplies in the soil and thus becomes a successful **soil inhabitant.** It survives adverse periods in infested soil as well as in infected plants. Thus, the bacteria overwinter in the soil, as soil inhabitants, and in infected perennial hosts.

Control. Avoidance of infected plant material and production of clean inspected nursery stock are first in importance for control of the disease. Second in importance is the denial of entrance of the bacteria by reduction of wounds and bruises, the protection of graft unions, and the control of chewing insects, especially subterranean ones. Other useful control measures include proper disposal of infected materials and disinfestation of soils in green houses. Fungicidal sprays and dusts have not been effective in field trials.

Diagnosis. A lesson in diagnosis is provided by a comparison of these galls with those of similar outward appearance caused by 2,4-D, an herbicide; *Plasmodiophora brassicae,* a fungus; and, *Rhizobium* sp., another bacterium (Table 9-1).

SUGGESTED READING

Braun, A. C., "Studies on Tumor Inception in the Crown Gall Disease." *American Journal of Botany,* Vol. 30 (November, 1943), pp. 674–677.

—— A. C., "Recovery of Crown-Gall Tumor Cells." *Cancer Research,* Vol. 11 (November, 1951), pp. 839–844.

—— A. C., "Studies on the Origin of the Crown-Gall Tumor Cell." *Brookhaven Symposia in Biology,* Vol. 6 (April, 1954), pp. 115–127.

—— A. C., "A Physiological Study on the Nature of Autonomous Growth in Neoplastic Plant Cells." Vol. 11 *Symposia of the Society for Experimental Biology.* 1957.

Lipetz, J., "Crown Gall Tumorigenesis: Effects of Temperature on Wound Healing and Conditioning." *Science,* Vol. 149 (August, 1965), pp. 865–867.

Experiment X

Crown Gall

Crown gall is an autonomous, nonself-limiting tumor initiated by infection of wounds by *Agrobacterium tumefaciens.* Once the abnormal growth is initiated, the bacteria are no longer required. In fact, bacteria are relatively infrequent in gall tissues compared to soft rotted tissues and have not been isolated from some secondary galls.

EXPERIMENT

Demonstrate the growth of galls and the role of wounds and causal bacteria by inoculating plants of sunflower (*Helianthus* sp.), tomato (*Lycopersicon esculentum*), or Jimson weed (*Datura stramonium* or *Datura tatula*). Jimson weed produces larger galls sooner than tomato and, unlike tomato, is devoid of the bumpy areas where adventitious roots emerge. These structures are sometimes mistaken for young galls by inexperienced observers.

1. Reserve one plant without wounding and inoculating as a control. Wound two plants just above the soil line with a cool, sterile scalpel or needle. Make wounds about ⅛ inch deep but not so deep nor so long as to cause the plants to be without support and collapse. Inoculate one wound with *Agrobacterium tumefaciens* but not another. Carefully smear bacteria on the stem of another plant which has not been wounded. Fig. 9-7 illustrates the kinds of results you should achieve.

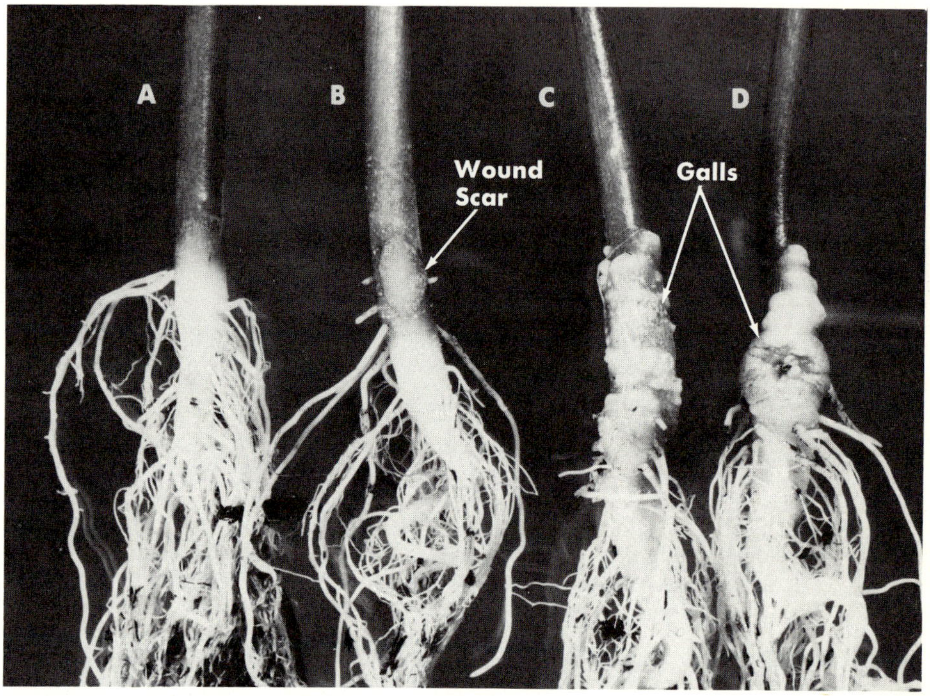

Fig. 9-7. Results of experiments with crown gall. (A) nonwounded, noninoculated plant; (B) wounded plant; and, (C, D) wounded and inoculated plants.

2. Fill the remaining space in the top of the pots with vermiculite and water thoroughly. If the wounds are not kept moist for about 5 days, the infections may fail. Slight swellings occur in 9–10 days and considerable swellings in 14 days.

3. After 6 days the vermiculite may be gently washed out.

Plant Material. Plants should be small, 5 to 6 in. high, when used for good results. Plants of Jimson weed with one set of true leaves are satisfactory. Plan two weeks from seeding until transplanting time and another two weeks from transplanting until ready for class use. Keep the soil low in the pots when transplanting so that there are $1\frac{1}{2}$ inches of clearance. This is to leave room for the wounds below the level of the rim of the pot so that the moist vermiculite covers them.

4. If desired by the instructor, crown gall is a very satisfactory disease with which to follow Koch's Postulates.

Expendable Supplies × Total Number of Students or Groups

1 culture of *Agrobacterium tumefaciens* on PDA slant
4 plants of Jimson weed, tomato, or sunflower
4 pot labels

Supplies Available

vermiculite
wax pencils or other waterproof marking pencils

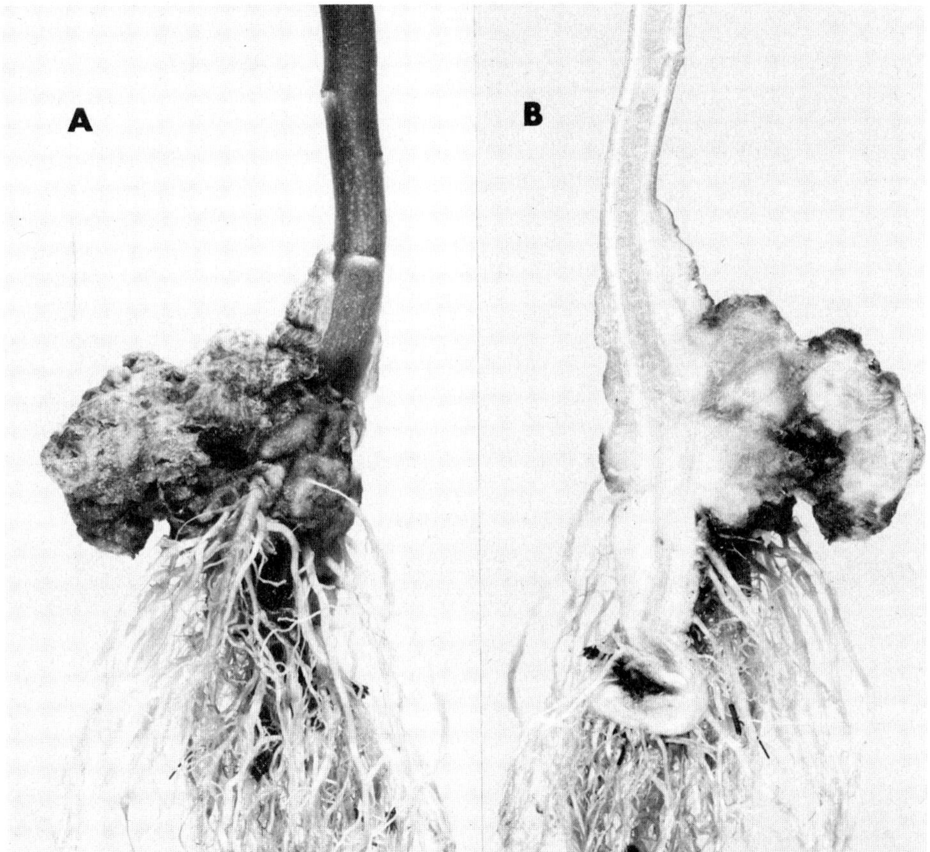

Fig. 9-8. Continued proliferation of experimentally induced galls on Jimson weed; (A) exterior view, and (B) interior appearance.

DISCUSSION QUESTIONS

1. Why do we wound but not inoculate one plant in the experiment?
2. What reasons might there be if no galls should form?
3. What is the significance of autonomous, nonself-limiting tumors? Are they like cancer?
4. How is crown gall controlled?

SUGGESTED READING

Braun, A. C., "Growth is Affected," in J. G. Horsfall and A. E. Dimond, eds., *Plant Pathology*, Vol. I (New York, 1959), pp. 189–240.

10

Root Nodules of Legumes

The root nodules of leguminous plants are caused by nitrogen-fixing bacteria belonging to the genus *Rhizobium*. Root nodules are well-organized and differentiated structures, a striking contrast to the cancerous growth of crown gall.

Nodules first become visible on susceptible plants in the upper portions of the primary root below the hypocotyl about two weeks after seedlings emerge. By the time a plant is several weeks old, typically, it has both a well-defined root system and conspicuous nodules everywhere except on the peripheral portions (Fig. 10-1). Nodules on herbaceous hosts are fragile and short-lived. On woody plants they may persist for four or five years and look like a string of beads.

Nodule Development. The bacteria (Fig. 10-5A) most frequently penetrate the root hairs but may also penetrate young epidermal cells (Fig. 10-5B). No other plant pathogenic bacteria are known to possess this ability of direct penetration! The infected root hairs curl and some controversy exists as to whether or not the bacteria produce hormones which can bring this about. No more than 4% or 5% of the root hairs become infected. Is there a mechanism restricting numbers of infections? It would seem so but no answers are available.

A gummy mucoid infection thread, called a **zoogloeal strand**, issues from the inside of the cell wall of the root hair immediately upon penetration (Fig. 10-5B). Electron microscopy reveals that the strand is a double-walled, tubular sheath within which the bacteria are embedded (Fig. 10-5C). Current thought is that the sheath is an invagination of the host cell wall because (1) it originates at the cell wall; (2) it is composed of cellulose, hemicellulose, and pectic substances like plant cell walls; (3) it ends at the next cell wall; and, (4) no sheath of cellulose composition has been found surrounding bacteria in the intercellular spaces. Perhaps the bacteria penetrate the cell walls by passing through plasmodesmata pores and then through the protoplast by invagination. Many of these details have been illuminated by electron microscopy. If you want to understand magnification beyond the capability of the light microscope refer to the paper by Jordan and colleagues cited at the end of the chapter.

In two days the strands extend from the inside tip of the root hair to the inner wall of the epidermal cell. Branching of the strand rarely occurs before this stage but almost always after penetration into the cortex. The strands penetrate into the third to fifth layers of cortical cells by the ninth to tenth day, but are not known to penetrate through the endodermis (Fig. 10-5D).

Bacteria are never found free of the strand until it ruptures in the deeper cortical cell layers. Although growth of the cortical cells appears to stretch and fragment strands on occasion, the more common means of release of the bacteria is by rupture of terminal and lateral vesicles which form on the strands (Fig. 10-5E). Once the bacteria are released, their nuclear material disperses, their cells enlarge, and they become enclosed in sacs with double membranes (Fig. 10-5F). These are now called **bacteroids.** The double membrane may be part of the endoplasmic reticulum of the host.

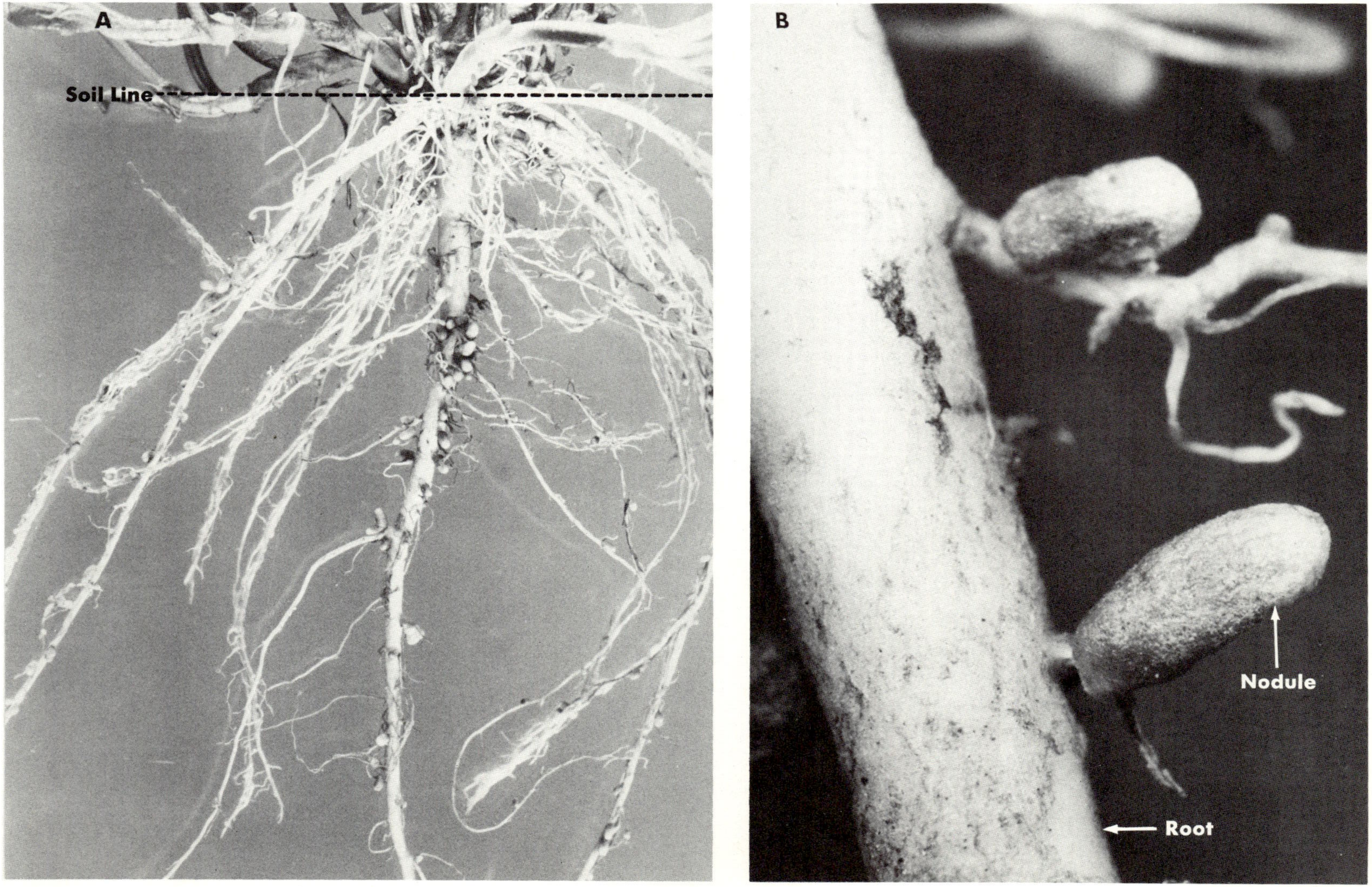

Fig. 10-1. **(A) Nodules of nitrogen-fixing bacteria on clover and (B) enlargement ($\times$ *8*).**

The host responds quickly to the invasion. Cortical cells surrounding the immediate area of invasion are stimulated to meristematic activity as soon as the cortex is invaded. By the time the threads reach the second cortical layer, *de novo* provascular differentiation is evident near the stele. Usually this occurs opposite the radial arms of xylem in the root stele but may be in between them.

The nodule continues to increase in size as a result of continued meristematic activity at the tip of the nodule (Figs. 10-5H, 10-2, 10-3B). The individual, infected cells increase in size once the bacteroids begin to develop. With the increase in nodule size, there is continued differentiation of vascular tissues such as pitted tracheids, scalariform vessels, and phloem cells. Usually two major vascular bundles develop which branch and surround the bacteroid area.

The chromosome number of the infected host cell is doubled. Whether this causes the enlargement of the host cell or is a result of enlargement is still a question.

The mature nodule is a complex organ constructed of several differentiated tissues (Figs. 10-2, 10-3, 10-5). It is surrounded by an epidermal layer formed from the cortical cells of the host. Near the tip there is a zone of meristematic cortical parenchyma which is the nodule meristem. Behind this is an area of infected enlarged cortical cells of fairly uniform size which contain nitrogen-fixing bacteroids. Finally, there is an older area of disintegrating bacteroids in the aging nodules. The bacteria become intercellular as the disintegrating cortical cells collapse, following vacuolation, hypertrophy, and disintegration of nuclei. Thus, the nodule is composed of a central core of bacteroid tissue surrounded by an epidermal layer, a layer of uninfected cortical cells, an endodermis, and a vascular system.

The nodules are not similar to secondary lateral roots (Fig. 10-4) for several reasons: (1) There are no root hairs, no stele, and no root cap. The structure is derived from cortical cells and are cortical in nature except for the differentiated vascular system. (2) The vascular system is not always opposite the protoxylem points as they tend to be in lateral roots. (3) The nodules do not digest their way out, as lateral roots do, but remain covered with cortical tissues. (4) The endodermis and vascular strands are differentiated from the cortex and not the mother-root stele. (5) Finally, they develop from cortical parenchyma and not the endodermis of the cortex.

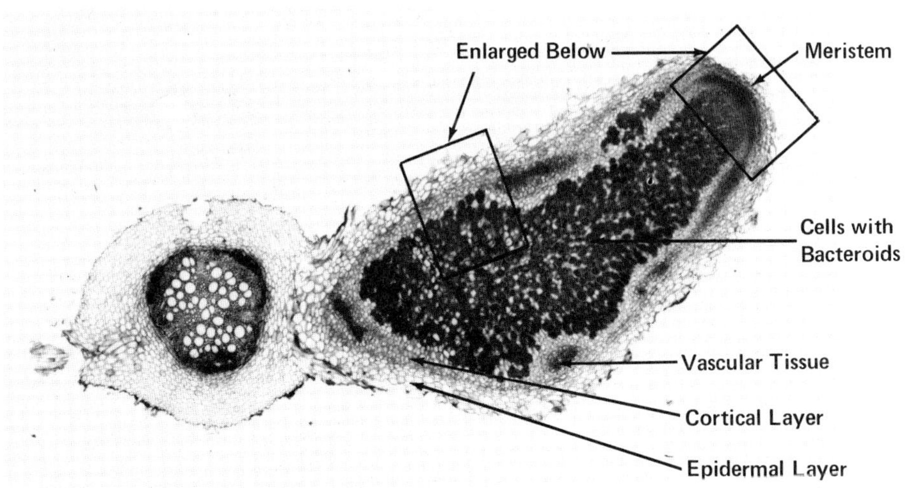

Fig. 10-2. Section through a root and nodule at the point of attachment.

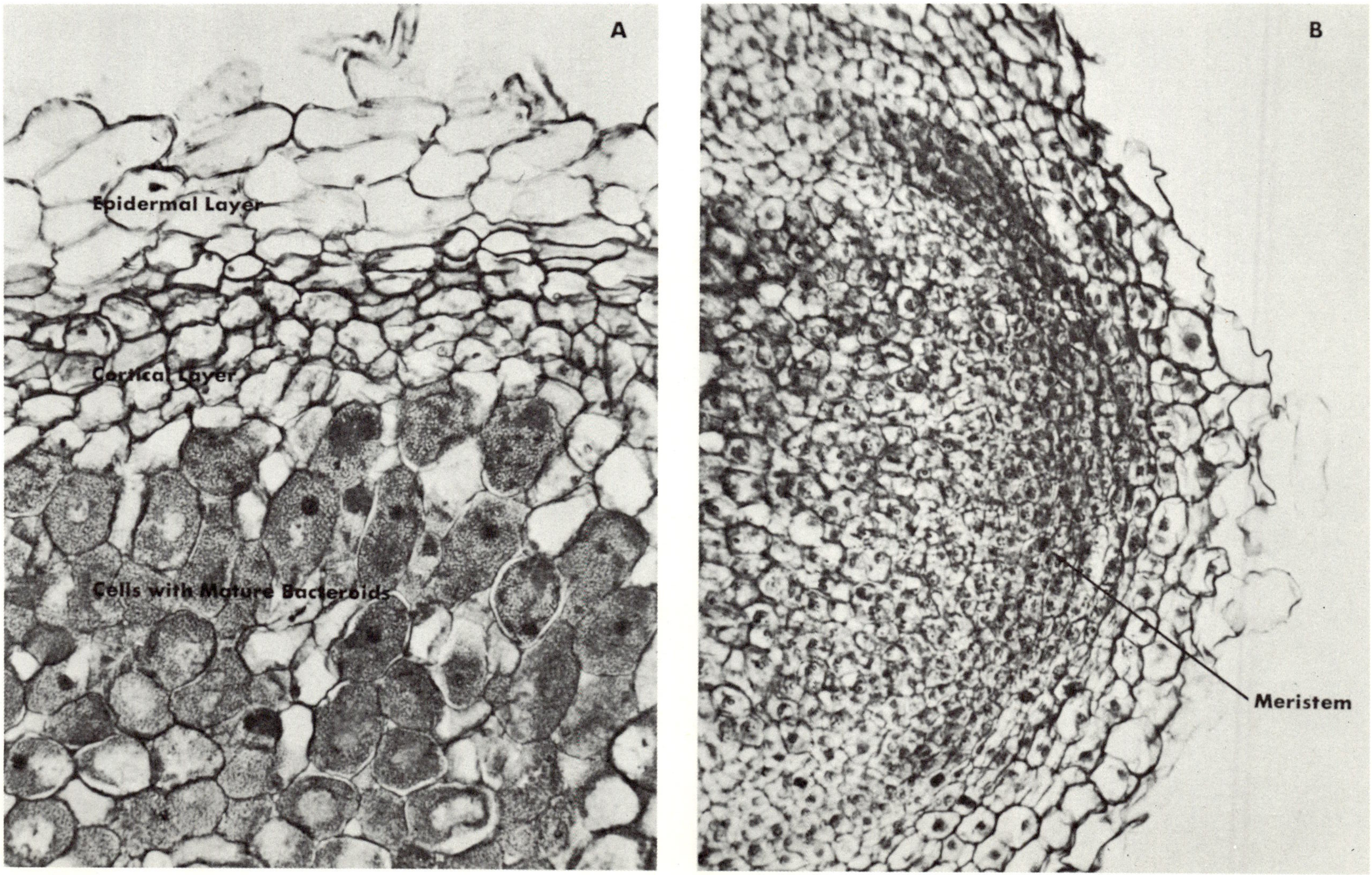

Fig. 10-3. Enlargement of a section through a nodule in an area of (A) mature bacteroids, and (B) nodule meristem ($\times$ *250*).

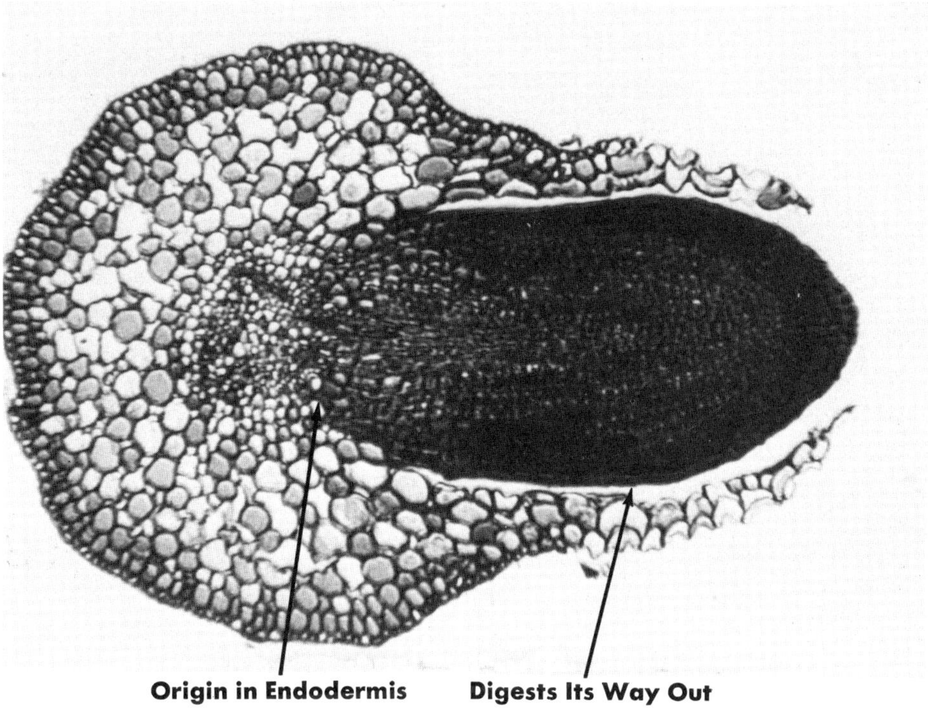

Fig. 10-4. Cross section of a secondary lateral root (× *10*).

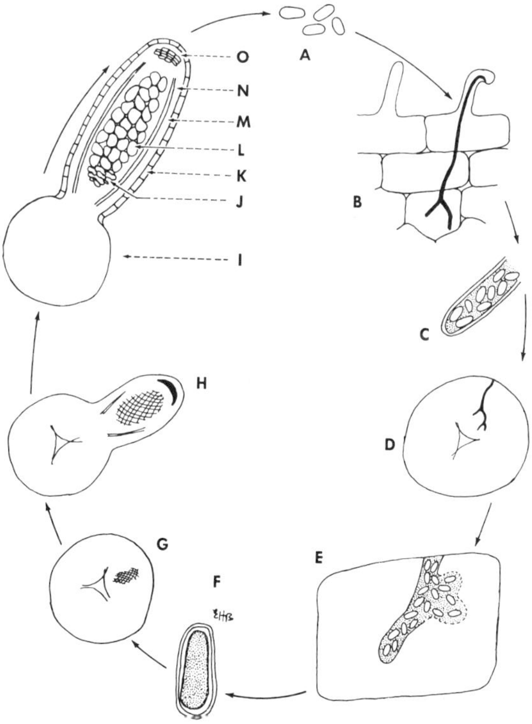

Fig. 10-5. Life cycle of *Rhizobium* and development of a nodule.

SUGGESTED READING

Bieberdorf, F. W., "The Cytology and Histology of the Root Nodules of Some Leguminosae." *Journal of the American Society of Agronomy*, Vol. 30 (May, 1938), pp. 375–389.

Bond, L., "Origin and Developmental Morphology of Root Nodules of *Pisum sativum*." *Botanical Gazette*, Vol. 109 (June, 1948), pp. 411–434.

Jordan, D. C., Ginyer, I., and Coulter, W. H., "Electron Microscopy of Infection Threads and Bacteria in Young Root Nodules of *Medicago sativa*." *Journal of Bacteriology*, Vol. 86 (July, 1963), pp. 125–137.

11

Virus Diseases

Shortly after the Germ Theory of Disease was firmly established, Iwanowski, in 1892, and later Beijerinck, in 1898, passed sap of tobacco plants infected with mosaic virus through bacterial-proof filters. They demonstrated that filtrates, although apparently sterile fluids, were infectious. Plant pathologists were well-pleased with the new and rewarding Germ Theory of Disease which stated that for each infectious disease there as a causal organism. Many assumed that the fluids contained very small bacteria which they could not see in the microscope but which could pass through the filters. This organismal theory was not abandoned until the 1920's, even though Beijerinck emphasized the noncorpuscular nature of the infectious filtrates and called them contagious living fluids.

Tobacco mosaic virus disease was the first virus disease to be recognized as infectious by Mayer, a Dutchman, in 1886. About this same time, peach yellows and peach rosette were serious diseases in Michigan and Erwin F. Smith of the U. S. Department of Agriculture pointed out the similarity and demonstrated transmission by grafting. Walter Reed gained notoriety in 1900 with the demonstration that the *Anopheles* mosquito transmitted the causal agent of yellow fever, a virus disease. Meanwhile, Japanese workers, from 1894 to 1904, demonstrated insect transmission of virus diseases of rice. In 1935, Stanley crystallized tobacco mosaic virus for which he was later awarded the Nobel prize. Shortly thereafter, Bawden and others demonstrated the nucleic acid content of the virus.

Most viruses that we know of today attack insects, vertebrates, bacteria, and flowering plants. (The viruses which attack bacteria are called **bacteriophages.**) There are only a few reports of viruses which attack algae, fungi, and Gymnosperms, and no reports of viruses which attack Pteridophytes. It is premature to believe that the ferns and mosses and their relatives are free of viruses.

Diagnosis of Virus Diseases. Positive diagnosis of virus diseases cannot be made on symptoms alone. This is so because the symptoms are either similar or absent, or because they are masked, or latent. Provisional diagnosis can often be made with experience but they are educated guesses. Some of the common symptoms include aberrations of chlorophyll distribution and density called **mosaics** (Figs. 11-1A, 11-1B, 11-3, 13-2A), necrotic local lesions, (Fig. 13-1), systemic ring spots (Figs. 11-1C, 11-2), tumors, sterility, malformations, and dwarfing (Figs. 11-4, 11-6). Symptoms can be quite variable (Figs. 11-1A, 11-1B, 11-2, 11-3). Less obvious effects may include necrosis of internal tissues (Fig. 11-5).

Aberrations of chlorophyll distribution can also be caused by nutritional deficiencies (Fig. 11-7), genetic anomalies (Fig. 11-8), and herbicides (Fig. 11-9). Some malformations resembling virus symptoms can be induced by growth regulators and related compounds such as 2,4-D (Fig. 11-10).

Fig. 11-1. (A, B) Systemic mosaic symptoms of cucumber mosaic virus in tobacco and (C) systemic ring spot symptoms of tobacco necrosis virus in tobacco.

Classification of Virus Diseases. So little is known about so few viruses that they cannot be classified with certainty on the basis of natural relationships. It is more convenient and less confusing to classify the diseases they cause. Virus diseases can be grouped into three artificial and sometimes ambiguous groups: mosaic, yellows, and ring-spot diseases. (Remember that there are exceptions to the following generalizations about them.)

The **mosaic diseases**, such as potato mosaic, characteristically produce chlorotic mottling and are usually transmitted by aphids. In most cases they are readily transmitted by mechanical means. Axillary buds are not induced to grow and blooming does not cease.

The true **yellows diseases**, such as aster yellows, do not produce spotting symptoms; instead, a sickly yellow coloration or even a bright and striking yellow coloration is produced. Auxillary buds are induced to grow and witches' brooms and rosettes are common. Very often flower parts are green and leaf-like and cessation of blooming is common. These diseases are generally transmitted by leafhoppers and are not readily transmissible by mechanical means.

The **ring spot diseases**, like tobacco ring spot, produce chlorotic spotting but have a tendency to regain normal appearance. These diseases are described as soil-borne but are, in reality, transmitted by fungi and nematodes. They can be transmitted mechanically but are not aphid- or leafhopper-borne. This group has the highest tendency for seed transmission. Although pollen may carry the virus and the embryo may be infected thereby, the plant bearing the flower that has pollinated does not become infected as a rule. Self-pollination tends to increase the rate of seed transmission when it does occur because the virus may be introduced into the embryo either through the egg or the pollen.

Morphology of Viruses. Viruses are composed of nucleic acids and proteins which are similar in chemical composition to other nucleic acids and proteins found in nature. The nucleic acids are of two kinds depending on the type of 5-carbon sugars they contain. Some contain DNA, (deoxyribonucleic acid), and others contain RNA (ribonucleic acid).

Fig. 11-2. Distortion, chlorotic splotching, and green ring spots of green ring mottle of sweet cherry.

Fig. 11-3. Chlorosis caused by apple mosaic.

Fig. 11-4. (A) Normal appearance of a cucumber plant compared with (B) a cucumber plant infected with cucumber mosaic virus.

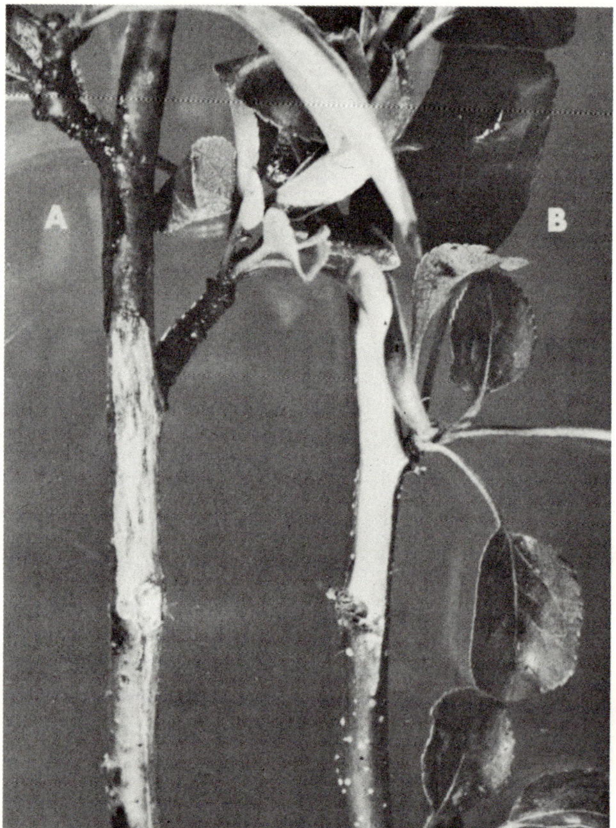

Fig. 11-5. (A) Necrotic symptoms of apple bark necrosis in a peeled apple twig compared to (B) the smooth white appearance of a peeled healthy apple twig.

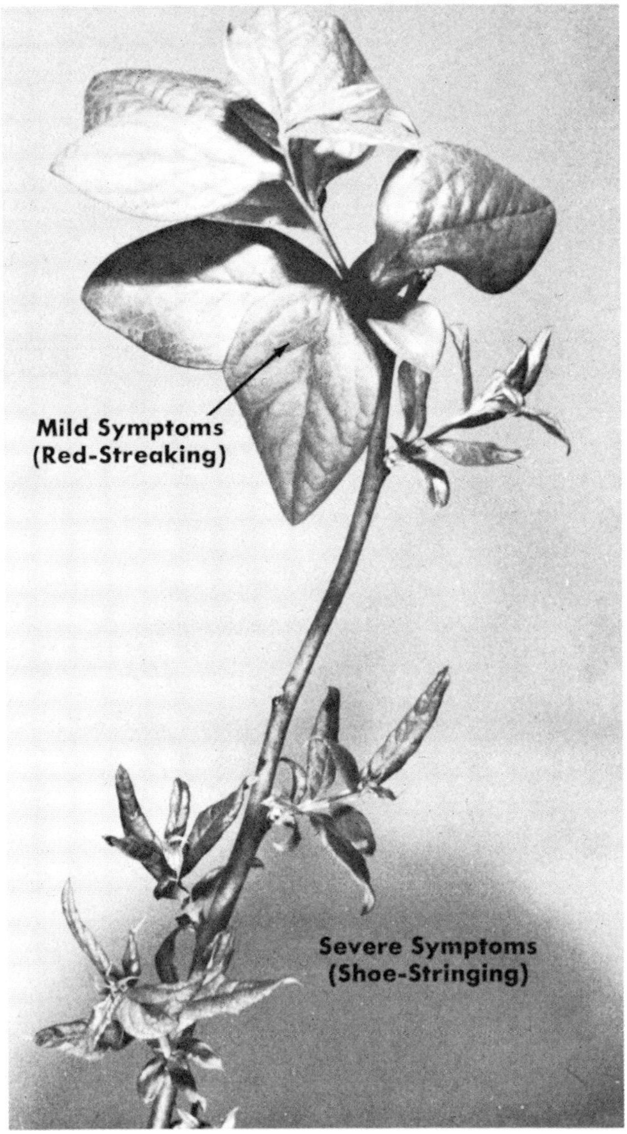

Fig. 11-6. Red streaking and stunting symptoms of shoe-string virus of blueberry.

Insect, animal, and bacterial viruses are mostly DNA viruses but some are RNA types. No plant virus has yet been found to contain DNA. Because so few have been analyzed, it is premature to conclude that all plant viruses are of the RNA type.

Of the plant viruses presently studied, two morphological types are the most common: (1) They are long and narrow, tubular rods or threads, 10-20 mμ $\times$ 200-1200 mμ and are composed of similar subunits arranged in a helical or coil-spring symmetry. There are, however, some rods (such as the alfalfa mosaic virus) which are short. (2) They are isometrical polyhedrons 20-60 mμ in diameter and of many (*poly*) sides (*hedrons*). The sides are equal (*iso*) measurements (*metric*) along each dimension. These viruses tend to be icosahedral, or twenty-sided. Each side of an icosahedron with isometrical symmetry is an equilateral triangle.

A virus particle is, essentially, a package of nucleic acid with a protein wrapper. The wrapper is called the capsid or **protein capsid** and the central package of nucleic acid is the **nucleocapsid.** Virus particles are 63%-95% protein and 5%-37% nucleic acid. It is not surprising that the polyhedrons, which are near-spheres, are 73% protein and 27% nucleic acid on the average, whereas the rods are 95% protein and 5%

Fig. 11-7. Iron chlorosis (due to iron deficiency) of pin oak.

Fig. 11-8. Genetic variegation in leaves of plants of various species.

Fig. 11-9. Herbicide injury to plants in three plant families. (The herbicide was probably amino triazole.)

Fig. 11-10. Injury to tobacco with 2,4-D.

nucleic acid. The sphere is the geometric configuration with the least amount of surface area per unit of volume.

The protein wrapper is constructed of hundreds to thousands of identical subunits. The subunits are composed of 100 to 200 amino acid molecules linked together in a long chain called a **polypeptide.** The subunits appear round or elongate because the polypeptides are folded back and forth upon themselves.

The nucleic acid is a long, very thin strand made of several thousand nucleotides. Each nucleotide is constructed of a 5-carbon sugar with a phosphate group and a nitrogen base. The nucleotides are made into a polymer, or chain, by linking the sugars together between phosphate groups. Four nitrogen bases occur in the nucleotides: adenine, cytosine, guanine, and uracil (or thymine for DNA).

There are only a few efficient designs for a biological container constructed of identical subunits. Efficiency is dictated by the distribution of energy into a stable configuration. Such configurations are (1) the helical or coiled arrangement of identical subunits in a tubular rod and (2) the isometrical or equally-spaced distribution of subunits in a polyhedron.

The protein wrapper, or capsid, protects the package of nucleic acid which carries the coded information for the construction of a virus particle. The nucleic acid is also the infectious portion of the molecule because purified nucleic acid without the protein can cause infection, induce typical symptoms, and result in the production of complete virus particles.

The **coding hypothesis** is the central theme of molecular biology today. In its simplest form, it is the concept that a sequence of nucleotides in a specific nucleic acid determines the sequence of amino acids in a specific protein. It appears that the code is universal. Thus, the code for the amino acid, histidine, is the same in rutabaga, rabbit, mice, and man. The coding unit is believed to be a triplet of three nucleotides and is called a **codon.** In general, the code is (1) **degenerate** in that more than one codon may be involved in making a single amino acid, (2) **ambiguous** because more than one amino acid may be made by a single codon, and (3) **nonoverlapping.** The code is read in groups of three nucleotides from a fixed starting point. Only a portion of the nucleic acid, about seven per cent, is involved in the protein code for the viruses that have been studied.

Transmission and Spread. The spread of most viruses depends on either (1) the distribution of infected nursery plants and propagating stock or (2) active insect vectors. These, for the most part, are sucking insects such as aphids and leafhoppers. Vectors, however, also include grasshoppers, earwigs, beetles, white flies, mealy bugs, and thrips as well as Eriophid mites, nematodes, fungi, birds, small mammals, and man. Would you consider pollen a vector too?

Spread of viruses by insect vectors depends on four factors: (1) the *virus*—numbers, distribution, and strength of sources; (2) the *host*—numbers, distribution, and susceptibility; (3) the *vector*—numbers, distribution, and efficiency of transmission; and, (4) *movement of the vector.*

The size of the vector population and the time of its entry into a crop area are important. Vectors may infect plants with viruses whether or not they colonize the plant. Therefore, the life history of the vector is critical for understanding how a virus is spread. Some insects, for example, may overwinter on a perennial plant and move to an annual crop species for colonization in the spring, carrying viruses with them. On the other hand, a flight of large numbers of insects, as with some aphids, may alight in a crop field, infect large numbers of plants by incidental probing, and move on.

Recognition of fundamental differences between virus diseases of annual and perennial plants must be made in regard to the efficiency of vectors and the spread of viruses. Inefficient, infrequently occurring, slow-moving vectors can cause serious spread and losses from viruses of perennial crops over a period of years. No matter how few the new infections are each year, they accumulate and the sum total of infections increases year after year, so long as the infected plants live for several years. The seriousness of swollen shoot of cacao in Africa and tristeza of orange in the United States are testimony to this.

Most annual crops begin each season free of viruses because only a few viruses, primarily in the ring spot group, are seed-borne. Infections are additive in the nonseed-borne viruses of annual crops only within the growing season and not year after year. Even with most of the seed-borne virus diseases, the level of virus infection fluctuates up and down because only a small percent of the seeds usually carry the virus. With perennial plants, the number of infected individuals tends to increase constantly so long as there is a source of virus, an active vector, and the infected plants remain alive for some years.

Virus and Vector Relationships. Relationships of virus and vector vary in specificity. Some viruses are transmitted by 50 different aphids while others require a specific strain of aphid or leafhopper and none other will do. Some aphids, such as *Myzus persicae,* can transmit nearly fifty distinct viruses and yet some aphids transmit none at all!

Two major kinds of relationships exist, **persistent** and **nonpersistent.** There are various intermediate types which are discussed later. The relationship is determined by the virus and not the vector because a vector may transmit one virus in a persistent manner and another virus in a nonpersistent manner. However, a given virus always behaves in only a persistent or a nonpersistent manner, regardless of how many different vectors are able to transmit it.

Persistence and nonpersistence are differentiated on two bases: (1) the length of time required after feeding before the vector is capable of transmiting the virus, and (2) the length of time the vector retains the capacity to transmit the virus without further access to the virus source.

Persistent viruses are primarily leafhopper-borne and generally cause the yellows type of disease. They usually cannot be transmitted mechanically. The vectors can acquire the virus in a short feeding time but cannot transmit the virus immediately. The period of time from acquisition to transmission is called the **latent** or **eclipse** period and may last 16 days. Once **viruliferous,** or containing the virus, vectors remain so for long periods of time, often for the remainder of their lives.

Most leafhoppers feed on the contents of the phloem and the viruses they transmit are found there also. Some exceptions do occur, as in the leafhoppers which transmit Pierce's disease of grape feed in the xylem. The same vectors transmit the virus to alfalfa and the disease caused by the virus in this plant is called alfalfa dwarf.

Some persistent viruses also are called **circulatory viruses** because the virus enters the circulatory system of the vector. The virus passes through the stylet, into the gut, through the gut wall, into the blood stream, and eventually reaches the salivary glands. When the insect extrudes saliva, it inoculates the plant on which it feeds with the virus. This circulatory route might be called the "blood and guts route."

The ability of an insect to transmit a virus cannot be determined by morphologic identification alone. A species of *Cicadulina,* a leafhopper, transmits maize streak. One leafhopper strain, which is morphologically indistinguishable from the vector, is unable to transmit the virus. The virus can be found in the blood of the vector but not in the blood of the nonvector. Surgical puncture of the gut wall permitted entry of the virus into the blood stream with the subsequent ability of the insect to transmit the virus. These two strains of leafhopper were found to differ by a single Mendelian gene which is sex-linked. Thus, a single gene difference, which was not expressed morphologically, determined which strain was capable of transmitting the virus.

Some persistent viruses are also called **trans-ovarian viruses** because they are carried (*trans*) in the eggs (*ova*). Fukushi, in 1935, worked with the rice dwarf virus and its vector, the green rice leafhopper, and found that progeny from a cross with a viruliferous female and a virus-free male were viruliferous. The progeny were not viruliferous when the female was virus-free and the male was viruliferous. He observed passage of the virus through the egg for six consecutive generations. He used 1200 rice plants to raise the insects and calculated the dilution of the original virus content to be 10^{-8}. He considered this to indicate the multiplication of the virus in the vector because the virus probably would have been diluted too much for transmission without multiplication.

The skeptical world did not accept Fukushi's triumph. The clover club leaf virus was used later by Black in a repeat of Fukushi's experiment. The vector can live on both clover, which is susceptible to the virus, and alfalfa, which is totally immune to the virus. By maintaining leafhoppers on alfalfa continuously, the possibility of further acquisition of the virus was eliminated. In each generation, a portion of the progeny was kept on the alfalfa and a portion was placed on clover to observe if symptoms would develop. Symptoms on clover would indicate that the insect carried the virus. If they carried the virus, they must have acquired it from their parents because the alfalfa, on which their parents lived, is totally immune to the virus. Twenty-one generations and five years later, leafhoppers which lived only on alfalfa and which never had contact with clover still were viruliferous. If there had been no multiplication the original quantity of virus would have been diluted 10^{-26}. This would have required a quantity of virus which weighed more than an adult leafhopper. Viruses which multiply in their vectors are called **propagative viruses.**

Multiplication was demonstrated with aster yellows virus in its insect vector by a more direct method. Infected leafhoppers were ground, diluted, and some of the diluted juice was injected into virus-free leafhoppers with a very fine micropipette. Later these injected leafhoppers were ground also and their diluted juices were injected into still other leafhoppers. Dilution beyond the original number of virus particles was demonstrated in this manner, indicating that the virus had multiplied.

Nonpersistent viruses are primarily aphid-borne (Fig. 11-11A), usually transmitted mechanically, and generally cause mosaic diseases. The aphids acquire the virus rapidly, can transmit it immediately, and then rapidly lose the ability to transmit it. The route of stylet insertion is quite pronounced and is called the **aphid tract** or feeding tract (Fig. 11-11B). It seems that the virus is carried on the distal 15 mμ of the stylet in those tested. These are, therefore, called **stylet-borne viruses.** Experimentors demonstrated this by carefully drawing the stylet out of its sheath and subjecting the tip to ultraviolet irradiation to inactivate the virus.

About 250 species of aphids have been tested as vectors of plant viruses. That is ten per cent of the known species of aphids. How many more are vectors?

Summary of Virus and Vector Relationships. Between the extremes of the defined

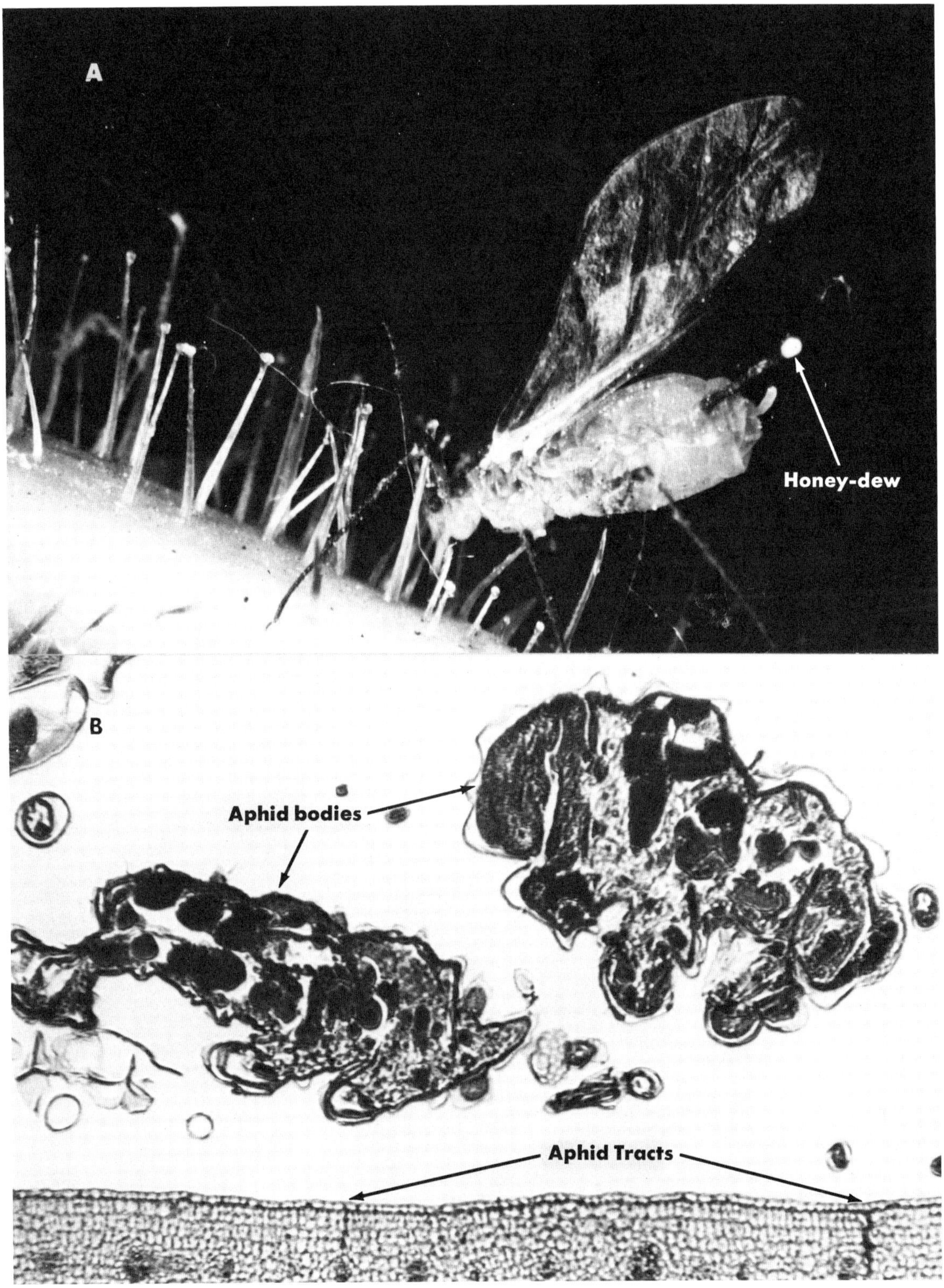

Fig. 11-11. (A) An adult, winged aphid (*Amphorophora rubi*) feeding on the stem of a wild black raspberry. Note the gleaming drop of honeydew. (B) Section through feeding aphids showing aphid feeding tracts in the plant section as well as cross sections of the aphid bodies (× *100*).

groups of persistent and nonpersistent viruses is an increasing number of intermediate types. For this reason, some authors are now discarding the persistent and nonpersistent classification and are referring to viruses as stylet-borne, circulatory, propagative, or trans-ovarian to indicate a relationship of the virus to the vector. These are useful and meaningful terms. Circulatory and propagative are not exclusive terms because one denotes the route of movement and the other multiplication, if it occurs. In addition, two other specialized terms have been introduced: NEPO and NETU viruses. The NEPO viruses are NEmatode-borne POlyhedral viruses and the NETU viruses are NEmatode-borne TUbular rod-shaped viruses.

In general, it still holds that most aphid-borne viruses are stylet-borne and nonpersistent. When we know more about a larger number of viruses and their vectors, a more meaningful and rewarding system for classification can be devised.

Control. Regulatory measures for most virus diseases usually depend on knowing what the virus is, what its vectors are, and from where both virus and vector are coming. In essence, a knowledge of the disease cycle.

The **first** control measure of importance is to plant virus-free seed, virus-free nursery, or propagating stock.

The **second** most important measure of control is to prohibit or restrict the introduction of viruses. A low level of infection by harvest time in annual crops can often be tolerated without much loss in quality or quantity of the crop. This is particularly so if the virus is introduced late in the season. Very few infections can be tolerated in perennial crops without reductions in yield, because the infections are additive, year after year. This is especially significant with the fruit, nut, lumber, and shade trees. Exceptions to this occur in the absence of vectors and with those crops which, although perennial, are harvested in one to three years before replanting.

Restriction of entry and spread of viruses within a planting involves the following practices, depending on the disease and locality:

(1) Roguing off-type plants is useful if the frequency of introduction is low and the virus-infected plants exhibit symptoms.

(2) Removal of volunteer plants is especially significant if such plants are the major source of virus.

(3) Eradication of alternative hosts is sometimes used to eliminate a source of the virus and sometimes the overwintering host for the vector.

(4) Decontamination of tools can be important with mosaic diseases caused by viruses which are easily spread mechanically.

(5) Chemical control of vectors can be important with nematode and leafhopper-borne viruses. Chemical control of migratory aphids has been of little value because the insects have adequate time to probe and transmit the virus before coming in contact with the chemical or before the chemical has time to kill them. Application of systemic insecticides has been very useful.

(6) Cultural practices, such as increasing the size of fields to reduce the proportion of border area, breaking-up crop cycles, and altering planting dates to miss insect movements, may reduce the incidence of virus infections depending on the disease and locality.

(7) Resistance to vectors will be an important control measure for some virus diseases in the coming years.

(8) Crop immunity to the virus will be a major control practice for some crops in the future.

Severe problems with virus diseases can be expected when new crops are introduced into an agricultural area, or when virgin lands are developed into a new agricultural area. In both cases, agricultural practices bring new hosts into contact with viruses. It is for this reason that we should be especially alert for virus infections where intensive agricultural assistance is offered to developing nations.

SUGGESTED READING

Plant Virology, Corbett, M. K., and H. D. Sisler, eds., Gainsville, Florida, University of Florida Press, 1964.

Bawden, F. C., *Plant Viruses and Virus Diseases*, 4th ed. New York, Ronald, 1964.

Black, L. M., "A Plant Virus that Multiples in its Insect Vector." *Nature*, Vol. 166 (November, 1950), p. 852.

Smith, K. M., *Textbook on Plant Virus Diseases*. London, Churchill, 1957.

12

Tobacco Mosaic

Tobacco mosaic virus (TMV) is not a typical plant virus. It remains infectious in dried plant material for many years—longer than any other plant virus, can be diluted to 1:1,000,000 and can be heated to 90° C. for ten minutes and still remain infectious. Because of these characteristics, it is easily handled and is thus an excellent virus for the beginner. It has served well as a tool for virus investigations.

Disease Cycle. TMV overwinters in infected plant debris in the soil and in tobacco products and refuse. The chief means of spread is through mechanical transmission such as the handling of transplants and contact of plants by man, machinery, birds, and chewing insects. The transmission of the virus by the false potato aphid is probably of little consequence. Seed transmission occurs in tomato on rare occasions.

Control. Avoid infected soil and debris. Wash hands thoroughly with soap to avoid contamination when handling susceptible plants such as tomato and tobacco. Do not smoke and keep tobacco products away from tobacco and other susceptible plants.

13

Potato Latent Mosaic

The disease of potatoes caused by potato virus X (PVX) is called latent mosaic because many potato cultivars, especially the old standards, exhibit no symptoms when infected. Other cultivars exhibit a chronic mosaic and still others react with an acute systemic reaction known as top necrosis.

The virus overwinters primarily in the potato tubers. Distribution of infected seed tubers disseminates the virus over long distances, whereas spread within a field is through mechanical contact by man and implements. Although some aphids do transmit some potato viruses such as virus Y, no sucking insect has been found to be a vector of PVX, whereas certain chewing insects are. The fungus, *Synchytrium endobioticum,* has been reported as an infrequent vector.

The disease is controlled by the use of seed tubers certified to be relatively free of virus. Complete resistance is known and someday will provide control in new cultivars.

Experiment XI

Biological Properties of Viruses—Local and Systemic Infections

Virus diseases, as all plant diseases, can generally be divided into two categories: systemic and local lesion diseases. Systemic infections result from a spread of the pathogen from the point of penetration to other areas of the host. In cases such as systemic virus diseases, FUSARIUM wilt of tomato, Dutch elm disease, and bacterial wilt of cucumber, the invasion is extensive. **Local lesion** infections are restricted to limited areas and do not spread into other areas of the host tissues. Legume nodules, local lesion infections of viruses, black rot of grape, and apple scab are examples of restricted infections. There are some diseases which are not extensive and yet are not small in size. Thus, there is a gradation between the two groups. Regardless of the variation, it is the distance of spread within the host tissues and not the rate of spread that defines systemic infection. Some diseases, such as dead arm of grape, exhibit both phases.

Local lesion and system reactions are dependent on the host and the virus (Fig. 13-1) as well as the environment.

EXPERIMENT

Inoculate plants of the species listed in Table 13-1 with TMV and PVX. Some extra plants of *Nicotiana tabacum* can be provided for those who wish to inoculate with tobacco products they may be carrying such as cigarette tobacco or snuff.

1. **Preparation of Inoculum.** Grind 1 g. of TMV-infected tobacco leaves in 10 ml. of distilled water with a mortar and pestle and a small pinch of fine, washed sand. Dilute this crude juice 1:100 for use as inoculum. Prepare the PVX inoculum in the same manner but do not dilute it. Two different students should prepare the inoculum for the TMV and PVX to avoid cross-contamintaion.

2. **Inoculation of Plants.** Lightly dust the leaves to be inoculated with carborundum, 400 to 600 mesh. The eye will hardly notice a light deposit. If gray deposits are visible, you have used too much. Dip a cotton swab or glass spatula into the inoculum and gently swab the leaf. Do not rub over any area more than once since this only serves to increase the amount of injury. Q-Tips are handy because they are sterile, inexpensive, and disposable, whereas glass spatulas must be cleaned and sterilized. While swabbing the leaf you may wish to support it underneath with a paper towel. Do *not* inhale carborundum.

3. **Observations.** Local lesions (Fig. 13-1) first appear as incipient infections in 3 to 4 days on *Gomphrena globosum* and in 2 days on *Nicotiana glutinosa*. Obvious systemic infections take longer to appear. Good growing conditions, especially adequate light and temperatures under 80° F., are required for the expression of optimum symptoms.

Examine the petioles (Fig. 13-2A) of leaves from *Nicotiana tabacum* for epidermal hairs (Fig. 13-2B). Plants which are used for preparing inoculum can be used prior to their grinding as well as plants which are inoculated after they exhibit systemic symptoms. Scrape some of the hairs and mount them in water on a microscope slide. Crystals of TMV are easily seen in these colorless hairs (Fig. 13-2 C-G). These are signs of disease. The crystals are flat and views from the top and the side are quite different (Fig. 13-2 E, F). Some crystals are irregular and not as easily recognized. Oblique illumination (Fig. 13-2 D) accents the three dimensions of the crystal. Holmes' strain of TMV from rib grass produces numerous and well-formed crystals.

Plant Material. Plan 2 weeks from seeding to transplanting for all plant material required. Plants of *Nicotiana tabacum* and *Datura tatula* which are used for sources of virus should be inoculated, respectively, with TMV and PVX 2 weeks after transplanting and may be used in class 3 to 4 weeks after inoculation. They can be used sooner but older infections give the class an opportunity to observe advanced stages of the diseases. Plan to use plant material for inoculation by 3 weeks after transplanting.

Expendable Supplies × Total Number of Students or Groups

3 plants of *Nicotiana tabacum*
(include extras for inoculation with tobacco products)
2 plants of *Nicotiana glutinosa*
2 plants of *Gomphrena globosum*
2 plants of *Datura tatula*
2 Q-Tips
9 pot labels

Nonexpendable Supplies × Total Number of Students or Groups of Students Which are Seated in a Laboratory Section

2 mortars and pestles (steamed between laboratory sections)

Fig. 13-1. **(A) Local lesions of PVX on *Gomphrena globosum* and (B) TMV on *Nicotiana glutinosa*.**

General Supplies × Number of Laboratory Sections

1 plant of *Datura tatula* infected with PVX
1 plant of *Nicotiana tabacum* infected with TVM

General Supplies Available

carborundum, 400 to 600 mesh, in a dispenser or salt shaker with 2 layers of cheesecloth
fine washed sand

TABLE 13-1. Symptoms of TMV and PVX infections in 4 different plants.

Suscept	Symptoms on infection with: TMV	PVX
Datura tatula	brown local lesions	systemic mosaic
Gomphrena globosum	brown local lesions	red local lesions
Nicotiana glutinosa	brown local lesions*	ringspots* or mottle
Nicotiana tabacum	systemic mosaic	ringspots or mottle

* Systemic with certain strains

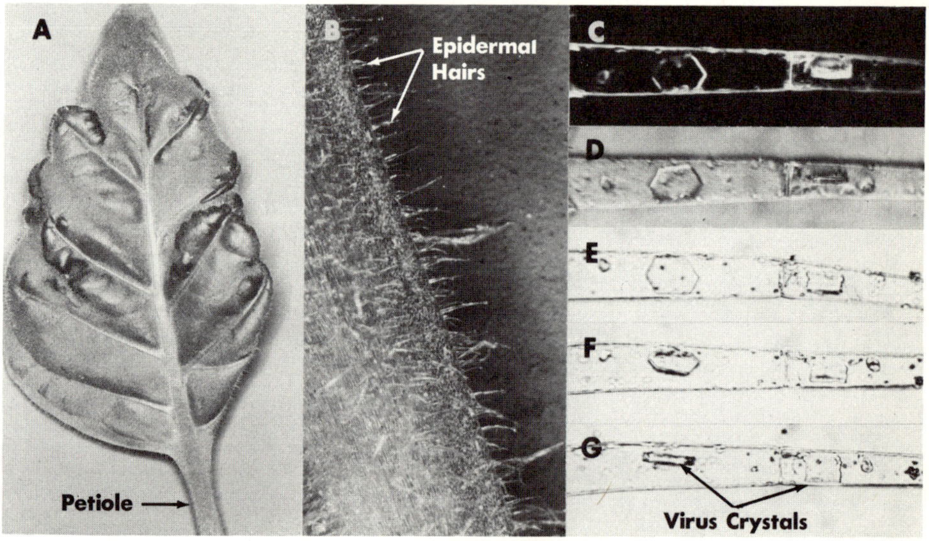

Fig. 13-2. (A) Puckering and mosaic symptoms on leaf due to systemic TMV infection on *Nicotinana tabacum;* (B) enlargement of a petiole showing capitate epidermal hairs; and, (C) crystals of TMV in an epidermal hair as shown by dark-field illumination; (D) oblique, bright field illumination, and (E-G) ordinary bright-field illumination. The hair was rolled on the microscope slide for the last three pictures in order to show the 3-dimensional configuration of the crystals.

DISCUSSION QUESTIONS

1. How can mosaic symptoms of nutritional deficiencies, herbicide injury, and tobacco mosaic be distinguished?
2. How can infections of a systemic fungus, systemic bacterium, systemic virus, nutritional deficiencies, injuries, and genetic anomalies be distinguished?
3. Why is carborundum used in inoculations? (Consider how viruses gain entry into their hosts.)
4. What is the value of local lesion and systemic hosts to a plant pathologist in the study of viruses and the diseases they cause?

Experiment XII

Biological Properties of Viruses—Synergism

Synergism is the phenomenon in which the effect of two things acting together is greater than the sum of the effects of the two acting alone. Synergism, it is well known, occurs with drugs and also with certain combinations of virus infections. TMV and PVX each cause a mosaic in separate infections of tomato but in combined infections they cause sudden wilting and death of young tomato plants (Fig. 13-3). PVX and PVY (potato virus Y) give a similar synergistic effect in combined infections in tobacco (*Nicotiana tabacum*). The occurrence of synergism depends on the host and the environment as well as the two viruses (Table 13-2).

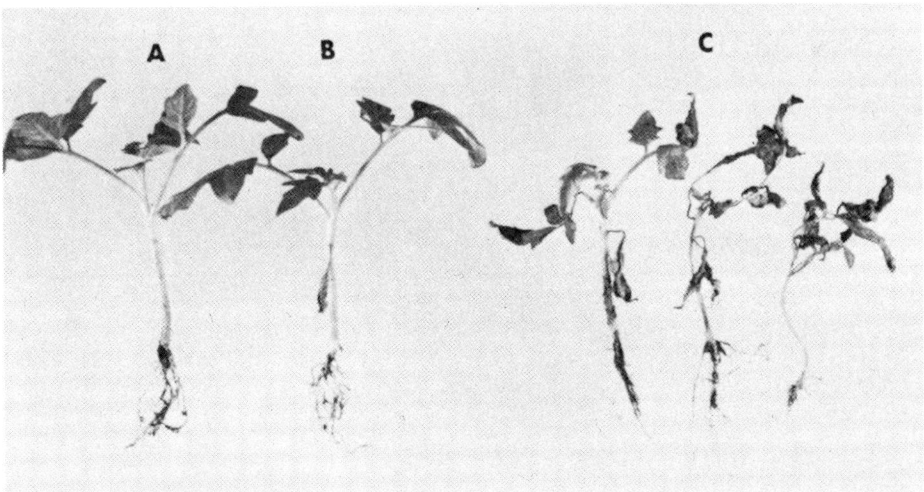

Fig. 13-3. (A) Symptoms of PVX single infection; (B) TMV single infection; and, (C) PVX and TMV combined infections resulting in synergistic action.

TABLE 13-2. Reactions of tobacco and tomato to multiple infections with TMV, PVX, and PVY. Synergism is designated with a (+) and its absence with a (—).

Virus Combination	Suscept	
	Tobacco	Tomato
PVX + PVY	+	—
PVX + TMV	—	+
PVY + TMV	—	—

Information as to the identity of unidentified viruses can be secured in some cases by determining if there are synergistic reactions with infections of plants in which known viruses have also been inoculated. Healthy uninoculated plants and plants inoculated with known mixtures serve as controls. Other plants, all known to be virus-free, are inoculated with a known virus and also the unidentified virus. An examination of Table 13-3 reveals the utility and some of the limitations of such a technique.

TABLE 13-3. Identification of TMV and PVX by inoculation of tomato (*Lycopersicon esculentum*) with combinations of TMV, PVX, and PVY. Synergism is indicated with a (+), no infection with a (—), and mottling or ringspots with (I) for infection other than synergism.

Virus to be Identified	Reaction of Test Plants Inoculated with the Virus to be Identified and:			
	no other virus (healthy control)	TMV + PVX (Synergism Control)	TMV	PVX
None	—	+	I	I
PVY	I	+	I	I
PVX	I	+	+	I
PVX + PVY	I	+	+	I
TMV	I	+	I	+
TMV + PVY	I	+	I	+
TMV + PVX	+	+	+	+

EXPERIMENT

Inoculate young tomato plants with (1) PVX alone, (2) TMV alone, (3) PVX and TMV on different leaves, (4) TMV and an unknown virus, and (5) PVX and an unknown virus. The instructor will provide the unknown inocula from portions of the TMV and PVX inocula prepared by the students, or by grinding uninfected leaves, or mixtures thereof. All inocula should be labelled with code numbers. This experiment is run conveniently with Experiment XI because the same inocula for TMV and PVX can be used.

Plant Material. See Experiment X for the preparation of infected plants for inocula. Tomatoes are transplanted 2 weeks after seeding and used in the experiment 2 weeks after transplanting. For quick, dramatic results, very young plants should be used.

Expendable Supplies × Total Number of Students or Groups of Students

5 tomato plants
5 pot labels
2 Q-Tips

DISCUSSION QUESTIONS

1. Can synergistic tests on tomato with PVX and TMV distinguish if other viruses, such as PVY, are present in unidentified materials?
2. Can both viruses which cause synergism in a suscept be natural parasites of the same plant in nature? Under what conditions? Of the plant in single infections in nature?
3. If a control plant inoculated with both TMV and PVX failed to show synergism, are the results of other tests run at the same time with the same inocula valid? What could the matter be?

Experiment XIII

Physical Properties of Viruses

Viruses are, in part, proteins and exhibit the physical properties and characteristics of proteins. They are crystallizable and yet retain infectivity; are inactivated by heat; are particulate and can be diluted to an end point; are antigenic and can induce specific antibodies. Proteins and viruses, because they are proteins, differ in tolerance to heat, dilution end point, shape, size, and other features.

EXPERIMENT

Determine the dilution end point (Fig. 13-4) and temperature inactivation (Fig. 13-5) of TMV. This experiment requires a good deal of miscellaneous equipment, careful attention to detail, and cooperation of class members. Each laboratory section should perform the experiment as a single group, providing replicates from section to section. The assay for each experiment requires 10 plants. Provision of this quantity for each small group or each student in a large enrollment class is prohibitive.

Very low concentrations of infectious virus particles are handled in order to measure the dilution end point and the high percentage of inactivation at the higher temperatures used. Contamination with the very infectious, highly concentrated crude juice would overshadow the low concentrations. In order to avoid contamination, class members should be assigned specific tasks. Each should wash thoroughly with soap

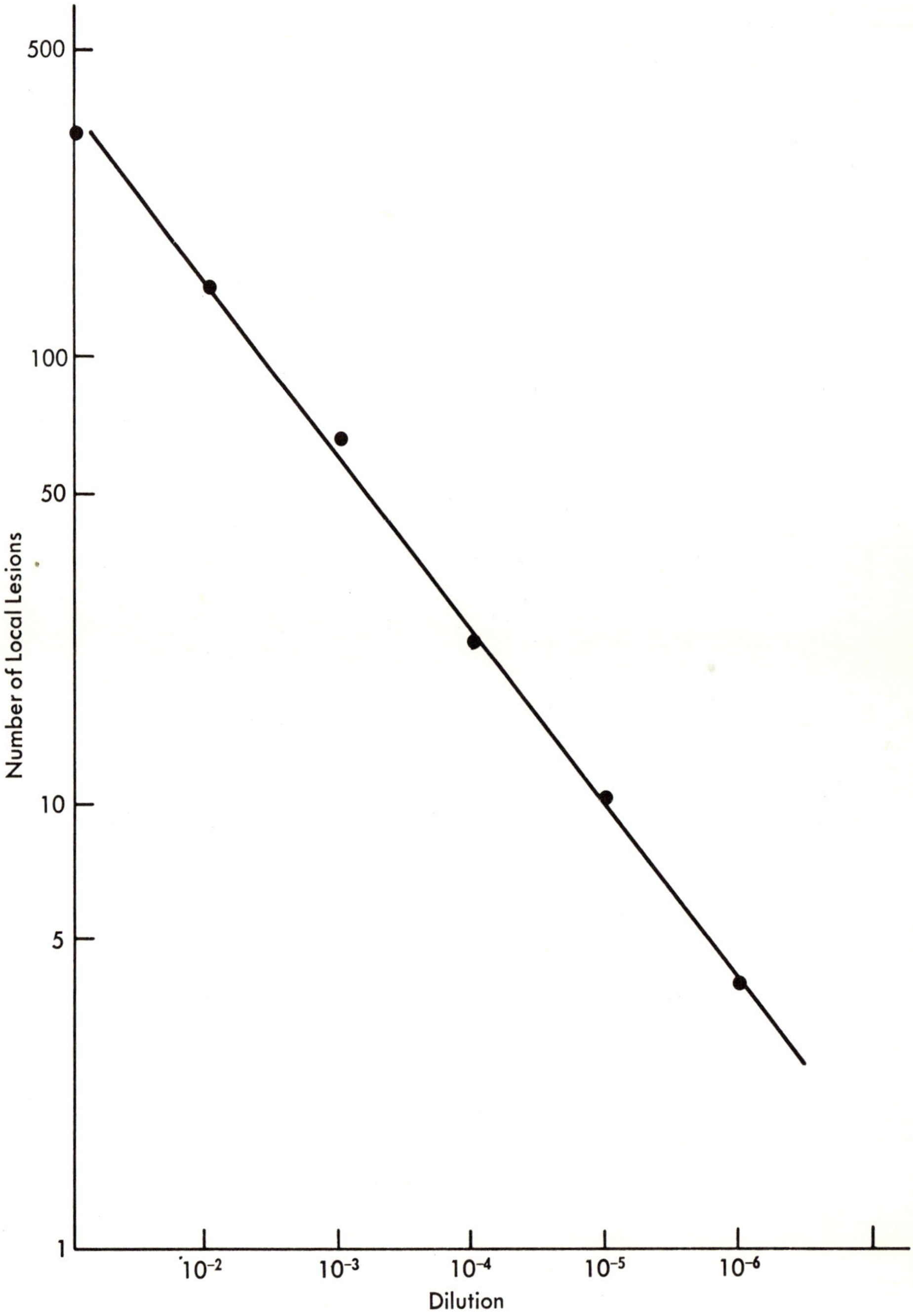

Fig. 13-4. Dilution curve of TMV.

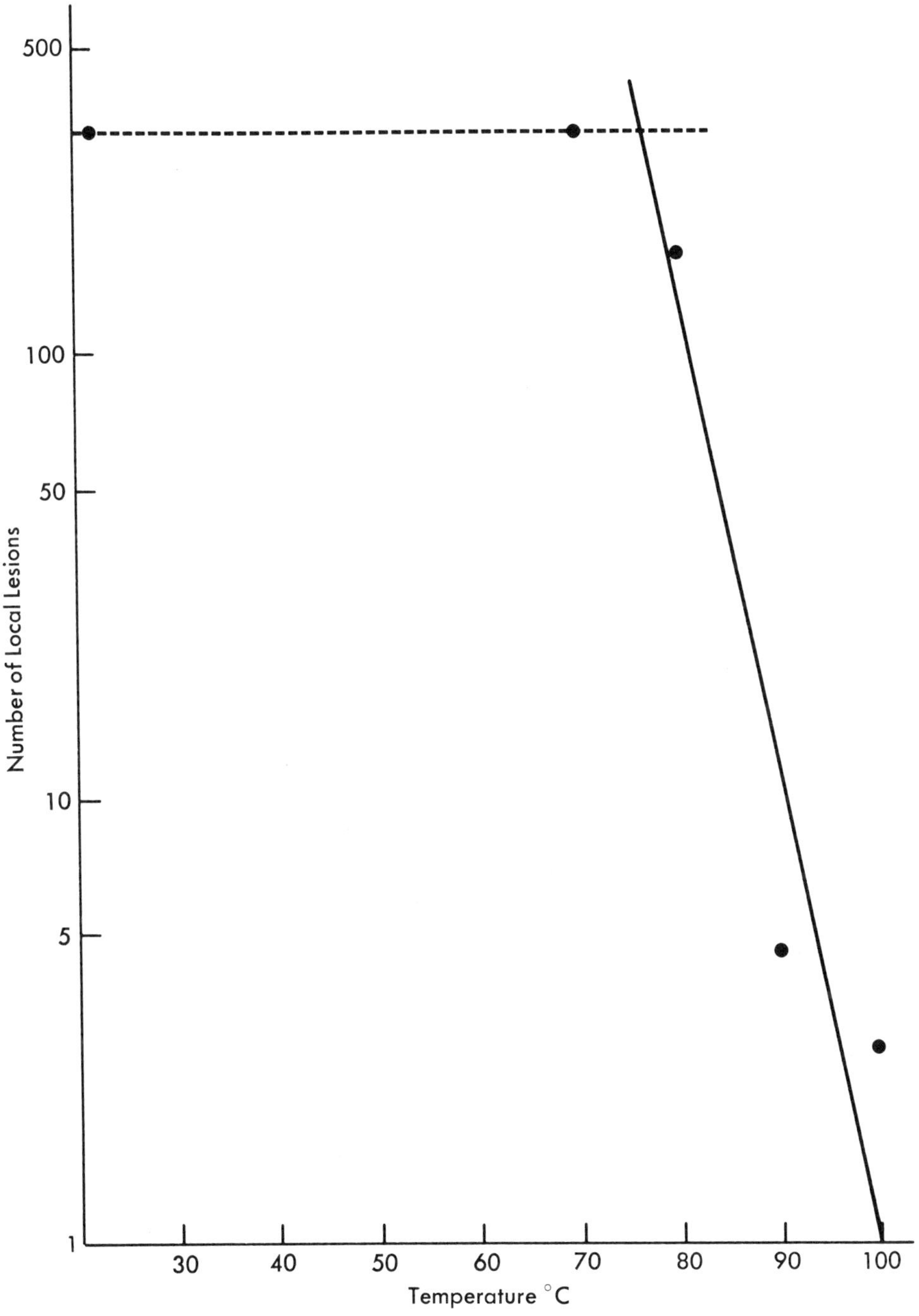

Fig. 13-5. Inactivation of TMV as a function of temperature.

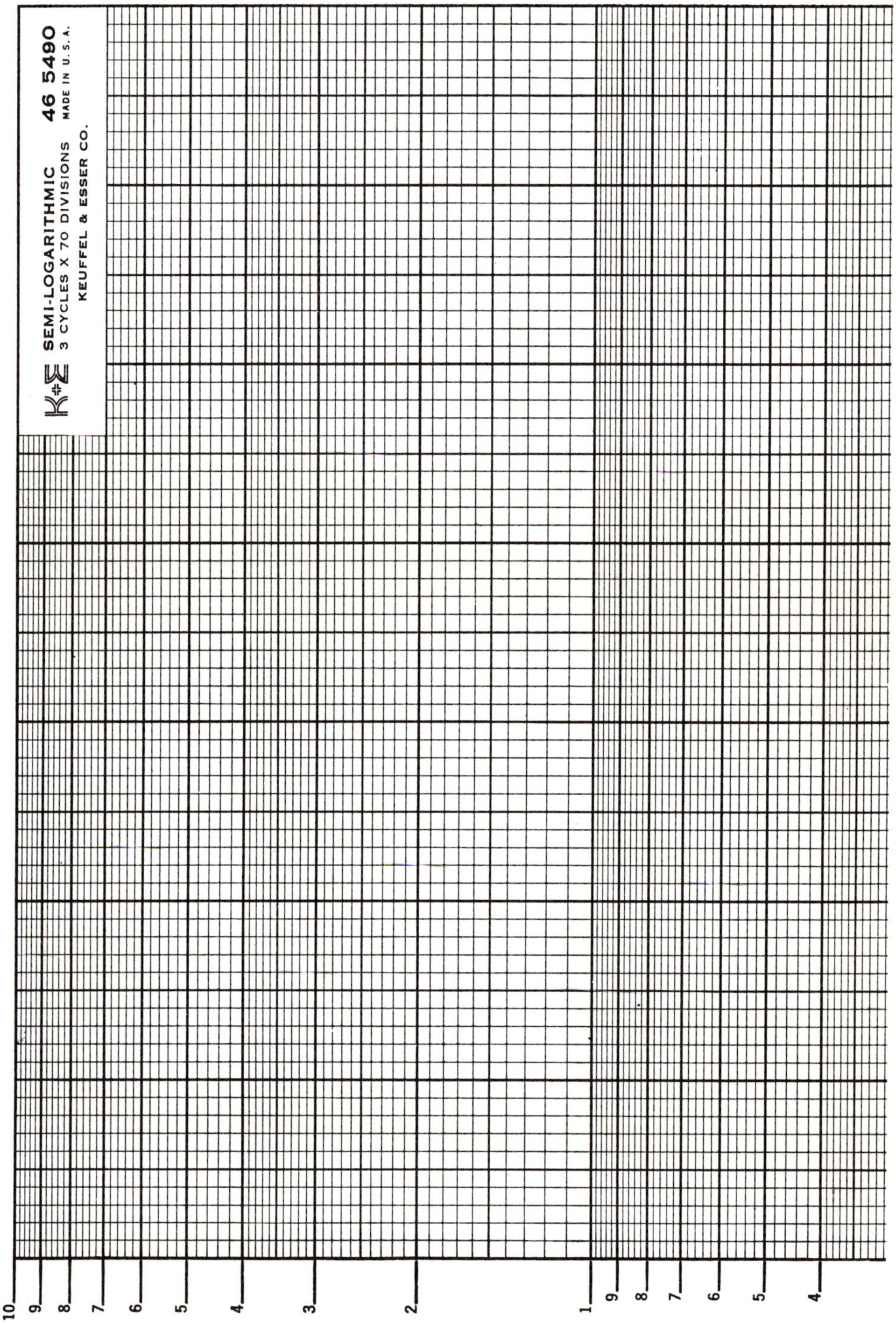
K+E SEMI-LOGARITHMIC
3 CYCLES X 70 DIVISIONS
KEUFFEL & ESSER CO.
46 5490
MADE IN U. S. A.
10 9 8 7 6 5 4 3 2 1 9 8 7 6 5 4

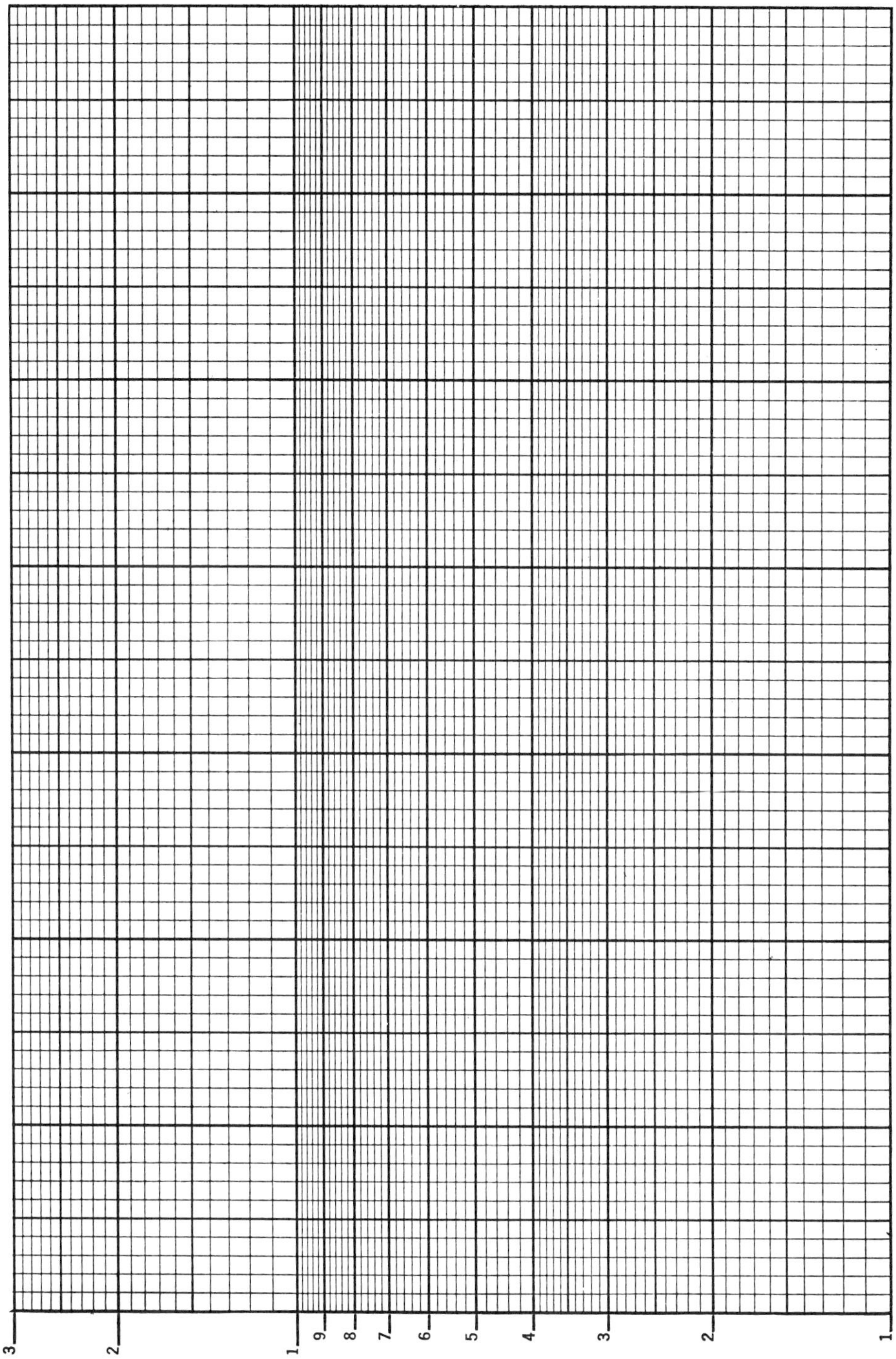
3
2
1
9
8
7
6
5
4
3
2
1

and not smoke or touch tobacco products, infected plants, or the areas where crude juice is prepared except for the grinders.

1. **Grinders** prepare the crude virus extract. Grind 2 infected leaves in a mortar and pestle with a little fine, washed sand. Filter the crude juice through 2 layers of cheesecloth into a graduate cylinder of 50 ml. capacity. Add enough distilled water to make a total of 30 ml. and mix. This is to insure that there is enough inoculum from a single source for the experiment.

2. **Manager** holds a flask while a grinder pours the crude virus extract into it. This is carried to a different table for the remainder of the experiment in order to avoid contamination with the crude sap or spills from it in the vicinity of its preparation.

3. **Pipetter** distributes 4 ml. of extract into one tube and 2 ml. into each of 5 others. Run the extract down the side of the tube with the trademark. The Inoculator (see Step 9) will tip the tube the other way to avoid contamination. The extract which remains on the side of the tube near the top does not receive the same amount of heat as the bulk of the extract in the bottom. A tenfold dilution series is made from the tube with the 4 ml. by serial transfer of 1 ml. of extract to 9 ml. of distilled water. With each transfer, rinse the extract up and down in the pipette to mix the contents of the tube and flush the pipette.

4. **Manager** distributes the tubes with 2 ml. of extract to the Firemen. He keeps one tube and labels it "no dilution and room temperature."

5. **Firemen** heat 2 ml. of extract in test tubes for precisely 10 min. at 70°, 80°, 90°, and 100° C. as follows. Place the tubes in water (which has been heated to the desired temperature) in a beaker on a ringstand. Swirl the tube to keep the contents stirred for uniform treatment. Constantly keep the water stirred. During the 10 min. treatment, keep the temperature constant. Remove the tubes after precisely 10 min. and promptly run cold water over the outside of the tube or immerse it in crushed ice. Be sure not to allow any water to run into the tube and dilute the virus extract.

To keep the temperature constant, place the Bunsen burner beneath the beaker when the temperature drops ½° and remove it after a moment. If the temperature does not rise, repeat. If the fire is held under the beaker until the thermometer registers a rise, it will rise too far. Practice holding the temperature constant before placing the tube of extract in for the exact 10 min. treatment. The accuracy of the experiment depends on the correct temperature for the correct length of time.

6. **Timer** is an assistant to the fireman. He watches his watch and times the tubes in and out of the bath.

7. **Labeler** prepares the plants for inoculation while the others are preparing and treating the aliquots of virus extracts. Set 11 plants of *Nicotiana glutinosa* in a row and remove, with a small blade or scissors, all the leaves (on 10 of them) except for 3 which are approaching maturity and are almost fully expanded. Label 10 pot labels as indicated in each vertical column, and insert one in each pot.

1	2	3	4	5	6	7	8	9	10
0	70	80	90	100	10^{-2}	10^{-3}	10^{-4}	10^{-5}	10^{-6}
10^{-6}	0	70	80	90	100	10^{-2}	10^{-3}	10^{-4}	10^{-5}
10^{-5}	10^{-6}	0	70	80	90	100	10^{-2}	10^{-3}	10^{-4}

8. **Keeper of the Swabs** organizes the tubes of treated and diluted extracts and sees that contamination does not occur. He keeps track of which leaf on which plant is inoculated with which extract.

9. **Inoculator** is a single individual who does all the inoculating for uniformity because the amount of pressure and number of swabbings per leaf affect the number of local lesions that result. He should practice on the 11th plant provided in order to perfect his touch and stroke technique. Many find it easiest to stroke in one direction rather than back and forth. As much of the leaf area should be stroked as possible without overlapping the strokes. Dust lightly with 400 to 600 mesh carborundum before inoculating.

10. **Calculate** the average number of lesions and dilution for each treatment when the lesions appear. Construct a temperature inactivation curve on a semilog scale. This curve is really the intersection of two lines (Fig. 13-5). The horizontal line represents a static situation with no effect of the temperature (room temperature) over the 10 min. time span. The slope of the steep intersecting line is the rate of protein denaturation. It's a first order chemical interaction with a Q_{10} of about 600. The larger the Q_{10}, the greater the increase in the rate of the reaction with an equal rise in temperature; or, in this case, the greater the rate that the protein would be denatured. An ordinary chemical reaction has a Q_{10} on the order of 2! In other words, the rate doubles with a 10° (Kelvin) rise in temperature.

Construct the dilution end point curve on a semilog scale. By placing the tenfold dilutions at equal spacings along the horizontal axis, a log scale is made creating a log–log plot of the dilution.

Expendable Supplies × Number of Laboratory Sections

11 potted plants of *Nicotiana glutinosa*
with at least 3 leaves in a near-expanded stage
10 pot labels
11 test tubes
2 sterile pipettes
1 piece of double cheesecloth about 8 inches square

Equipment Available

4 ringstands
4 ringclamps
4 asbestos screens
4 beakers (600-800 ml.)
4 thermometers
carborundum, 400 to 600 mesh

Equipment Available—Steamed Between Laboratory Sections

1 large mortar and pestle
2 test-tube racks
1 graduate cylinder, 50 ml.

DISCUSSION QUESTIONS

1. Why is the location of the 3 leaves inoculated with any one extract so designed that the 3 leaves are of different ages and are on 3 different plants? Why not inoculate all 3 leaves on each plant with only one extract?
2. Sometimes the dilution curve is flat at the higher concentrations. Why?
3. Construct hypothetical inactivation curves for viruses which are more tolerant and less tolerant of heat than TMV.
4. Why is the local lesion host preferred over a systemic host for assay?
5. How can you assay the number of infectious virus particles in an extract with a systemic host? (It can be done!)

14

Aster Yellows

Aster yellows is a classic example of a virus disease in the yellows group. Symptoms include a general yellowish color. Often a plant is affected unilaterally, one branch at a time, while the rest of the plant appears healthy. Dormant buds are stimulated and yellow, and thin branches with long internodes appear in profusion to produce the characteristic witch's broom. Roots may also proliferate. Flower parts are often leaf-like in shape and color, and flowering may cease altogether. Plants may be stunted.

Although species in 150 genera are susceptible, economic hosts are primarily ornamentals in the Compositae, such as China aster, and vegetables in the Umbelliferae, such as carrot, celery, and parsley.

The virus overwinters in perennial weed hosts and the introduction of the virus into crops is dependent, in most cases, on the leafhopper vectors (Fig. 14-2). The leafhoppers transmit the virus in a persistent manner. The insects are unable to transmit the virus for about 10 days after acquisition by feeding on infected plants but are able to transmit the virus without contact with an infected plant for as long as three months afterward. The virus is not mechanically transmitted.

Fig. 14-1. Yellow, spindly, unilateral symptoms of aster yellows of celery.

Control. High value crops, such as the aster, may be grown under screens to exclude the leafhoppers and, therefore, to exclude the virus. With some crops, such as carrots, the population of leafhoppers can be kept low with repeated insecticide sprays. This reduces the incidence of disease to acceptable levels. With lettuce and other crops, which are sown in seed beds and later transplanted, the disease is reduced by the eradication of weed hosts in a border area and with a program to reduce the population of the vectors until transplanting time.

Fig. 14-2. A species of *Cicadulina*, a leafhopper, which is a vector of aster yellows virus.

15

The Fungi

The mushrooms, mildews, smuts, rusts, and puffballs are fungi (sing. fungus) and do not have stems, leaves, roots, fruits, or vascular systems. They are usually defined as simple plants with filamentous, branched, thread-like structures having true nuclei and no chlorophyll. Hence, fungi do not photosynthesize and must obtain reduced carbon compounds, such as sugars, as saprobes or parasites. Most fungi reproduce asexually by spores as well as sexually. They are generally nonmotile although some produce motile reproductive cells.

The individual filaments (Fig. 15-1) that make up the fungus body or **thallus** (pl. thalli) are **hyphae** (sing. hypha) and are, collectively, called a **mycelium** (pl. mycelia). The hyphae of the Oomycetes and Zygomycetes are **coenocytic** or without cross walls, whereas the hyphae of the Ascomycetes, Basidiomycetes, and Deuteromycetes are **septate**, or with cross walls, septa (sing. septum).

There are pores in the septa through which living protoplasmic strands interconnect the protoplasts on either side. The pore is simple in the Ascomycetes and most of the Deuteromycetes but complex in the Basidiomycetes. This complex pore is called the **dolipore** and consists, essentially, of a short piece of tubing inserted through the septum at right angles (Fig. 15-1E); the pore is a short tunnel. Two membranes, which look like parentheses, face the ends of the tube and are called **parenthesomes.**

Hyphae produce specialized branches and structures for (1) *absorption* (haustoria), (2) *survival* in adverse conditions (chlamydospores, sclerotia), and (3) *reproduction* (gametangia, sexual and asexual fruiting structures, spores).

Haustoria (sing. haustorium) are specialized hyphal branches which penetrate the cell of the host and absorb nutrients (Fig. 15-1F). They may be small and knob-like, bladder-shaped, or elongate. They contain nuclei and a concentrated number of mitochondria. Haustoria do not rupture the protoplasmic membrane of the host cell but invaginate it, as though poking a finger into a balloon. This greatly increases the absorptive surface that the fungus has in contact with the host protoplast and the rate of absorption is directly proportional to the amount of absorbing surface. (See the paper by Ehrlich and Ehrlich listed at the end of the chapter for electronmicrographs of haustoria.)

Haustoria are commonly produced by some of the Oomycetes, Ascomycetes, and Basidiomycetes. Although some facultative parasites, such as species of *Phytophthora*, produce haustoria, they are produced more frequently by obligate parasites such as the fungi which cause the white rusts, powdery mildews, downy mildews, and rusts.

Sclerotia (sing. sclerotium) are produced by such fungi as *Claviceps* (Fig. 28-1), *Sclerotinia* (Figs. 33-1, 33-2), and *Botrytis* (Fig. 40-3); are ⅛ inch to 1 inch or more in length; are generally dark purple to black; and, are composed of hyphal cells in nonlinear, close-packed order. These cells resemble parenchyma and are called **pseudoparenchyma.** Some fungi, such as *Verticillium*, produce minute black resting bodies less than 1/32 inch which are called **microsclerotia.** (Fig. 42-4).

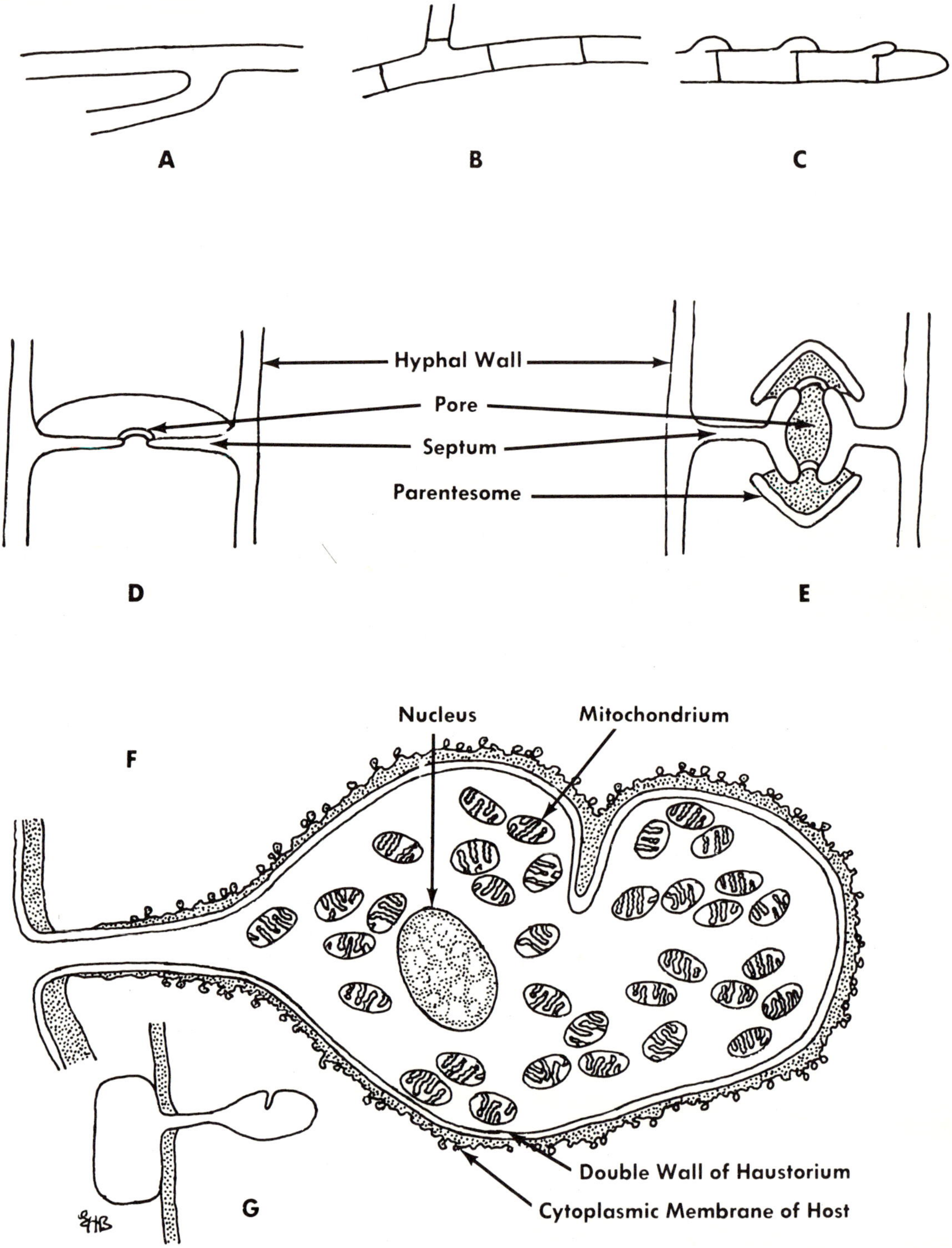

Fig. 15-1. Hyphal structures: (A) hypha without septations (coenocytic); (B) hypha with septations; (C) hypha with clamp connections; (D) simple pore; (E) dolipore with stipuled areas to indicate 3 dimensions; (F) haustorium; and (G) cross section of an intercellular hypha from which the haustorium represented in (F) issued.

Spore is a general term used to describe nearly all detachable units of the fungus thallus which function like seeds even though none are true seeds in structure. They are produced as a result of both asexual and sexual reproduction. The term spore would generally not be used to describe unmodified hyphal fragments. Spores and all other parts, fragments, or other structures of the fungus thallus which can grow and reproduce the organism are called **propagules**, because they propagate the organism.

Spores may be variously colored or colorless; one-celled, 2-celled, or many-celled; round, pointed, long, or coiled; uninucleate, binucleate, or multinucleate; and thin-walled, thick-walled, or variously sculptured with warts, ridges, spines, or other protrusions. The dimensions of most spores fall in the range of 2 to 100 μ although the length of the spores of a few species may reach 400 to 500 μ.

Asexual Spores. (Fig. 15-2). The several kinds of asexual spores may be classified as (1) **oidia** (sing. oidium), which are formed by fragmentation of hyphae into individual cells (**arthrospore** is an older term); (2) **conidia** (sing. conidium), which are borne on the tips or sides of specialized branches of hyphae (**endoconidia** are conidia which are formed within the specialized branch); and (3) **sporangiospores**, which differentiate in indefinite numbers from the protoplast of a cell called a **sporangium** (pl. sporangia). If the sporangiospores are motile they are **zoospores**, and if the sporangia always produce zoospores they are **zoosporangia**.

Sporangia and spores, such as oidia and conidia, are borne on a specialized

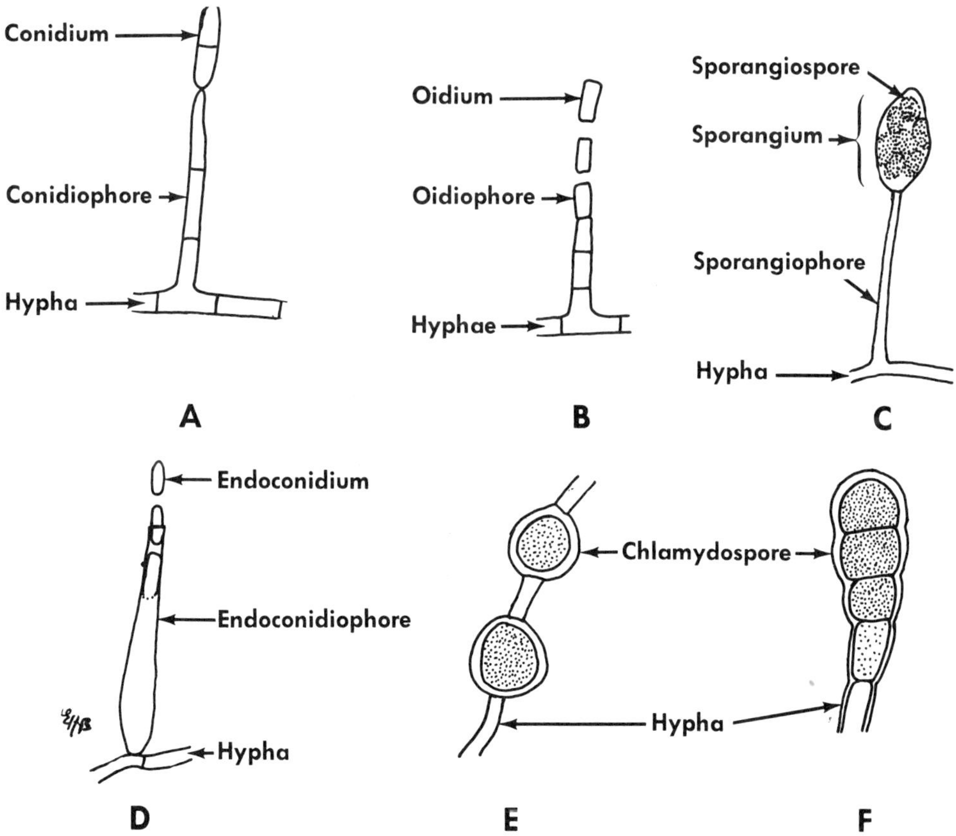

Fig. 15-2. Asexual spores: (A) conidium, (B) oidium, (C) sporangium, (D) endoconidium, (E) intercalary chlamydospore, and (F) terminal chlamydospore.

branch of the hypha called the **sporophore** (*phoreus*, bearer). Thus, the sporophore bearing a hypha which fragments into oidia is an **oidiophore**; if it bears sporangia, it is a **sporangiophore**; and if it bears conidia, it is a **conidiophore.** The sporophore may be short or long, branched or unbranched (simple), and determinate or indeterminate in growth. The tip continues to differentiate more branches of the sporophore if growth is **indeterminate** whereas differentiation ceases if growth is **determinate.** The sporophore may differ markedly in appearance from the hypha, or it may appear as an unmodified hyphal branch.

Asexual spores with thick walls are **chlamydospores** and act as resting spores capable of surviving adverse conditions. Chlamydospores may form from terminal hyphal cells or from intermediate (intercalary) cells in the hyphal strand (Fig. 15-2 E, F).

Sexual spores are formed as a result of sexual reproduction. The oospore, zygospore, ascospore, and basidiospore are sexual spores which are characteristic of the classes Oomycetes, Zygomycetes, Ascomycetes, and Basidiomycetes, respectively. Typically, the oospore and zygospore form after plasmogamy has occurred and require a rest period prior to germination. Ascospores and basidiospores form after meiosis and, as a rule, germinate promptly.

Life Cycle. The cyclic, repeating chain of events that occur when an organism reproduces is its life cycle (Fig. 15-3). The life cycle may be asexual, sexual, or a more complicated, multiple cycle involving both asexual and sexual reproduction. Understanding the life cycles of pathogenic fungi is important because a multiplication of disease requires a multiplication of the causal organism. The propagules of the organism are the primary, secondary, and overwintering inocula of the disease cycle.

The sexual cycle is the chain of events involved in reproduction by a mechanism which provides for an alternation of doubling and halving the number of chromosomes. This may be defined as the sex-function and usually involves the fusion of haploid nuclei to form diploids which are subsequently reduced to haploids. The specific mechanism by which an organism provides for the alternation of doubling and halving varies considerably but typically includes three acts—plasmogamy, karyogamy and meiosis.

Plasmogamy is the union (*gamos*) of two protoplasmic bodies and results in the formation of a dikaryon, or body with two (*di*) nuclei (*karyon*) per cell. The nuclear condition of the dikaryon is noted as (**n** + **n**). **Karyogamy** is the subsequent union (*gamos*) of the two nuclei to form a diploid cell which is the zygote. Its nuclear condition is noted as (2 **n**). **Meiosis** is a nuclear division by which a diploid nucleus is reduced to four haploid nuclei. Haploid nuclei are noted as (1 **n**). You may recall that meiosis in the higher plants occurs during megasporogenesis and microsporogenesis—the origin of the egg and sperm nuclei, respectively. However, in the higher plants, three of the four nuclei, resulting from the meiotic division of the diploid nucleus during megasporogenesis, disintegrate. In most fungi, all four haploid nuclei resulting from meiosis remain viable.

The three events which produce the dikaryon, diploid, and haploid phases do not occur in rapid succession. The three phases vary in their duration within the time span of the life cycle. In general, the haploid phase dominates the life cycles of the Oomycetes, Zygomycetes, and Ascomycetes, whereas the dikaryon dominates the life cycle of the Basidiomycetes. There is now limited evidence, however, that the asexual and dominant phase of the Oomycetes is diploid. In any case, it is the dominant phase of an organism that typically propagates by asexual reproduction. In contrast, you might recall that the life cycles of the seed plants are dominated by the diploid phase.

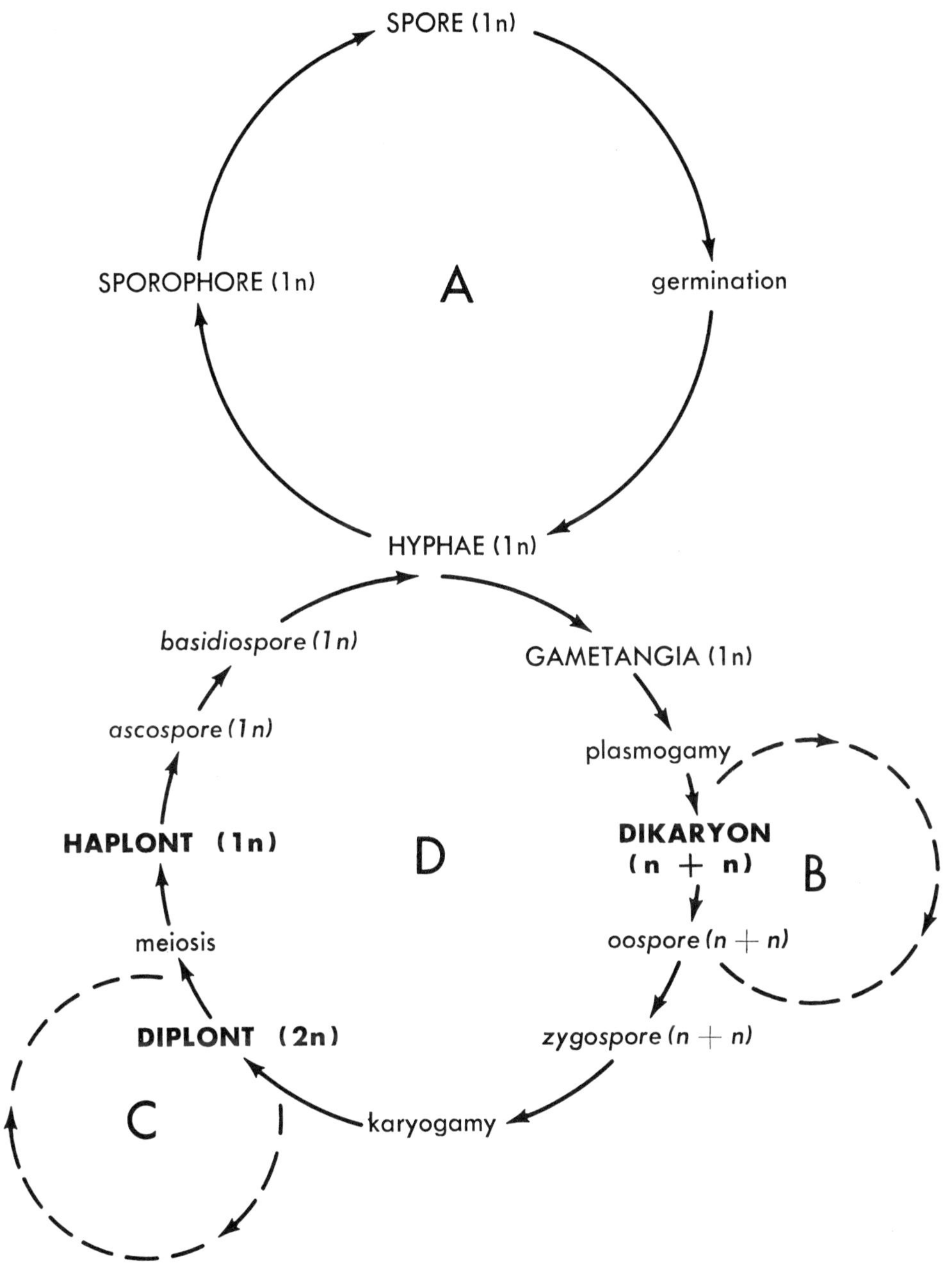

Fig. 15-3. Diagrammatic representation of the life cycle of a fungus with possible variations. Cycles A, B, and C are asexual cycles of reproduction of the haploid, dikaryotic, and diploid phases, respectively; and, cycle D is the sexual cycle. The cycle of a single organism would include only one of the sexual spores listed (oospore, zygospore, ascospore, or basidiospore). The 4 major sexual spores are listed to indicate where they typically occur in a life cycle.

Moreover, it is the dominant diploid phase of the seed plants that propagates asexually, such as the tubers of the potato. Sometimes there is asexual reproduction of both haploid and diploid phase of plants which exhibit a pronounced alternation of generations such as some of the ferns and mosses and their relatives.

Asexual reproduction is the propagation of the organism without nuclear doubling and halving. The haploid, dikaryotic, and diploid stages may be propagated by (1) fission, (2) budding, (3) fragmentation, or (4) spore formation. Asexual reproduction in the Oomycetes, Zygomycetes, and Ascomycetes is generally limited to the haploid phases, although asexual reproduction of some of the Oomycetes may be the diploid phase. In the Basidiomycetes, both haploid and dikaryotic phases are propagated by asexual mechanisms. Propagation of the diploid phase is more often found in advanced plants like potatoes, rather than with fungi.

For many plant-pathogenic fungi, the sexual cycle occurs only once each growing season. Those that are saprophytic tend to mature sexually late in the season whereas those that are parasitic tend to mature sexually when their host matures or dies. The asexual cycle, on the other hand, usually occurs again and again during the growing season and is, therefore, called the **repeating cycle**. It is this repeating cycle of the causal organism that gives a high birth rate to some diseases.

Classification. Many organisms have two names—a common or colloquial name and a scientific name or latin binomial. The scientific name is important because many creatures are obscure, unknown to the layman, and have no common name. Sometimes an organism has two or more common names but only one acceptable scientific name. The same common name may be used for two or more different organisms.

The scientific name is called the **latin binomial** because it is composed of two (*bi*) Latin names (*nomen*). The first name is the genus and the second name is the species. The genus should always be capitalized. Our standard, herein, for the species is the lower case although the rules permit proper nouns used as species names to be capitalized. The Latin binomial should be italicized when printed and underlined when written to indicate italics. Sometimes the binomial is followed by a man's name or by two men's names, the first of which is in parentheses. The last name indicates the author who most recently described the species and the name in parentheses indicates an author who previously described the same organism but used a different name for the genus or species. The apple scab fungus, for example, was named *Sphaerella inaequalis* by Cooke in 1866. Winter transferred the species to *Venturia* in 1875 and the recognized Latin binomial today is, therefore, *Venturia inaequalis* (Cke.) Wint. The designation of authors for the binomials is not to be used here. Often, the genus but never the species is abbreviated to the single initial letter when the binomial is repeated in a single paper. Thus, a second reference to *Venturia inaequalis* would appear as *V. inaequalis*. All binomials are spelled out since our interest is to become familiar with the binomial names.

Similar species are grouped into genera (sing. genus), similar genera into families, families into orders, and orders into classes. Each taxonomic division is a **taxon** (pl. taxa). Taxa are not unalterable divisions. Taxa are man-made categories and different authorities group organisms differently, depending upon which characteristics each believes to be most significant in indicating natural relationships. I follow the classification system adopted by C. J. Alexopoulos in his text on mycology listed at the end of this chapter. It is a well-organized, easily readable reference for the student who is beginning to learn about fungi. The terminology in his text is also adopted here in order to avoid ambiguity or confusion in referring to it.

There are five classes of fungi which are very important in plant pathology: the Oomycetes, Zygomycetes, Ascomycetes, Basidiomycetes, and Deuteromycetes—sometimes called the Fungi Imperfecti or imperfect fungi (Table 15-1). These words are Greek to the beginner—and indeed they are! All end with the Greek term *mycetes* which means fungus class. Hence, they are the *Oo*-fungus class, *Zygo*-fungus class, *Asco*-fungus class, and so forth. The first term describes the kind of sexual spore which is characteristic of each class. For example, Oomycetes characteristically produce oospores, Zygomycetes produce zygospores, Ascomycetes produce ascospores, and Basidiomycetes produce basidiospores. A sixth class of fungi with important plant pathogens is the Plasmodiophoromycetes. Members of this class have no hyphae, are without cells, and consist of naked bodies of protoplasm. Their relationship to other fungi is obscure. *Plasmodiophora brassicae,* one of this group, causes clubroot of crucifers and is discussed later.

TABLE 15-1. General characteristics of five classes of fungi which are important in plant pathology.

Class	Hyphae	Pore	Asexual Spores	Sexual Spores
Oomycetes	nonseptate	simple	sporangiospores motile and nonmotile	oospores
Zygomycetes	nonseptate	simple	sporangiospores nonmotile	zygospores
Ascomycetes	septate	simple	conidia, oidia	ascospores
Basidiomycetes	septate, clamp connections	complex	conidia, oidia	basidiospores
Deuteromycetes	septate	simple, some exceptions	conidia	not used in classification

GENERAL REFERENCES

Ainsworth, G. C., *A Dictionary of the Fungi.* Commonwealth Mycological Institute, Kew, Surrey, 1961.

—— G. C., and Sussman, A. S., *The Fungi,* Vols. I, II, and III. New York, Academic Press, 1965, 1966, 1967.

Alexopoulos, C. J., *Introductory Mycology,* 2nd ed. New York, John Wiley & Sons, Inc., 1962.

Arthur, J. C., *Manual of the Rusts of the United States and Canada.* Lafayette, Indiana, Purdue Research Foundation, 1934.

Barnett, H. L., *Illustrated Genera of Imperfect Fungi.* Minneapolis, Minnesota, Burgess, 1960.

Christensen, C. M., *Keys to the Common Fleshy Fungi.* Minneapolis, Minnesota, Burgess, 1946.

—— C. M., *The Molds and Man.* Minneapolis, Minnesota, University of Minnesota Press, 1951.

Clements, F. E., and Shear, C. L., *The Genera of Fungi.* New York, H. W. Wilson, 1931.

Cummins, G. B., *Illustrated Genera of Rust Fungi.* Minneapolis, Minnesota, Burgess, 1959.

Ehrlich, H. G., and Ehrlich, Mary A., "Electron microscopy of the Host-Parasite Relationships in Stem Rust of Wheat." *American Journal of Botany,* Vol. 50 (February, 1963), pp. 123–130.

Fergus, C. L., *Illustrated Genera of Wood Decay Fungi.* Minneapolis, Minnesota, Burgess, 1960.

Fischer, G. W., and Holton, C. S., *Biology and Control of the Smut Fungi.* New York, Ronald, 1957.

Ingold, C. T., *Dispersal in Fungi.* Oxford, England, Clarendon Press, 1953.

—— C. T., *The Biology of Fungi.* London, Hutchinson, 1961.

Large, E. C., *The Advance of the Fungi.* New York, Henry Holt and Co., 1940.

16

Club Root of Cabbage

Club root of cabbage is caused by *Plasmodiophora brassicae.* Only members of the Cruciferae, the mustard family, such as cabbage, turnip, and rutabaga are susceptible. Infected roots enlarge into club-shaped structures of irregular dimensions. Aboveground parts may be stunted, yellow, and wilted, reflecting the unhealthy structures below. Examine Table 9-1 for a comparison of the characteristics of various kinds of overgrowths caused by different pathogenic agents.

Causal Organism. *Plasmodiophora brassicae* is an obligate intracellular parasite. Since it lives within the cells of its host, it is an **endoparasite.** It cannot live as a saprobe in nature or grow on artificial media. The ungerminated, resistant spores, however, persist for years in the soil. The protoplast, called a **plasmodium,** is intracellular and in intimate contact with the host protoplast. As the host cells divide, portions of the parasite protoplast are frequently carried along with the dividing host protoplast so that both daughter cells of the host contain a plasmodium of the parasite. Continued divisions of infected host cells produce a group of neighboring cells, each enlarged and containing a plasmodium. This group of diseased cells is known by the German term, **Krankheitsherde,** meaning disease (*Krankheit*) group (*herde*) (Fig. 16-1).

Disease Cycle. The organism overwinters for one to several years as uninucleate, unicellular, haploid resting spores in decaying plant debris or soil. These spores are actually **resting zoosporangia.** Under favorable conditions, the spores germinates, or perhaps more accurately "hatch" since the cell wall ruptures and a single biflagellate cell called a **zoospore,** emerges (Fig. 16-2 A-C). This motile body swims in the film of moisture in wet soil and penetrates the root hairs of susceptible plants if contact is made (D, E). The flagella are lost, presumably by retraction, on penetration and the body of the fungus is now called a **myxamoeba** (E). The events at several stages in the life cycle of this organism are a matter of controversy. A general picture is presented for simplicity with one major controversial point indicated.

The myxamoeba enlarges and the nucleus divides repeatedly to give rise to a multinucleate haploid protoplast called the **haploid plasmodium** (F). So far, the organism is still within the root hair. Cell walls develop around each nucleus along with a portion of the protoplasm (G) and form **zoosporangia** (H). **Zoospores** are differentiated (I) and released (J). There is a controversy as to whether these zoospores pass through the root hair, into the soil, and reinfect the root hairs, or whether they remain within the root, act as gametes, fuse, and infect the cortex as zygotes (K). Presumably, if the cortex is invaded by zygotes, the diploid nucleus divides repeatedly and produces a **diploid plasmodium** (L) indistinguishable from the haploid plasmodium. If these presumptions are correct, then the plasmodia in the cortex are diploid whereas those in the root hairs are haploid. A stage called the **akaryote phase** (N) then occurs in which the organized nuclear material cannot be demon-

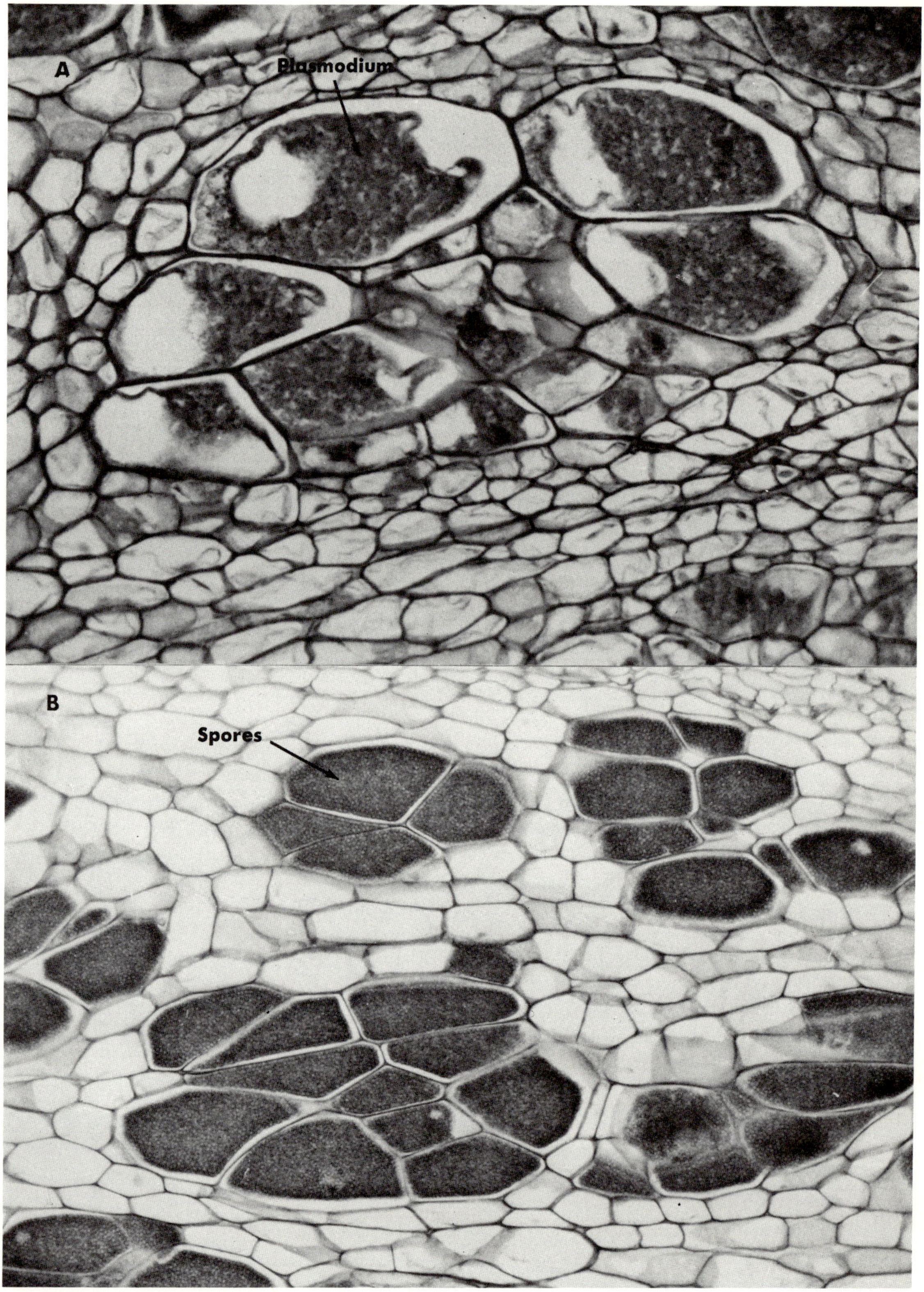

Fig. 16-1. (A) Plasmodia and (B) spores of *Plasmodiophora brassicae* in enlarged host cells. The groups of enlarged cells are Krankheitsherde (× *240*).

strated by staining techniques. This stage is presumably followed by meiosis and cell wall development (O) around each nucleus to form the resting zoosporangia (spores).

Plasmodia and spores can be readily distinguished in prepared microscope slides by their texture. A mass of spores appears as a finely-granular and uniformly-textured mass whereas plasmodia appear irregular in texture and may contain vacuoles of various sizes (Fig. 16-1).

The causal organism is disseminated by drainage water, farm implements, wind-blown soil, man, animals, and infected and infested transplants.

Control. The first step in control is the use of seedlings free from infection and infestation of the causal organism. These should be reared in noninfested seed beds. Soil fumigation can be used to disinfest seed beds. Not a single plant should be transplanted into a clean field from a field with a history of club root or from a seed bed showing a single seedling with club root. Liming to raise the pH of heavy soils and the use of fungicides in transplanting water are useful control measures. Resistance to the causal organism is known and resistant cultivars may become available.

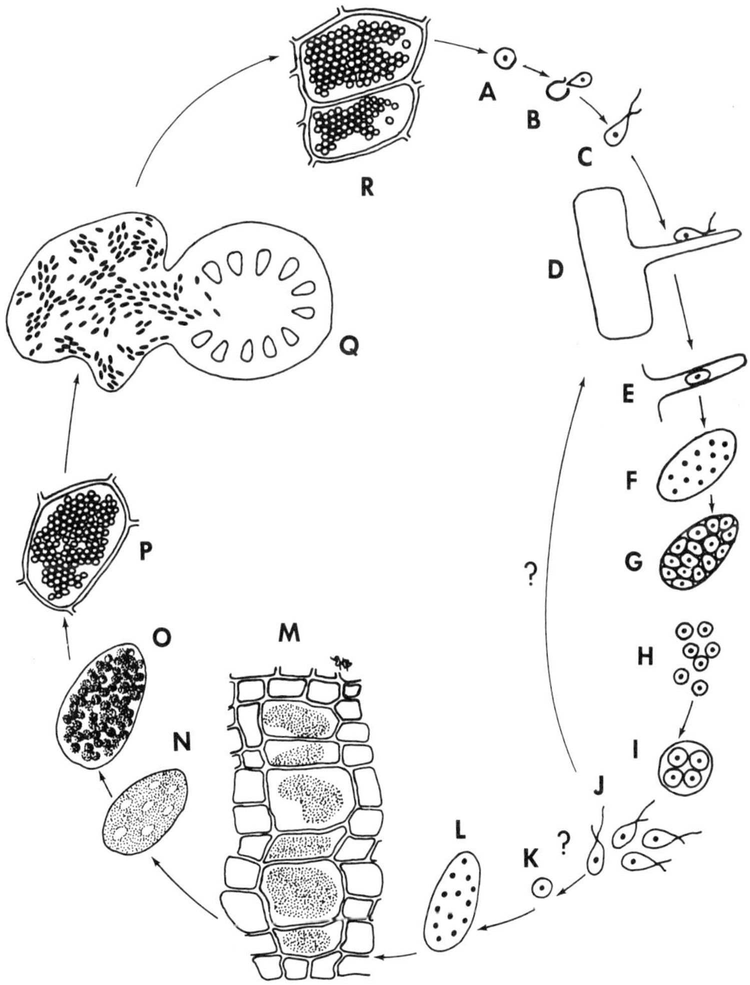

Fig. 16-2. Life cycle of *Plasmodiophora brassicae.*

17

The Oomycetes

The Oomycetes typically produce oospores, nonseptate or coenocytic hyphae, and sporangia. The sporangia may germinate by the production and liberation of zoospores with two flagella or they may germinate by the production of a germ tube. Zoospores are generally produced at lower temperatures than the germ tubes. Since zoospores are biflagellate and swim, moist conditions generally favor their dissemination. The sporangia, on the other hand, are deciduous and are disseminated by wind and splashing rain.

Taxonomic Outline

Oomycetes (*Class*)
- Peronosporales (*Order*)
  - Pythiaceae (*Family*)
    - *Pythium, Phytophthora*
  - Peronosporaceae (*Family*)
    - *Bremia, Peronospora, Plasmopara, Sclerospora*
  - Albuginaceae (*Family*)
    - *Albugo*

The three families, Pythiaceae, Peronosporaceae, and Albuginaceae, are distinguished by the characteristics of the sporangia and sporangiophores. The sporangiophores in the Pythiaceae are similiar to hyphae but are strikingly different in the other families. The sporangia are borne in chains in the Albuginaceae and singly in the other families. Haustoria are absent in *Pythium*, infrequent in *Phytophthora*, and common in species of the Peronosporaceae and Albuginaceae. *Pythium* and *Phytophthora* are distinguished by the differentiation of zoospores in the sporangium of *Phytophthora* and in a vesicle produced by germination of the sporangium in *Pythium*. Species of *Pythium* cause root rots and damping-off diseases (See Chapter 22, Damping-Off Diseases). Species of *Phytophthora* cause fruit rots, collar rots, root rots, and blights. Late blight of potato, caused by *Phytophthora infestans*, is examined in more detail in Chapter 18.

18

Late Blight of Potato

History has recorded the great famines in Ireland and elsewhere in Europe during the 1840's due to the failure, in part, of the potato crops. In Ireland, alone, during those years from 1845 to 1860, one million people died and one and one half million more emigrated.

The disease apparently originated in Mexico from where it spread, along with the potato, southward through the isthmus of Central America into the Andes and from the Andes into Europe and the United States. Distribution became worldwide and, as a result, today the disease is of major importance in areas where potatoes are grown in the cool, humid regions of the temperate zones.

Dieback of the growing point (Fig. 18-1) and blighting of the foliage (Fig. 18-2B) as well as discoloration of tubers are symptoms of the disease. Foliar symptoms are not clearly diagnostic as they are similar to symptoms of early blight caused by *Alternaria solani* (Fig. 18-2A), and very similar to those of *Botrytis* blight.

Life Cycle (Fig. 18-4). Although oospores are the usual structures of Ooomycetes which overwinter and perpetuate these organisms, oospores of *Phytophthora infestans* are exceedingly rare in the United States and insignificant in the life cycle. Oospores are important in the life cycle in some areas such as the mountains of Mexico. Mycelium in tubers, cull piles, and potato seed pieces are the chief overwintering habitats. The mycelium in infected tubers systemically invades the shoot of the tuber when it grows in the spring and the fungus sporulates to produce sporangia.

Sporangiophores emerge through stomates of the lower surface of leaves. A terminal sporangium is produced and turned aside as the sporangiophore continues to grow, branch and produce sporangia in succession (Fig. 18-3). This is **indeterminate** growth. The angle of the branches is acute and serves as a diagnostic characteristic. Sporangial production requires a high humidity and increases with moderately cool temperatures. With favorable conditions during sporulation, the undersurface of leaves becomes white and moldy.

The sporangia are deciduous, like leaves on trees, and are readily air-borne. They germinate in one of two ways, depending on the temperature. At low temperatures they **germinate indirectly** by producing zoospores which are rapidly spread by splashing rain, and at warmer temperatures **germinate directly** by the production of a germ tube. The sporangium functions like a conidium when it germinates to give rise to a germ tube and the sporangia are frequently called conidia for this reason. They are, nonetheless, sporangia.

The primary inoculum is the zoospores or the sporangium itself, if it germinates directly by a germ tube. Zoospores or sporangia also constitute the secondary inoculum.

Control. Although resistance among different cultivars varies, protectant chemical sprays are applied in many areas during periods of favorable weather for inoculum production and infection. Some areas of both Europe and the United States have worked out reasonably dependable schemes for the prediction of infection based on the accumulative moisture and current temperature trends. When dependable, such schemes for **disease forecasting** are worthwhile since the grower can save his money when sprays are not needed and they save his crop when they are.

Fig. 18-1. Dieback of the growing point of a potato caused by *Phytophthora infestans.*

Fig. 18-2. (A) Foliar symptoms of early blight of potato caused by *Alternaria solani.* (B) Foliar symptoms of late blight of potato caused by *Phytophthora infestans.*

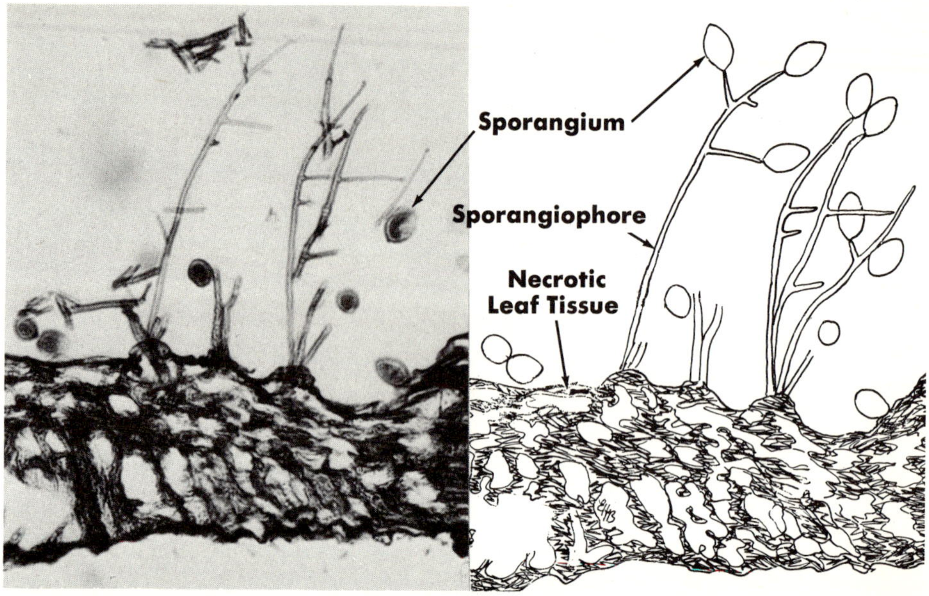

Fig. 18-3. Conidia and conidiophores of *Phytophthora infestans* with diagrammatic reconstruction (× *180*).

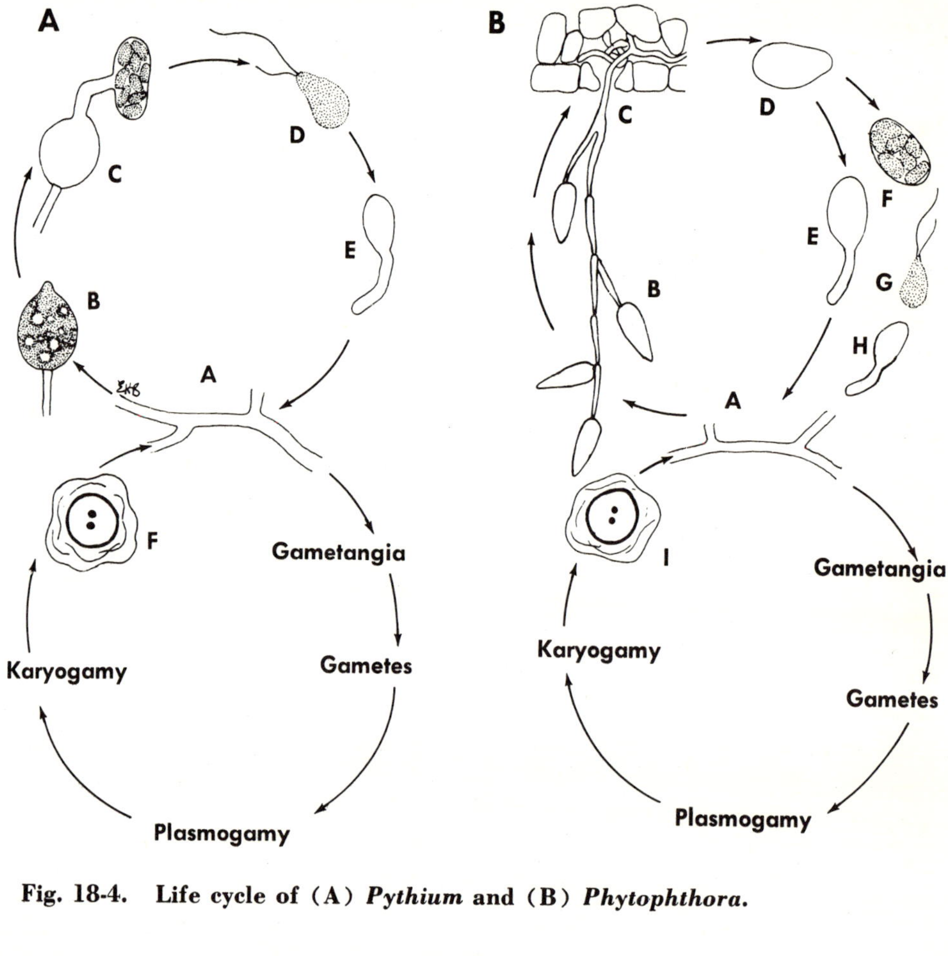

Fig. 18-4. Life cycle of (A) *Pythium* and (B) *Phytophthora*.

19

The Downy Mildews

Discovery is a word that excites in the minds of many, a romantic vision of some courageous explorer. More often discovery is a nondramatic, quiet business of alert observation and careful experimentation. Such a tale of discovery is the story of Professor Millardet and his exploration of the vineyards in the French countryside one day in 1882. . . .

He walked down a lane that led him along a vineyard. Seeing the grapes he mused upon just how long the wine industry would last. Ever since the downy mildew disease (Fig. 19-1) had been brought over from America, the blight had become progressively severe.

First, it had been the *Phylloxera,* a root aphid. It too had never occurred in Europe until it came from America. The French grapes were extremely sensitive to the *Phylloxera* and the wine industry seemed doomed. Wild American grapes, however, were resistant and it seemed a good thing to bring American rootstocks to France. The susceptible French grapes were grafted onto rootstocks of resistant American grapes and the root aphids were no longer a problem. The wine industry had been saved. Then, the mildew struck! Whole vineyards were devastated; the fungus having been imported into France on the rootstocks from America.

Suddenly Millardet stopped! He looked more closely at the vines bordering the path. The leaves appeared to be greener and much healthier than the rest of the vineyard. On still further and closer observation, he noticed that there was a bluish-white residue on the leaves. On inquiry, he found that the owner had dusted the vines along the lane with copper sulfate and lime so that the conspicuous and poisonous-looking substances would discourage pilfering of the grapes.

Millardet subsequently tested mixtures of copper sulfate and lime for the control of the mildew. He later collaborated with Gayon, his friend and an able chemist, in the study of the chemistry of the mixture that they called Bouillie bordelaise, or **Bordeaux mixture.** Bordeaux mixture came to be used throughout the world for control of downy mildew of grape and other diseases. It is still widely used for the control of many diseases.

Millardet applied the principle of chemical crop protection from disease by alert observation while on a quiet exploration along a country path. He was aware of the fungal spores of the downy mildew fungus and their role in causing the mildew disease and he had been working on chemical control of the disease prior to his chance discovery. Was he also aware of Prévost's findings of some 70-odd years earlier? Prévost had discovered that the black dust of bunt was composed of microscopic particles which functioned like little seeds in view of the fact that they grew. He discovered their role in causing the bunt disease and, by chance, noticed that water from a copper container inhibited growth of the little seeds or spores. He then proceeded to elucidate the principle of preventative chemical control with copper sulfate treatment of wheat seeds to prevent germination of the spores, but not the wheat seeds.

Genera of the Downy Mildew Fungi. All of the species in the Peronosporaceae are obligate parasites which cause diseases called downy mildews (Fig. 19-1). There is another group of diseases called powdery mildews which have a gross similarity in appearance. These are caused by a distinct group of species of the Erysiphaceae of the Ascomycetes all of which are also obligate parasites.

The various genera of the Peronosporaceae are distinguished primarily on the branching characteristics of the sporangiophores (Figs. 19-2, 19-3). Some of the common species and their hosts are:

Bremia lactucae	lettuce
Peronospora destructor	onion
Peronospora parasitica	crucifers
Peronospora tabacina	tobacco
Plasmopara viticola	grape, Boston Ivy
Pseudoperonospora cubensis	cucurbits

Hyphae are commonly intercellular (Figs. 19-4, 19-5) and haustoria are common and easily seen as branched, bulbous configurations in *Peronospora* (Figs. 19-5, 19-6).

Life Cycles: follow the general pattern of *Pythium* and *Phytophthora.* In most cases the sporangia germinate by zoospores or germ tubes, depending on environmental conditions. Sporangia of *Peronospora,* however, always germinate by a germ tube. The oospores usually germinate by a germ tube which produces a single, terminal sporangium.

Disease Cycles. Overwintering is usually through mycelium in biennial or perennial wild hosts, mycelium in crop refuse, or oospores. For downy mildews of lettuce, tobacco, and onion, mycelium is most important. Oospores are significant in the overwintering of the causal organisms of the diseases of grape and grass. An important source of inoculum for field-grown cucumbers are the sporangia carried by air currents from cucumbers grown in green houses. Inoculum carried by winds from southern areas is also important with the curcurbits. Although most downy mildew fungi grow best during cool weather, *Pseudoperonospora cubensis* grows best in warm temperatures.

Secondary inoculum is always zoospores from sporangia or the sporangia themselves if they germinate by germ tubes. The sporangia are deciduous (or detachable) and are blown by the wind. They desiccate and die over long distances if the air is dry. Germ tubes penetrate the stomatal openings.

Control. Chemical sprays, especially carbamates, are fairly effective in the control of these diseases of grape, onion, cucumber, and lettuce. Eradication of perennial wild onion, in the vicinity of commercial fields, is reported to be helpful. Control of weed hosts and seed treatments are useful, along with chemical sprays, for control of the cruciferous diseases.

Some cultivars of lettuce, muskmelon, and cucumber are resistant and provide control. Little resistance, however, is available in most commercial cultivars of crucifers, onion, and tobacco.

Fig. 19-1. (A) Enlarged view of symptoms of downy mildew on the upper surface of a grape leaf showing necrosis and chlorosis, and (B) the general symptoms of the disease on the lower and upper surfaces.

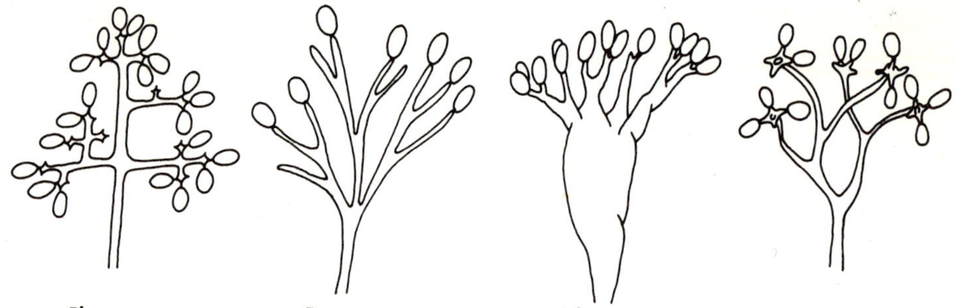

Fig. 19-2. Characteristic sporangiophores of some genera of downy mildew fungi.

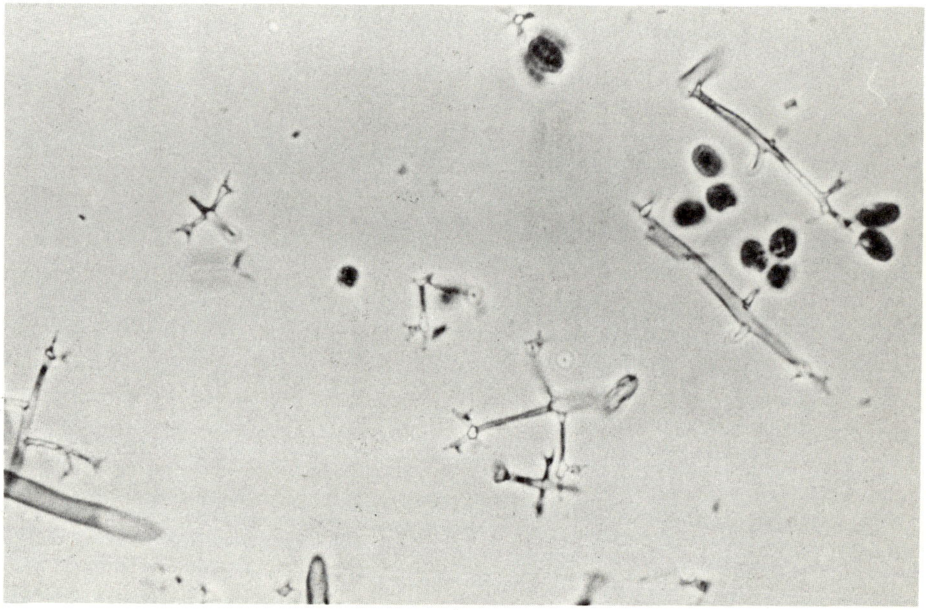

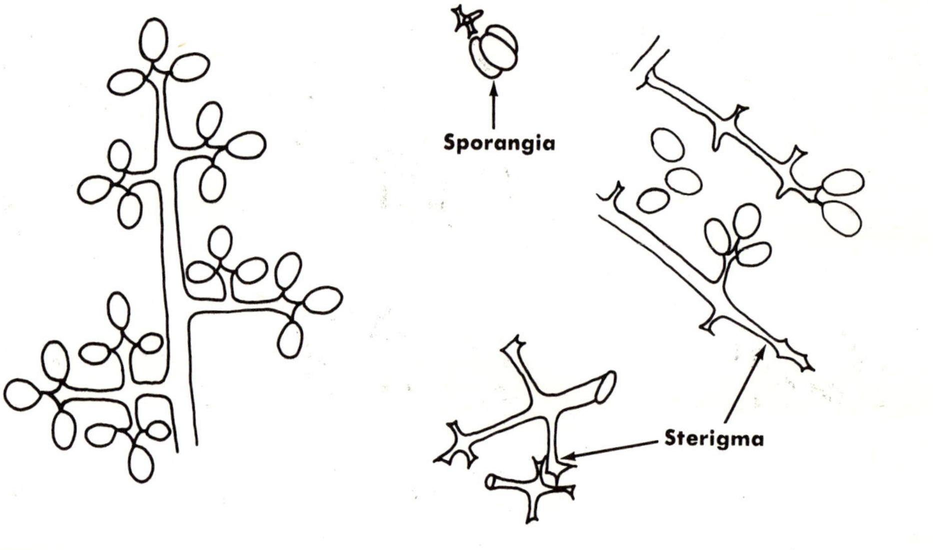

Fig. 19-3. Photomicrograph of sporangiophores and sporangia of *Plasmopara viticola* and diagrammatic reconstruction (× *225*).

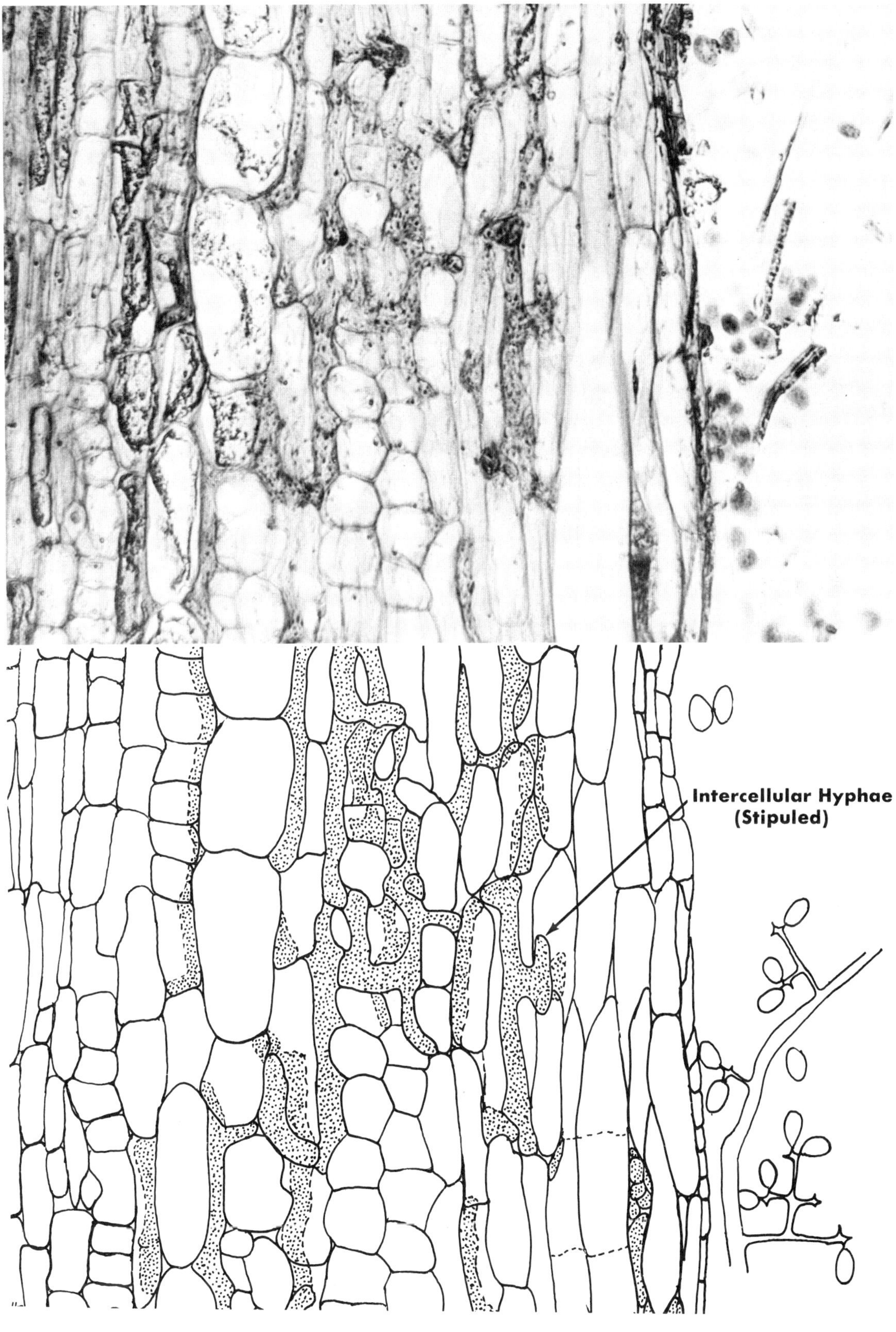

Fig. 19-4. Photomicrograph of intercellular hyphae of *Plasmopara viticola* and diagrammatic reconstruction (× *250*).

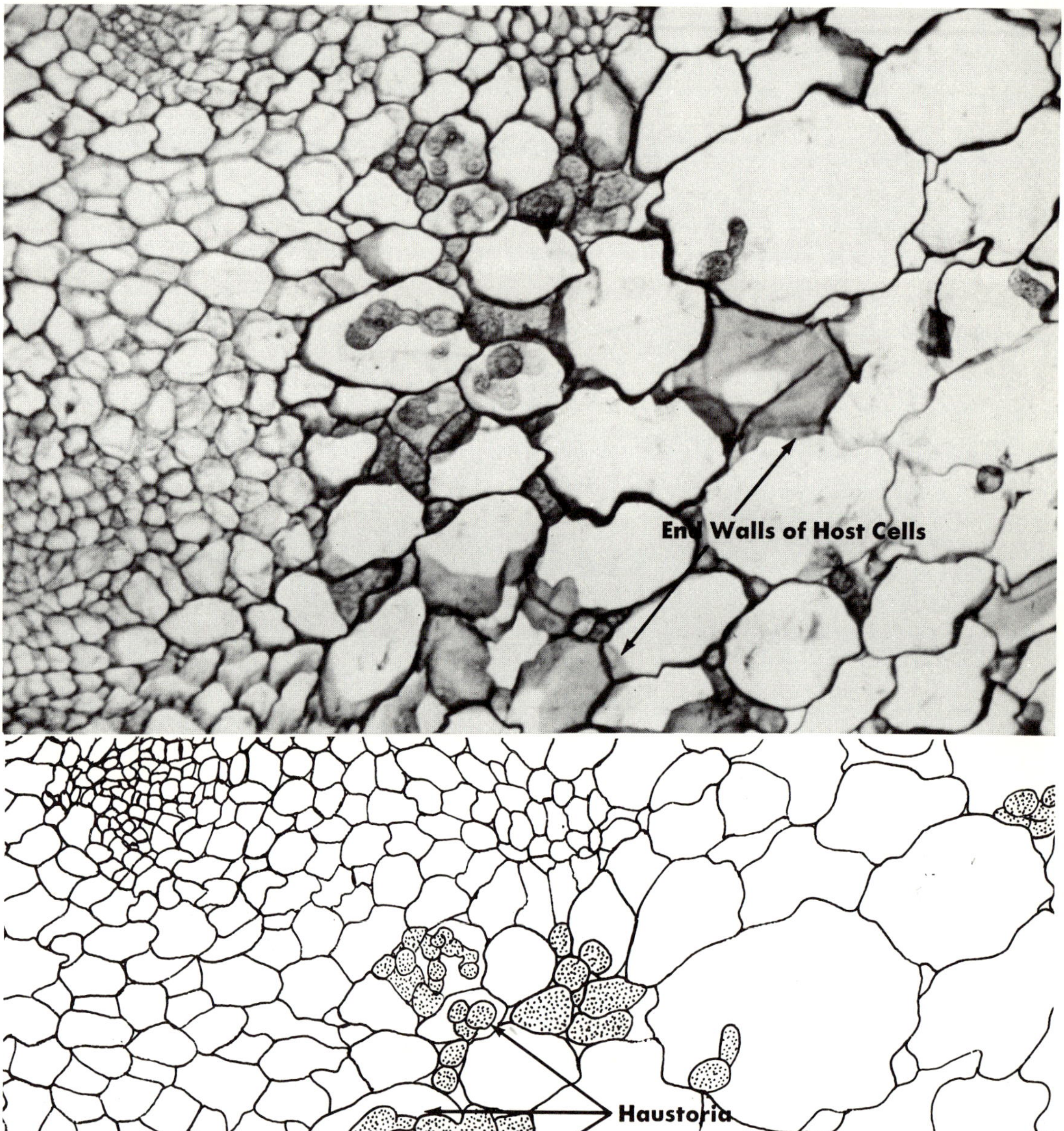

Fig. 19-5. Photomicrograph of intercellular hyphae and haustoria of *Peronospora parasitica* and diagrammatic reconstruction (× *250*).

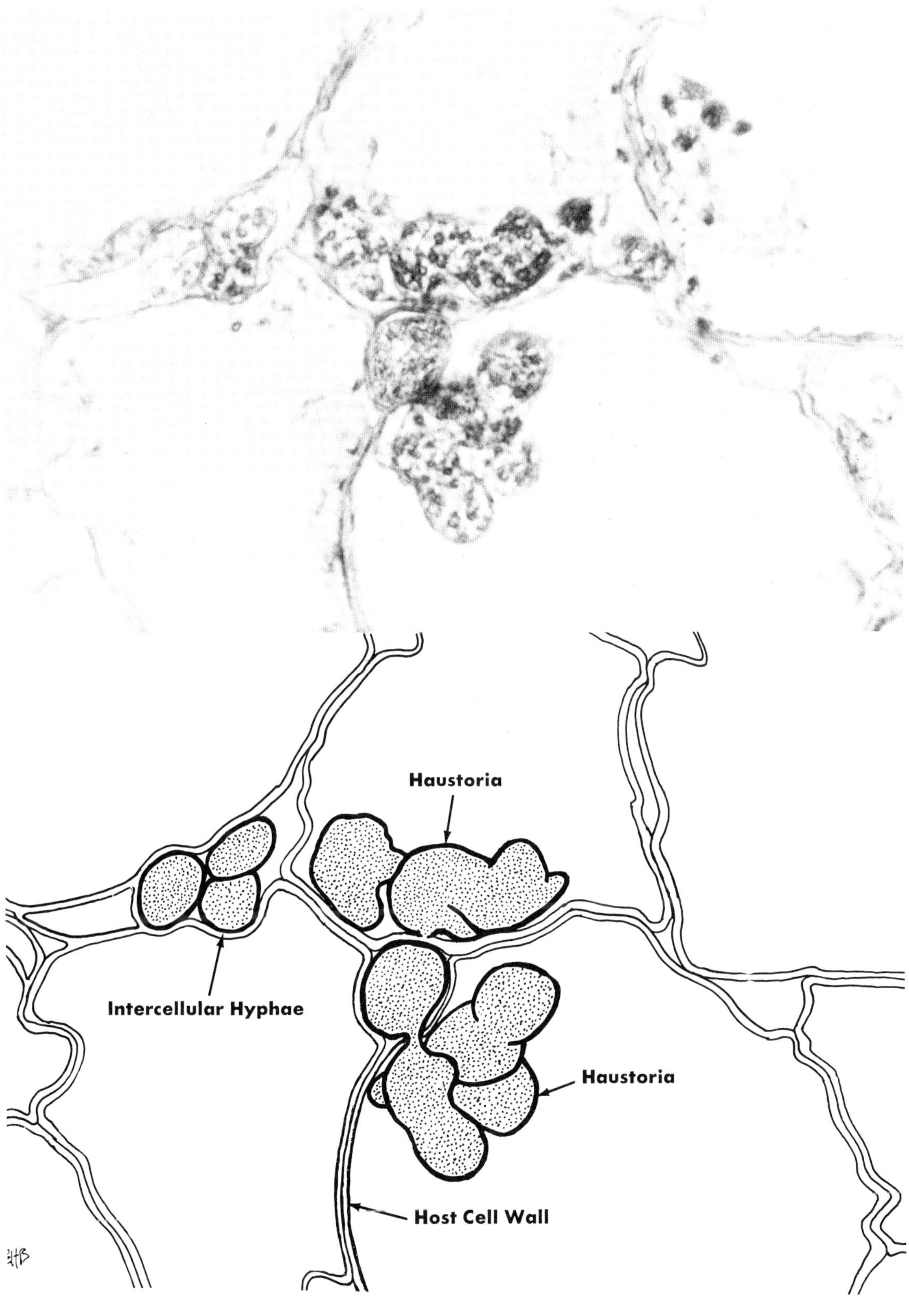

Fig. 19-6. Photomicrograph of intercellular hyphae and haustoria of *Peronospora parasitica* and diagrammatic reconstruction (× *480*).

20

White Rusts of Crucifers

The white rusts occur worldwide and are best known and most common on the Cruciferae. Spinach, radish, turnip, horseradish, and mustard are common economic hosts, whereas cabbage, brussels sprouts, salsify, and cauliflower are less affected. These diseases are also common on many cruciferous weeds but are less well-known on sweet potato and species of the Capparidaceae and Amaranthaceae.

The descriptive name, white rust, is due to the pustules which form on leaves, stems, and floral parts. Rupture of the epidermis exposes white powdery masses of spores (Fig. 20-2) as white dust in small patches to large, irregular, coalesced areas (Fig. 20-1). Local lesions and systemic infections develop in plants about to seed, but, frequently, local lesions predominate in vegetative plants. Systemically infected plant parts are often enlarged and distorted.

Causal Organisms. The white rusts are caused by species of *Albugo*, the single genus of the family Albuginaceae. All are obligate parasites of vascular plants. *Albugo candida* is found on crucifers, *Albugo occidentalis* on spinach, and *Albugo bliti* on species of the Amaranthaceae. The thick wall of the oospore may be ridged, warty, spiny, or variously sculptured. This characteristic is useful in distinguishing the species of *Albugo*.

Most species of *Albugo* occur only on a particular group of plants. For example, *Alubgo candida* attacks radish, cabbage, shepherd's purse, and sweet alyssum. Although the fungus is morphologically identical on all of these hosts and classified as one species, the strain from each host cannot infect any of the others. This specialization is called biological or physiological and the strains are called **biological or physiological races**. Some strains may attack a few closely related species of hosts in some species of *Albugo*.

Disease Cycle. Most species overwinter by means of mycelium in perennial parts of overwintering hosts and oospores in debris of infected plant parts. Most diseases cannot be controlled by eradication of weed hosts because the fungus on the weed host and economic host are different biological races. An exception is the fungus which causes white rust of salsify, since it is the same biological race of the fungus that occurs on goat's beard, a common weed that's a relative of the dandelion.

The mycelium of *Albugo* is intercellular and secures nutrition through knob-like haustoria. The sporangiophores are borne in solid layers just beneath the epidermis. Sporangia are differentiated in succession to produce chains with occasional interstitial spacer cells (Fig. 20-2). This characteristic of chains distinguishes the Albuginaceae from all the related families. Because of pressure under the epidermis from increasing numbers of sporangia, the sporangia are more often flattened and polyhedral than round.

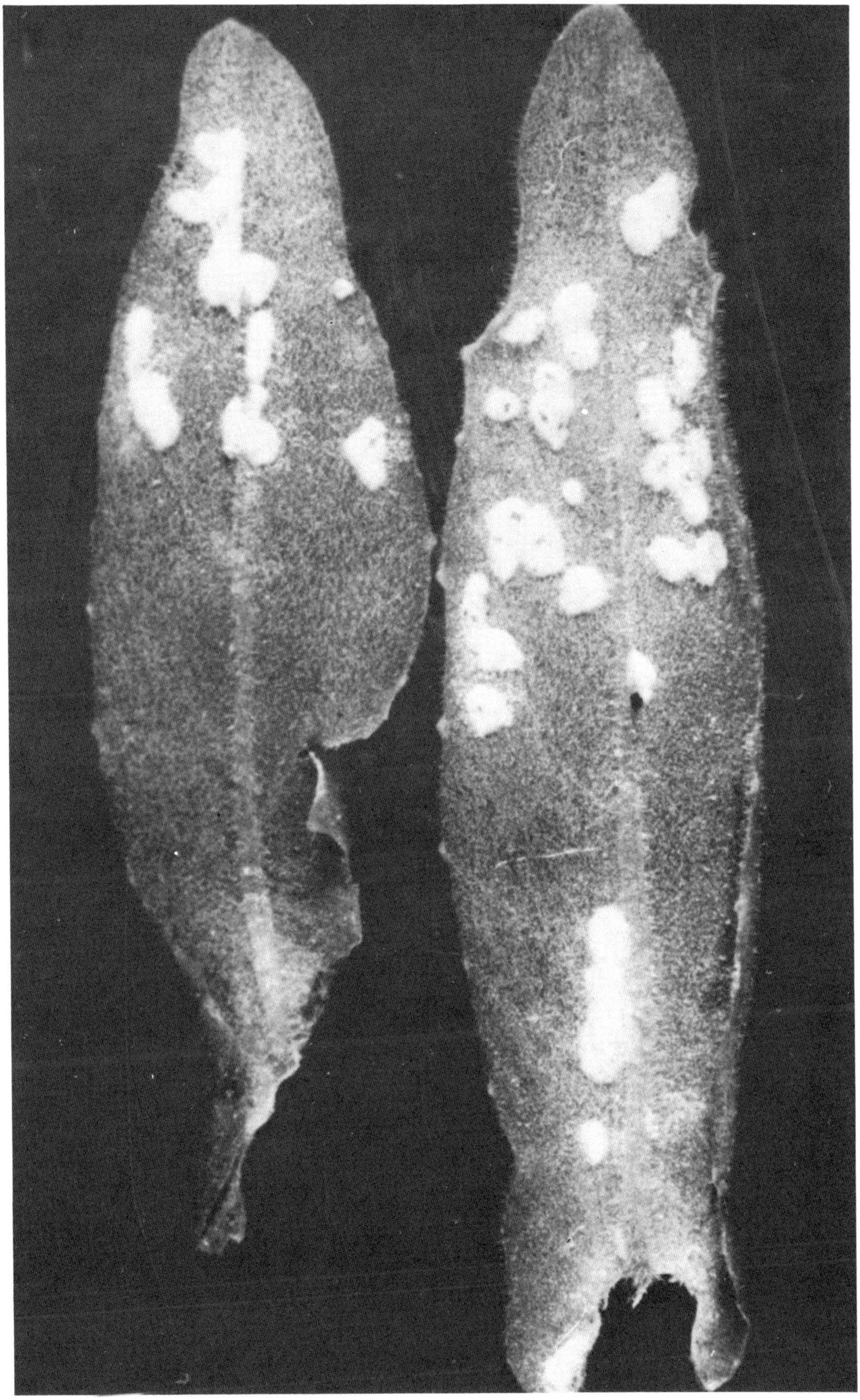

Fig. 20-1. Pustules of sporangia of *Albugo*.

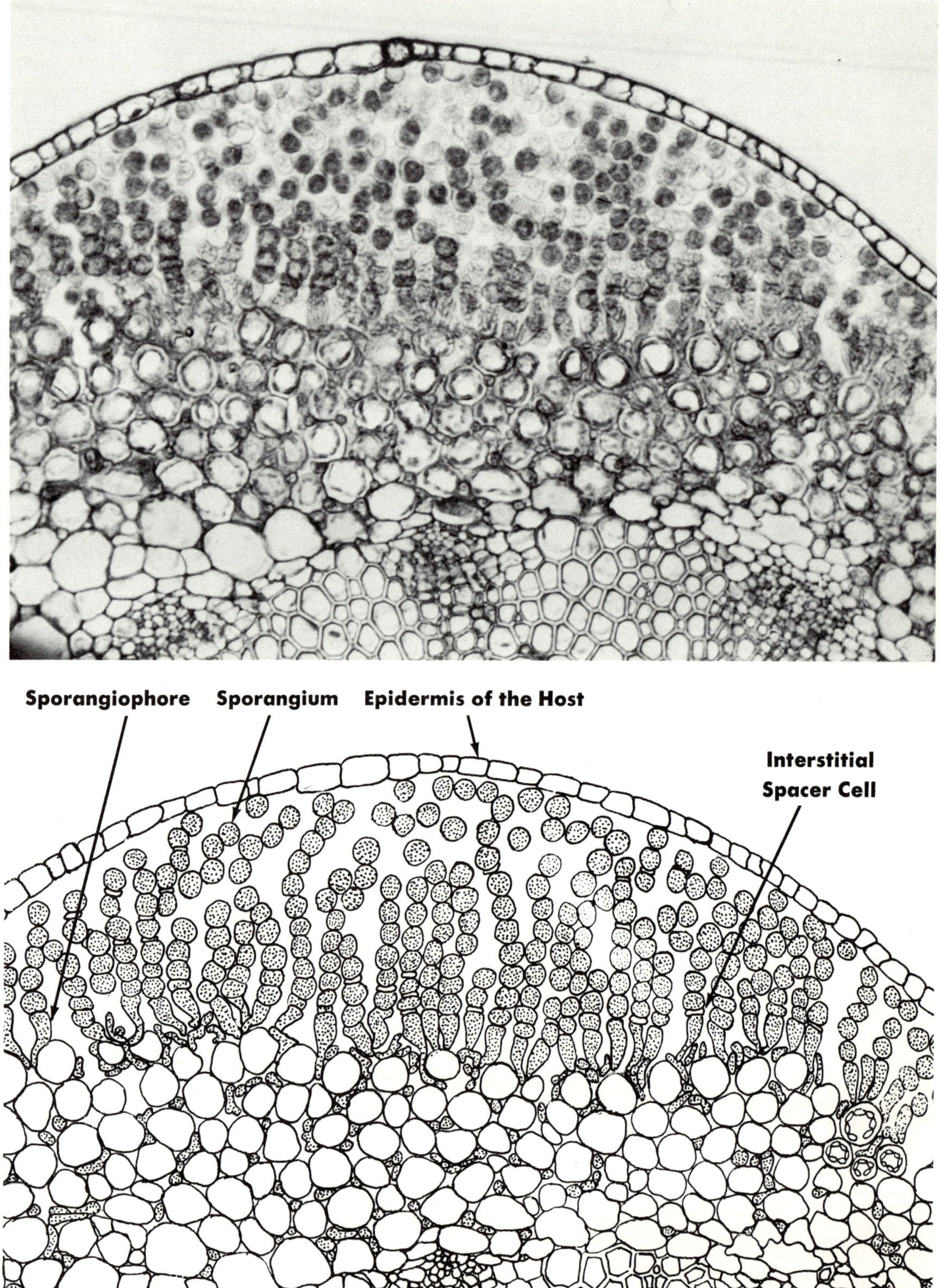

Fig. 20-2. Photomicrograph of an unruptured pustule of sporangia of *Albugo* and diagrammatic reconstruction (× 250).

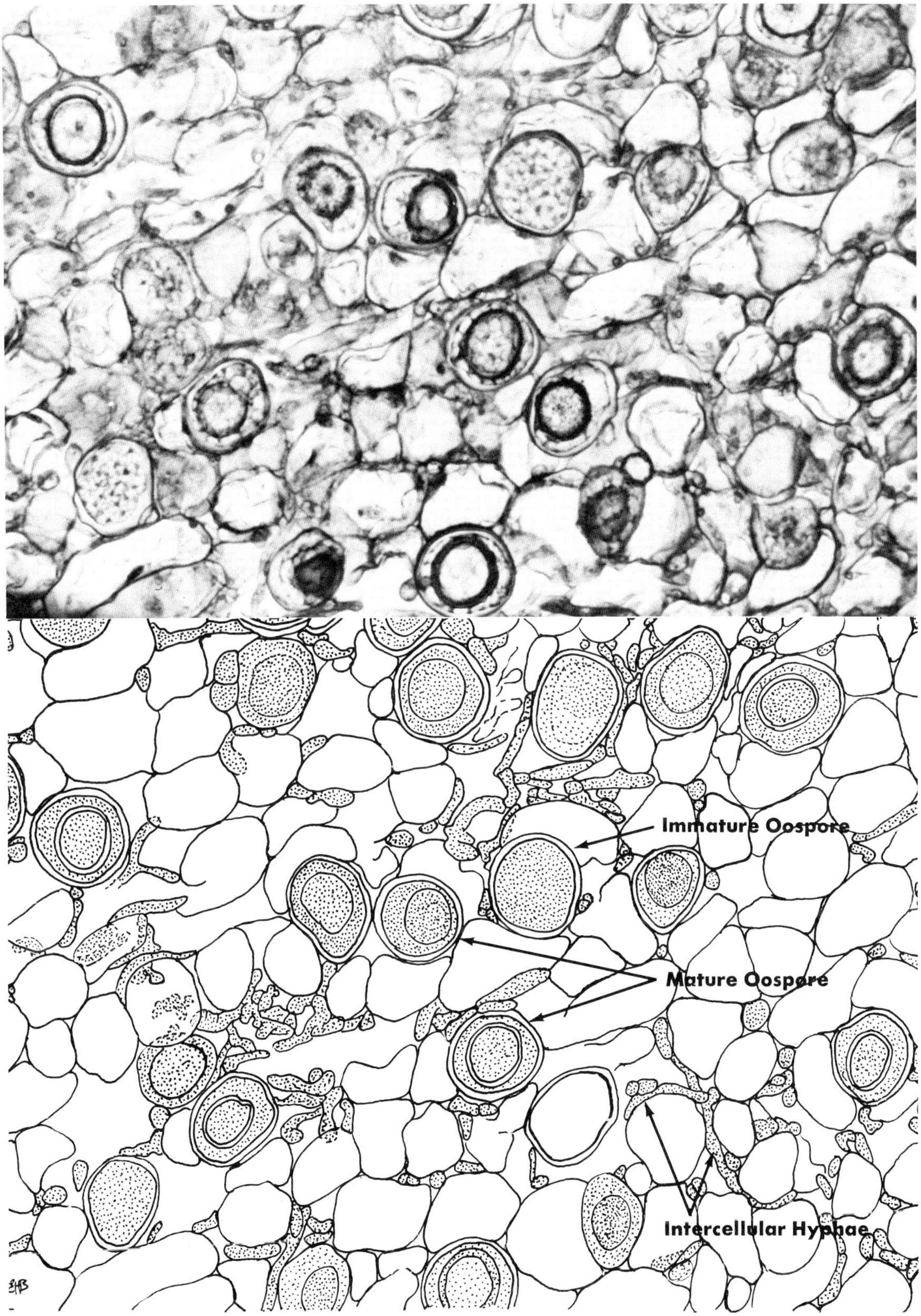

Fig. 20-3. Photomicrograph of immature and mature oospores of *Albugo* and diagrammatic reconstruction (× *250*).

The sporangia are disseminated by wind and rain. They germinate by production of either germ tubes or zoospores depending on the temperature. Sexual reproduction results in the production of oospores (Fig. 20-3) which are dark and may cause the host tissues to appear black. Oospores germinate by the production of zoospores and not by a germ tube. Mycelium is capable of overwintering in perennial hosts, especially in perennial roots such as those of horseradish.

Primary inoculum consists of overwintered mycelium and zoospores produced by germination of oospores. Moisture is required for the germination of oospores, sporangia, and zoospores and for penetration through the stomates. Sporangia which germinate directly and zoospores which are produced by indirect germination of sporangia constitute secondary inocula.

Control. The white rust diseases are of limited economic importance and control measures are essentially lacking. European horseradish is more resistant than our commercial types and may be utilized more in the development of new, resistant cultivars.

SUGGESTED READING

Endo, R. M., and Linn, M. B., *The White Rust Disease of Horseradish.* Ill. Agric. Expt. Sta. Bull. 655, 1960.

21

Zygomycetes

A common fungus that most people encounter in addition to the mushroom on their steaks is the bread mold, *Rhizopus stolonifera. Rhizopus* is a member of the Mucorales. The other two orders of this class are the Entomophthorales and the Zoopagales which contain parasites of insects and nematodes, respectively.

Our primary interest in *Rhizopus* is to acquaint ourselves with the characteristics of a weak parasite and the kind of disease it causes. A weak parasite is always a good saprobe, otherwise it would be unable to compete with vigorous parasites and would be selected against. *Rhizopus* is a good saprobe and is commonly found on decaying vegetable matter and old bread.

Diseases caused by weak parasites are, typically, diseases of senescent and dormant tissues. *Rhizopus* causes a soft rot of sweet potato roots and a rot of strawberry fruit called leak, because the fruit becomes soft, mushy, and leaks. Many of the diseases of produce in transit and in storage are caused by weak parasites. A weak parasite is not aggressive, as a rule, and penetration of host tissues is usually by a passive route such as wounds. *Rhizopus* is strictly a **wound-parasite**, that is, it can penetrate host tissues only through fresh wounds.

Causal Organism. Rhizopus produces sporangia and rhizoids on the hyphae where the sporangiophores occur (Figure 21-2 A, B). Continuation of the sporangiophore into the lumen of the sporangium is swelled and called the **columella.** The spores are haploid and nonmotile and are properly called sporangiospores (C). They always germinate by a germ tube (D).

A pure culture of *Rhizopus* derived from a single sporangiospore is a **strain.** A pure culture of a single strain of *Rhizopus* never reproduces sexually. If pure cultures of many sporangiospores are tested in pairs, some pairs reproduce sexually and others not. Two strains which mate and reproduce sexually are compatible in mating type and are designated (+) and (—). All strains of *Rhizopus* are either (+) or (—) and matings of two (+) or two (—) strains are infertile whereas matings between a (+) and a (—) strain are fertile. The (+) and (—) differences are not differences in sex structure but provide for a similar function of outbreeding.

Progametangia (E, F) develop as short, swollen hyphal branches in juxtaposition, or opposite each other, on the hyphae of two different mating types. The tips fuse and the cell walls dissolve. Two new cell walls are laid down which create a central cell now designated the **zygospore mother cell** (G, H) (Figure 21-1 A). The outer cells between the hyphae and the zygospore mother cell enlarge and are called **suspensor cells.** The zygospore matures with the development of a very heavy, warty wall (I) (Figure 21-1 B) and germinates with the production of sporangiophore and sporangium (J). Presumably, meiosis occurs just prior to or during germination so that the sporangiospores are haploid (K).

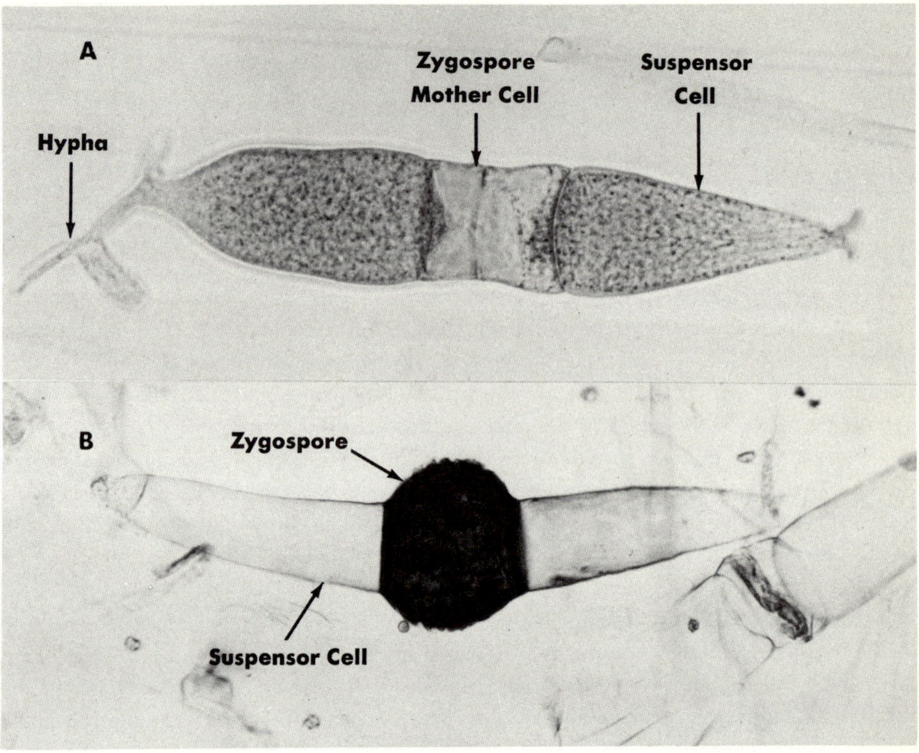

Fig. 21-1. (A) Early stage of zygospore mother cell showing remnants of the central cross walls, and (B) zygospore of *Rhizopus* (× 225).

Disease. The soft rot of sweet potato caused by *Rhizopus* seldom occurs on roots in the ground or on above-ground parts, but is almost entirely a disease of roots in storage. Penetration is through fresh wounds and bruises made by harvesting, handling, insects, rodents (in storage), and other diseases.

The sweet potato root is able to produce a corky layer of wax-impregnated cells around wounds rapidly under conditions of high temperature (80°–85° F.) and relative humidity (90%). *Rhizopus* cannot penetrate this corky layer and it appears that the corking-over proceeds more rapidly than the development of the fungus under these conditions and the disease is successively checked.

Control: includes sanitation, wound prevention, and proper storage conditions: (1) **sanitation** to clean-up debris-containing inoculum in order to avoid the disease; (2) **wound prevention** to prohibit the entrance of inoculum which may be present; and, (3) **proper storage conditions** to provide an environment unfavorable for the growth of the parasite but favorable for wound healing of the host.

Rhizopus soft rot of sweet potato is controlled by careful handling to avoid wounds, washing to remove debris and inoculum, rapid drying, and curing in storage at 80°–85° F. and 90% relative humidity for a week. This permits wound healing prior to dropping the temperature to 55° F. for continued storage. If the roots are moved they should be given another curing to permit healing of any new wounds.

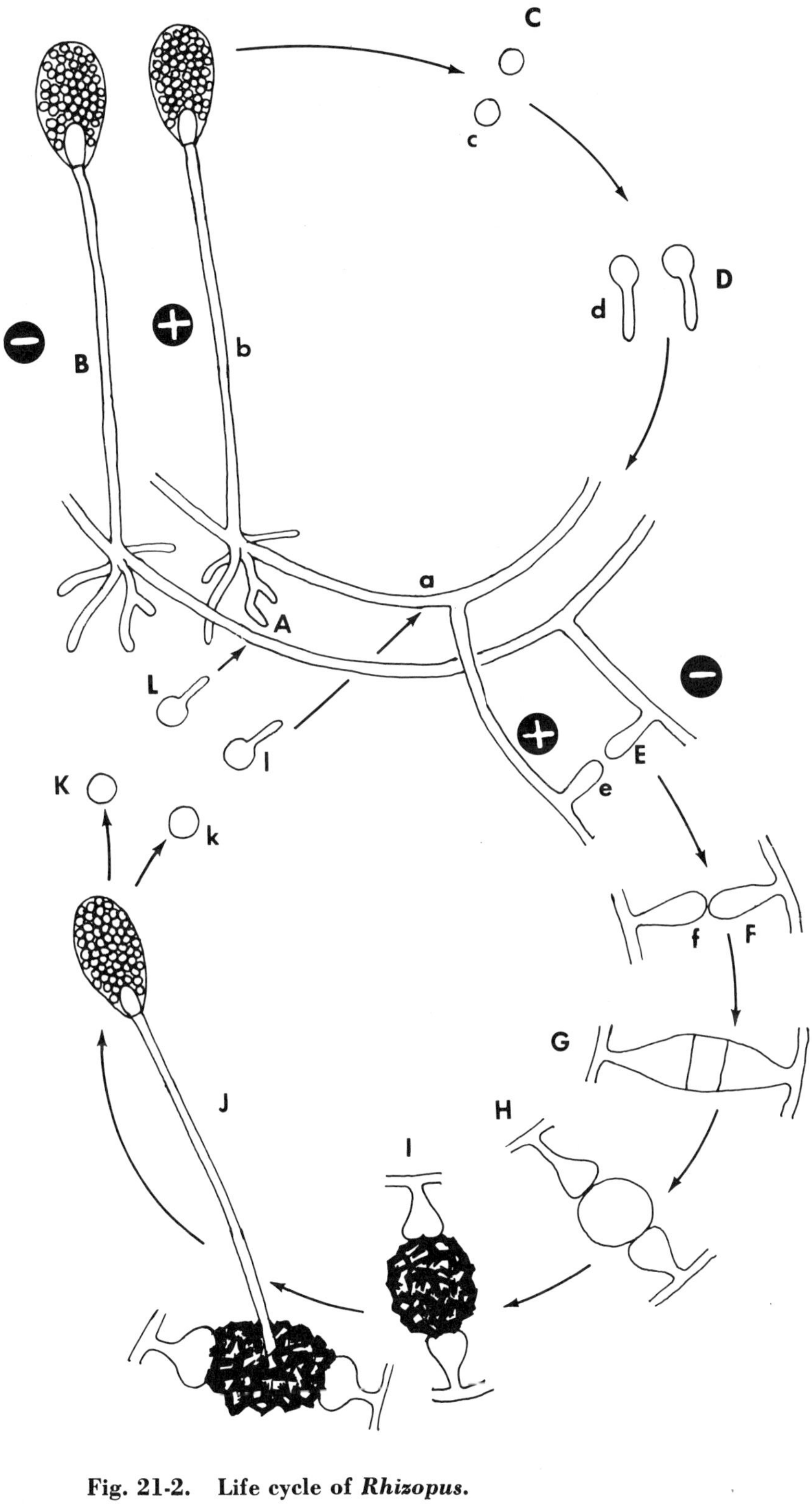

Fig. 21-2. Life cycle of *Rhizopus*.

22

Damping-Off

Damping-off is a disease of the seedlings of any plant, caused by any agent, in which the seedlings wilt, collapse, and die. **Pre-emergence** damping-off occurs before the seedlings have emerged through the soil and **post-emergence** damping-off occurs after emergence (Fig. 22-1).

Species of *Pythium, Rhizoctonia,* and *Fusarium* are most important as the causal agents of damping-off although species of *Alternaria, Botrytis, Rhizopus, Phytophthora, Sclerotinia,* and many other genera are also agents. Most all of these causal agents of damping-off are also responsible for root rots, collar rots, fruit rots, and other maladies of their hosts as well.

In general, these diseases occur most commonly in heavy soils and organic soils. They are favored by high moisture levels and frequent wet periods.

Control: effected with chemical seed treatment and soil sterilization, especially in greenhouses. Chemicals which provide reduction in damping-off caused by *Pythium* may not be very effective with damping-off caused by *Rhizoctonia* and *vice versa.*

Experiment XIV

Damping-Off: Inoculum Potential

One measurement of the **inoculum potential** of soil-inhabiting and soil-borne pathogens is an estimate of the population of infectious propagules. Under the most favorable conditions, the number of infections cannot exceed the number of infectious propagules. Presumably then, the larger the population of infectious propagules, the higher the inoculum potential and the higher the potential incidence of disease. What is the relationship of the number of pathogenic propagules per unit of soil and probable incidence of disease (Figs. 22-1, 22-2).

EXPERIMENT

1. Prepare a series of flats with mixtures of sterilized and nonsterilized muck soil which contain damping-off fungi such as species of *Pythium.* Prepare 1 flat of 0, 10, 25, 50, 75, and 100% nonsterilized muck soil. Also, prepare additional flats of 100% nonsterilized muck soil to provide space for 1 row of treated seeds per group. Plant 10 rows per flat. The sterilized muck soil should be prepared only a few days before use to prevent microbes from recolonizing. Mix the sterilized and nonsterilized muck soil very thoroughly. Water to the dripping point each day for two days before use to provide the *Pythium* with optimum growing conditions. Keep in a cool location. These preparations may be completed by the instructor or the students.

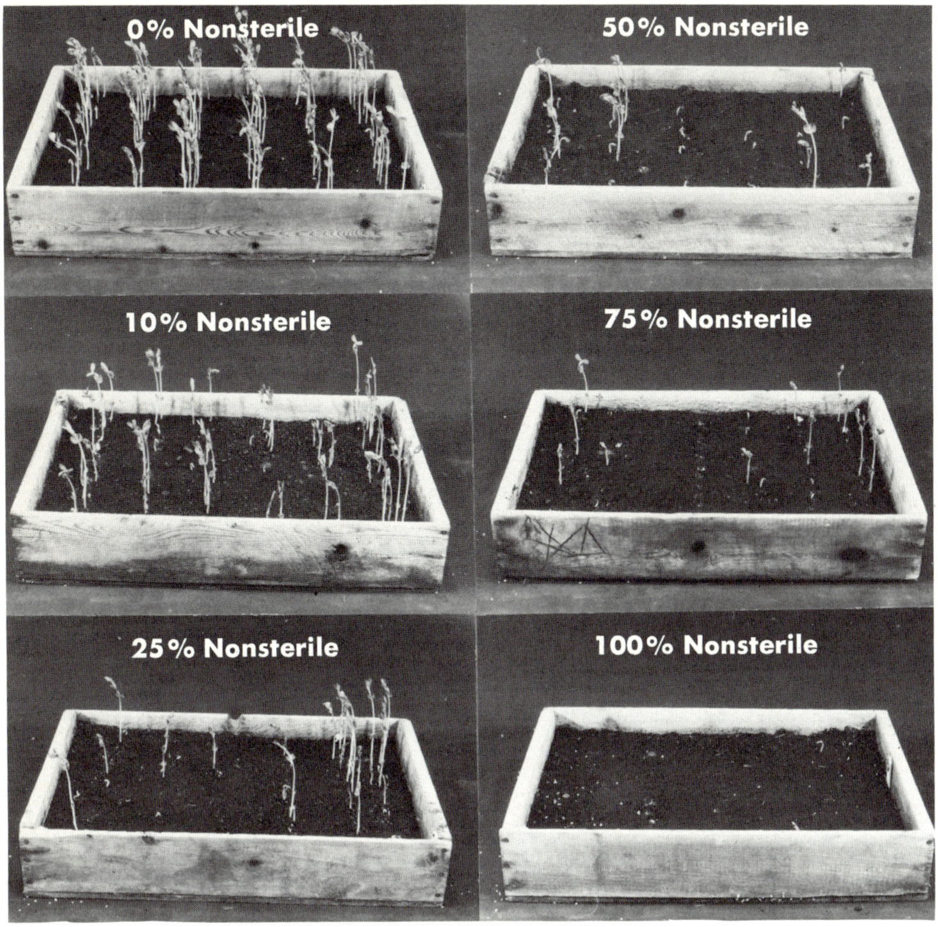

Fig. 22-1. Damping-off of pea seedlings in various mixtures of sterilized and nonsterilized muck soil containing *Pythium*.

2. Each group should treat 10 seeds with a slurry of a fungicide, such as Thiram or Arasan, and plant them in a single row in a flat of 100% nonsterilized soil. Allow the slurry treated seeds to air dry well before planting. Mark each row when planted with a stake, indicating date and treatment. The instructor should give some directions for depth of planting and compaction of the soil afterward.

3. The flats representing a series of dilutions of inoculum in the soil should be planted with 100 seeds each as 10 rows of 10 seeds. This totals 70 rows in the 7 flats. Divide 70 by the number of groups and have each group plant their share of rows—evenly distributed throughout all 7 flats so that differences in planting depths and compaction after planting can be averaged.

4. Keep the flats where they will not be too warm and water thoroughly once each day or every other day.

5. Count emergent seedlings and prepare a graph on the semilog scale (Fig. 22-2) plotting per cent emergence as a function of per cent nonsterile soil. If per cent damping-off is plotted, all figures have to be corrected for the germinability of the seed lot.

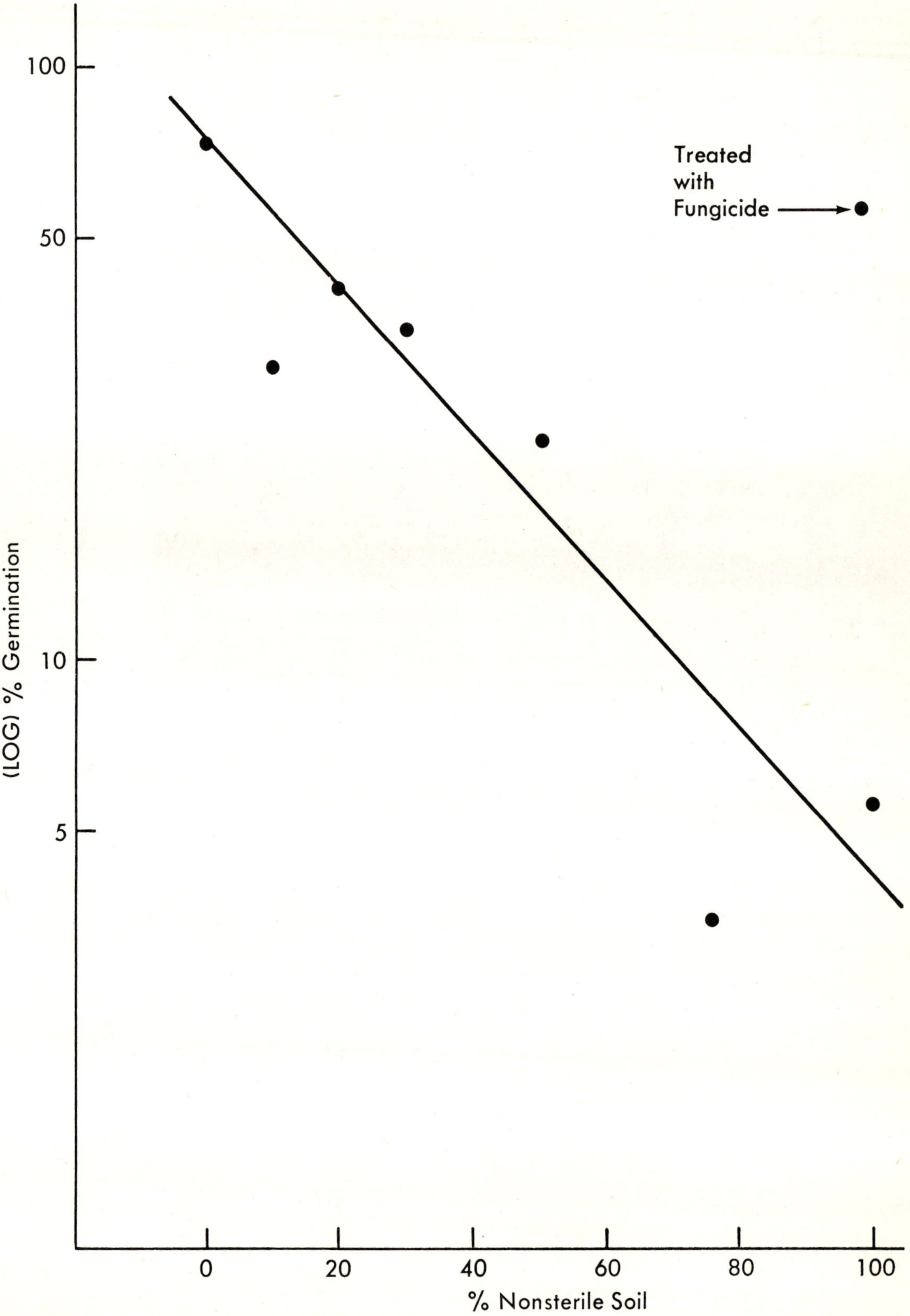

Fig. 22-2. Percentage of emergent seedlings as related to dilution of nonsterilized muck soil with sterilized muck soil.

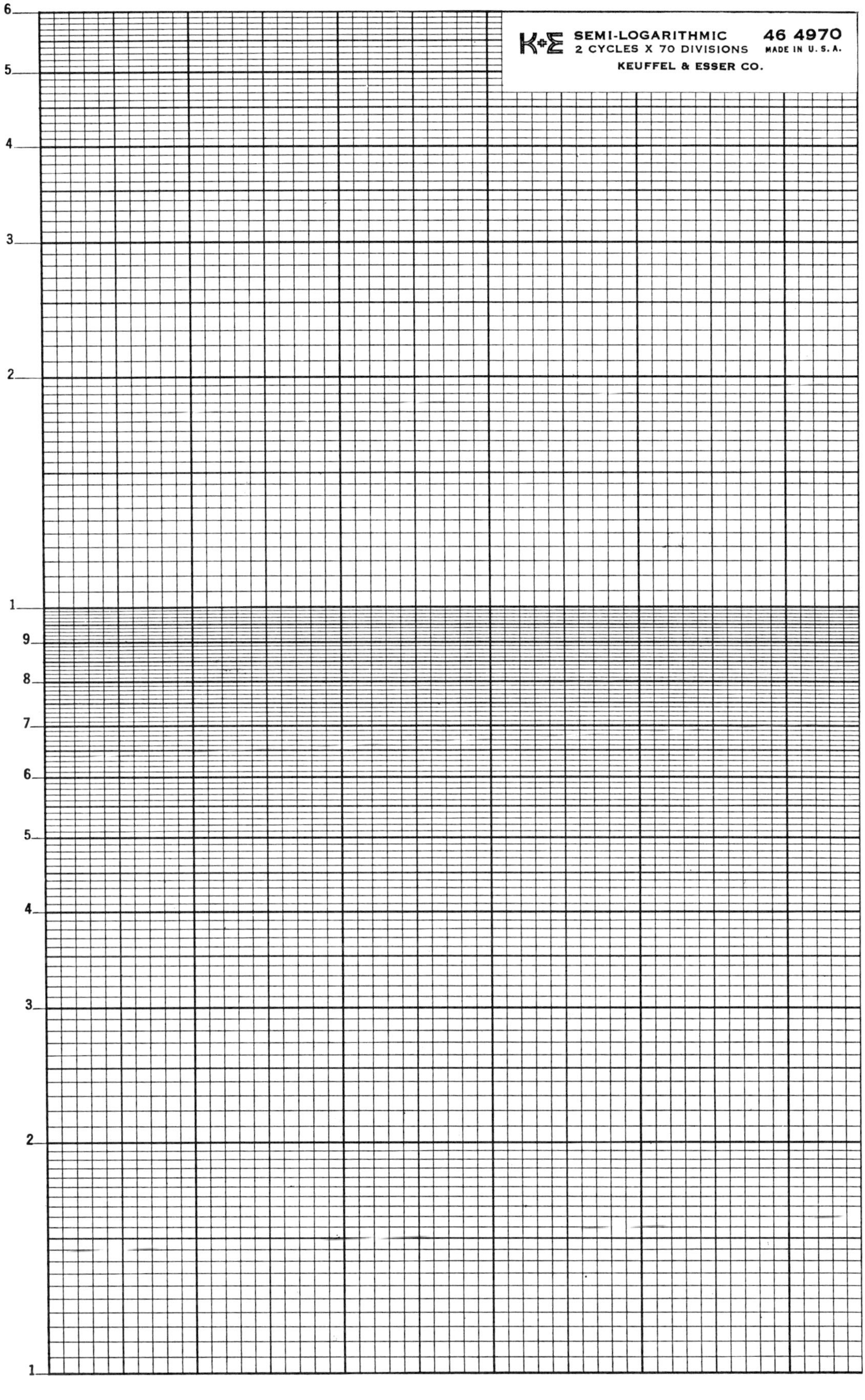
K+E SEMI-LOGARITHMIC 46 4970
2 CYCLES X 70 DIVISIONS MADE IN U.S.A.
KEUFFEL & ESSER CO.
6
5
4
3
2
1
9
8
7
6
5
4
3
2
1

Equipment and Materials

Alderman peas or seeds of other susceptible plants—700 plus 10 times the number of groups
pot labels—cut to 3¾ in. in length to prevent loss from flats—10 plus 10 times the number of groups (place in each row as planted)
sterilized flats
sterilized muck soil
nonsterilized muck soil
fungicide (Thiram, Arasan)
flasks for preparing slurries (as many as there are groups)
weatherproof pencils

DISCUSSION QUESTIONS

1. If damping-off affects 10%–15% of seeds planted in a particular field, would increasing the planting rate give higher yields?
2. Why is steaming of soil sometimes unsuccessful for control of damping-off?
3. Why is it not possible to control damping-off by crop rotation?
4. Would damping-off occur regardless of how little inoculum there is in the soil, or is there a required minimum concentration for the occurrence of disease?
5. Do you think that the slope of the curve in Fig. 22-2 will vary with environment, susceptibility of the host, and pathogenicity of the causal organism?

23

The Ascomycetes

The Ascomycetes are the sac fungi: the fungi that produce asci (sing. ascus, sac) in which ascospores are borne. This group includes the yeasts that leaven the bread and brew the beer; the molds that make penicillin; and, the morels of the dinner table. This group also includes great destroyers—the pathogens that cause Dutch elm disease, chestnut blight, apple scab, brown rot of stone fruits, the powdery mildews, and many, many others.

Characteristics. The hyphae of the Ascomycetes are branched and septate with simple pores in the septa. The hyphae are well-organized in some structures and look like parenchyma tissue since the cells do not appear to be in linear chains but in close-packed order. This is **pseudoparenchyma,** or false parenchyma. Sclerotia, stromata, and sexual and asexual fruiting bodies are generally composed of pseudoparenchyma.

Asexual Reproduction. Asexual spores are conidia or oidia; none are sporangia. Conidia are generally produced on conidiophores which vary from short, hyphal-like structures to complex, branched structures which may differ markedly in appearance from the hyphae. Conidiophores and conidia may be borne free and independent without organization (Fig. 23-1A) or they may be organized into definite asexual structures: the acervulus (D), synnema (B), sporodochium (C), or pycnidium (E).

The **acervulus** (pl. acervuli) is a pile or heap of hyphae bearing a compact layer of conidiophores (Fig. 23-1D, 37-3, 43-1). The **synnema** (pl. synnemata) is a columnar structure of united conidiophores bearing conidia principally at the apex (Figs. 23-1B, 25-3, 25-5B). *Syn-* is a combining form meaning together. A **sporodochium** (pl. sporodochia) is a cushion-shaped, seed (*sporos*) container (*dochium*) composed of a mass of hyphae covered with conidiophores (Figs. 23-1C, 31-3). Some sporodochia are small, microscopic masses as with *Fusarium* (Fig. 23-1C) whereas others are macroscopic and well-formed as with *Nectria* (Fig. 31-3). The **pycnidium** (pl. pycnidia) is a hollow, asexual spore-bearing structure lined with conidiophores. The walls are generally composed of pseudoparenchyma (Figs. 23-1E, 31-1, 35-2, 35-5). Pycnidia may be round to flask-shaped; with a short neck, very long neck, or no neck; and dark or sometimes highly colored. They may be almost completely immersed in the host tissues or other substrate, partially embedded, or on the surface.

Sexual Reproduction. The perfect or sexual fruiting structure is a **sporocarp.** In the Ascomycetes, the sporocarp contains asci and is properly called the **ascocarp,** regardless of the style. The sporocarp of the Basidiomycetes is called the **basidiocarp** and it contains basidia. Asci may be borne free, without an ascocarp, or in one of three basic kinds of ascocarps (Fig. 23-2A): the perithecium (C), cleistothecium (B), and

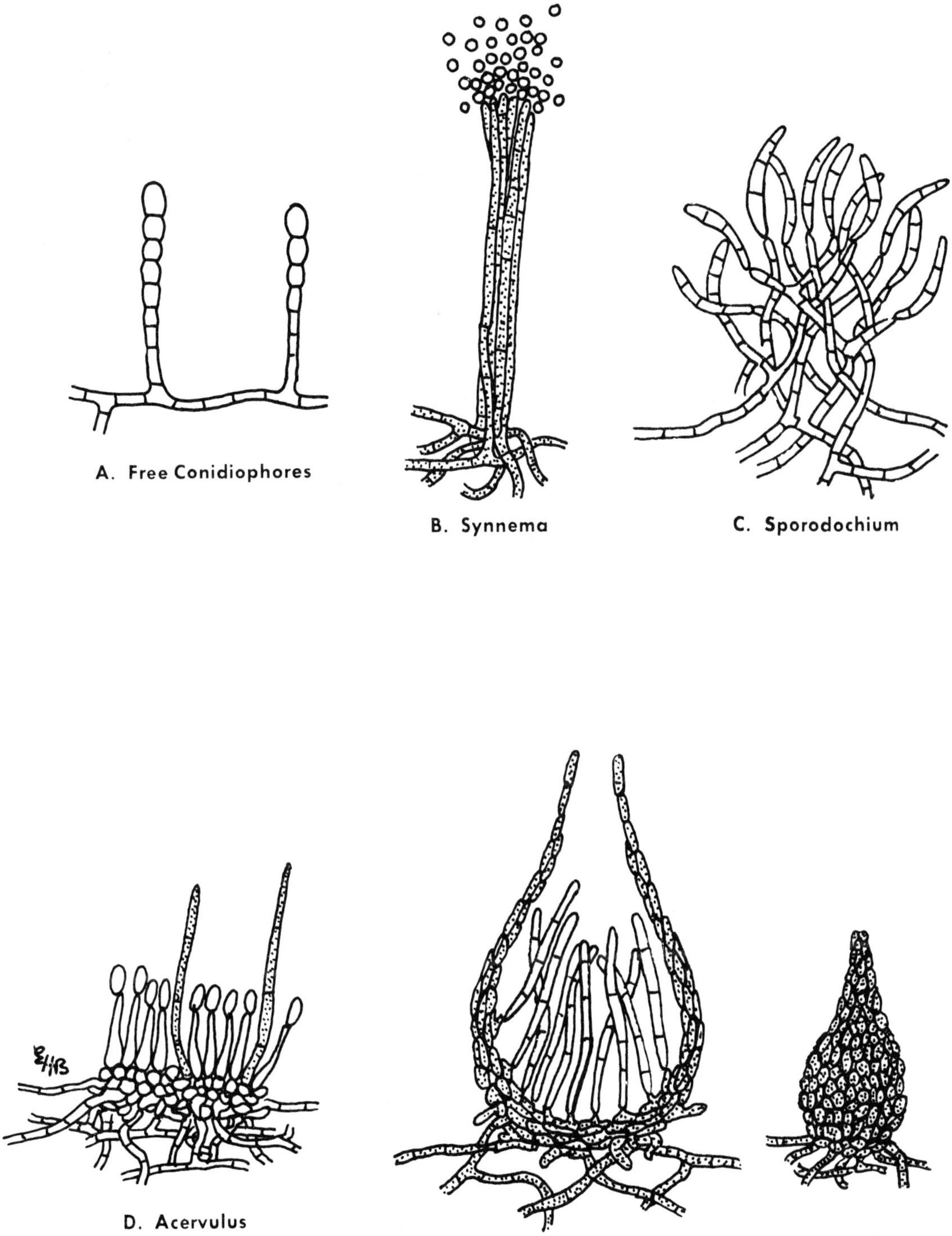

Fig. 23-1. Varous ways in which conidia are borne.

apothecium (D). The **perithecium** has a neck with an opening at the top. *Peri-* is a combining form meaning around, as in perimeter. The **cleistothecium** is a closed structure with no opening. *Cleisto-* is a combining form meaning closed, as in cloister. The **apothecium** is a cup-shaped structure on a short to long stalk called a **stipe.** The word apothecium is derived from the same root as apothecary, meaning storehouse. The apothecary of old times is today's drugstore. All ascocarps contain **asci,** the sac-like structures which contain **ascospores.** The textbook ascus is long and narrow with eight ascospores in linear array. Some asci are round; some contain fewer than eight or more than eight ascospores; and some deliquesce (or dissolve) within the ascocarp. Asci may be thin-walled or have a thickened apex and may or may not have some sort of natural opening at the apex. The ascospores may be hyaline (colorless) or dark-colored, one-celled, many-celled, short, long, or even twisted. All of these features are important diagnostic characteristics.

Ascogenous Hyphae and Crozier Hook Formation. Nuclei are brought together in the female gametangium, the **ascogonium,** by the transfer of a nucleus from the male gametangium by gametangial fusion, spermatization, or other mechanisms. The fertilized ascogonium produces **ascogenous hyphae**—so called because they produce the asci. The cells of the ascogenous hyphae are binucleate, or dikaryotic. Ascogenous hyphae develop within the ascocarp.

The tip of the ascogenous hypha (Fig. 23-3) curls and is called the **crozier hook** (B). The two nuclei, one from each parent, divide with their spindles so oriented (C) that a new cross call (D) between the pairs of nuclei makes the ultimate (last) and the antipenultimate (third from last) cells uninucleate and the penultimate (next to last) cell dikaryotic and positioned at the apex. The nuclei in the penultimate cell fuse and this cell is now a zygote with a diploid nucleus and is the ascus **mother cell** (E). A meiotic division occurs (F) followed by an equational division of all four haploid nuclei so that the elongating young ascus now contains eight haploid nuclei (G)—four sets of identical pairs. Cell walls are laid down around the nuclei to form eight ascospores (H). Meanwhile, the nucleus from the antipenultimate (third from last) cell migrates though a pore dissolved in the wall into the ultimate cell (I). This develops into another crozier hook (J).

Classification. Reproduction of Ascomycetes is typically both asexual and sexual. A number of these fungi have asexual spores which are very similar to each other and some have no known asexual spores. There are some other fungi that are known only by asexual spores which are similar to the asexual spores of the Ascomycetes. These sexless fungi are the so-called imperfect fungi, the Fungi Imperfecti (Deuteromycetes), in contrast to those that are sexual and supposedly perfect. The classification of the fungi is based on sexual characteristics so how can we classify imperfect fungi in a "perfect" scheme? We can't. Therefore, an artificial class, the Deuteromycetes, or Fungi Imperfecti, was organized. This group and all taxonomic subdivisions of it are called **forms.** Thus, there is the form-class Deuteromycetes with form-orders, form-families, and so forth. If an organism is referred to by a perfect name, the entire organism is properly the subject. If an organism, with a known sexual stage, is referred to by an imperfect name in the Deuteromycetes, then only the imperfect, or asexual, cycle should be the subject. The Deuteromycetes are discussed separately and in more detail later.

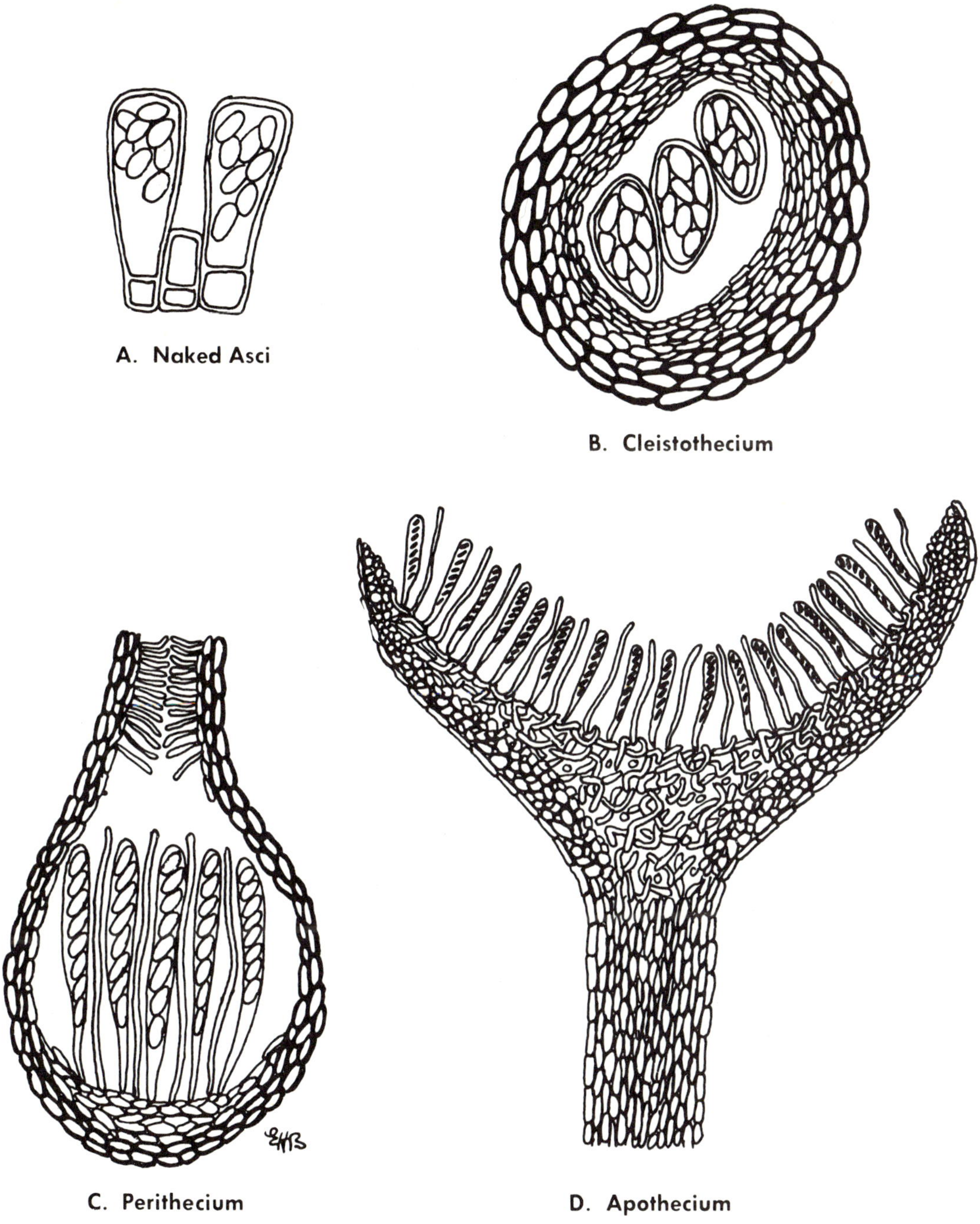

Fig. 23-2. Various ways in which asci are borne.

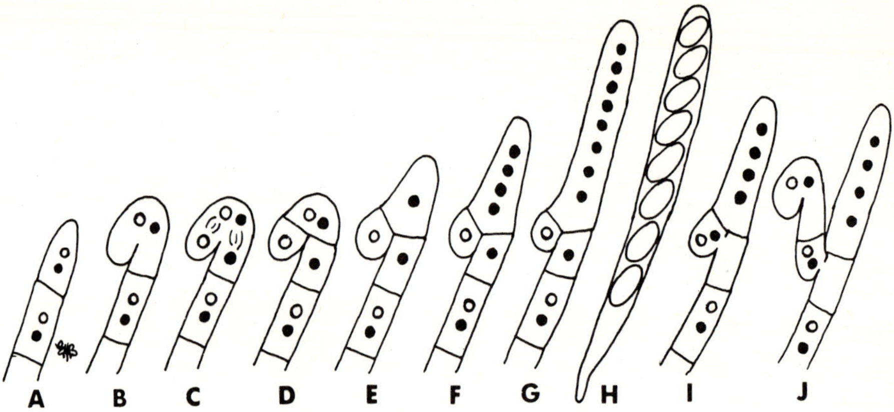

Fig. 23-3. Development of the crozier hook and ascus from an ascogenous hypha.

OUTLINES OF GENERA TREATED

Hemiascomycetidae (*Sub-Class*)

Taphrinales (*Order*) *Taphrina deformans, Taphrina pruni*

Euascomycetidae (*Sub-Class*)

Plectomycetes (*Series*) *Ceratocystis ulmi*

Pyrenomycetes (*Series*)

Erysiphales (*Order*) *Erysiphe, Microsphaera, Podosphaera, Uncinula, Sphaerotheca, Phyllactinia*

Sphaeriales (*Order*) *Hypoxylon pruinatum*

Clavicipitales (*Order*) *Claviceps purpurea*

Diaporthales (*Order*)

Gnomoniaceae (*Family*) *Gnomonia ulmea*

Gnomonia veneta

Diaporthaceae (*Family*) *Nectria coccinea*

Discomycetes (*Series*)

Helotiales (*Order*)

Phacidiaceae (*Family*) *Rhytisma*

Sclerotiniaceae (*Family*) *Sclerotinia*

Monolinia

Loculoascomycetidae (*Sub-Class*)

Dothidiales (*Order*) *Guignardia bidwellii, Dibotryon morbosum*

Pleosporales (*Order*) *Venturia inaequalis*

24

Peach Leaf Curl

Peach leaf curl, caused by *Taphrina deformans,* is relatively unimportant as an economic factor in peach production even though it occurs rather continuously throughout the significant areas of peach production in the world. Control, where and when necessary, is effected with a single application of fungicide before bud-break. Such sprays, applied while the host is dormant, are called **dormant sprays**. The disease is important primarily in the urban, back-yard habitat where the busy commuter has a single peach tree. Either the single dormant spray technique should be used here too, or the one cut prunning method—at the ground line!

Symptoms: appear in early spring as curled leaves with a puckering of the blade along the midrib (Fig. 24-1A). Infected areas turn reddish when the underlying layers of asci mature. Premature leaf drop is common in severe cases and the otherwise dormant lateral buds issue leaves and shoots. This lowers the general vigor of the tree, predisposing it to winter injury, insect attack, and increased disease problems from other causes.

Microscopic observation of the crinkled leaves reveals a layer of fungus structures on the surface of epidermal cells, between them, and between parenchymatous cells (Fig. 24-2). Plum pockets is a disease caused by *Taphrina pruni,* a related fungus (Fig. 24-1B).

Life Cycle. In the spring, during cool, wet periods, overwintered conidia and ascospores germinate and produce germ tubes (Fig. 24-3L) which infect only expanding peach leaves and young shoots. During elongation of the germ tube, the single nucleus divides and thereafter the cells are binucleate. Mycelium grows beneath the cuticle (M), pushes between the epidermal cells, and into the intercellular locations and spaces of the parenchyma (N). The mycelium, especially in the subcuticular area, becomes fragmented (O) and each cell (A) is capable of producing an ascus. The nuclei fuse, producing a zygote (B). This diploid nucleus divides (C) and a cell wall is laid down (D). The lower cell is the **foot cell** while the upper cell is the **ascus mother cell.** It elongates and meiosis occurs, producing four daughter nuclei (E). Usually an equational division occurs (F) so that the ascus contains eight ascospores when mature (G). The asci are naked, that is, they are not borne in an ascocarp such as a perithecium, cleistothecium, or apothecium. Buds (I) are formed by the ascospores (H) and are called conidia. The conidia may also bud to form secondary conidia.

Disease Cycle. The primary inoculum is the overwintered ascospores and conidia (buds). They lodge beneath the scales of the dormant buds of the host. There is little secondary infection. This is why a single application of fungicide just prior to bud-break can effectively control the disease.

Fig. 24-1. (A) Symptoms of peach leaf curl and (B) plum pockets.

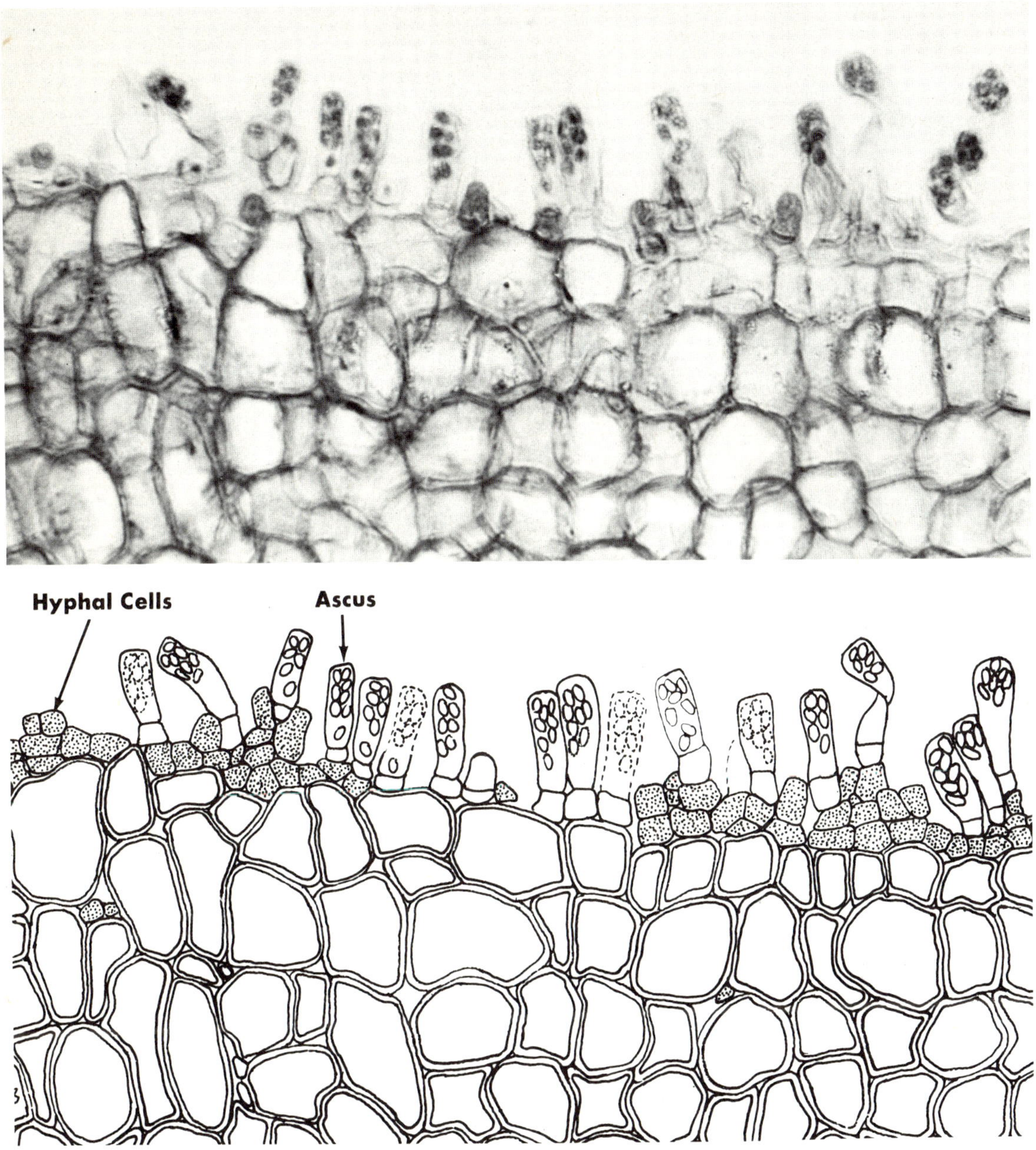

Fig. 24-2. Asci and ascospores of *Taphrina deformans* and diagrammatic reconstruction (× *250*).

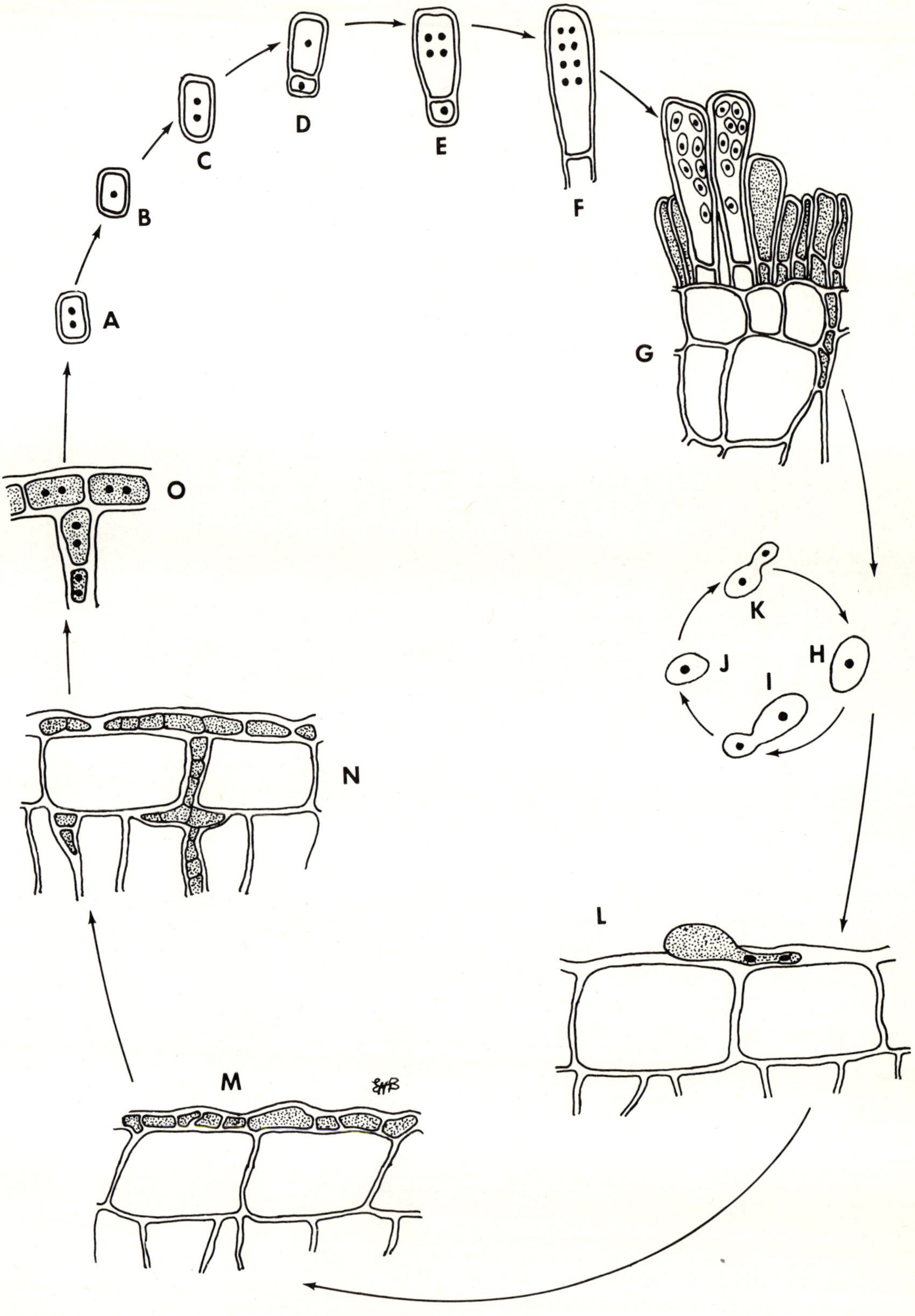

Fig. 24-3. Life cycle of ***Taphrina deformans.***

25

Dutch Elm Disease

The most well-known plant disease in the Eastern United States and Canada is the Dutch elm disease. It has laid waste to Main Street, U. S. A., and the countryside alike. Great groves of elm along our new superhighways stand stripped and bleached in the sun (frontispiece and Fig. 25-1 A).

Dutch elm disease is a misleading name since it describes not the country of origin but the country where pathologists intensively studied the disease. The disease was first observed in France, Belgium, and Holland during the First World War. Then it spread to all points of the compass: north to Sweden, east to Russia, south to Spain, and west to England. Although the origin is obscure, the disease probably came out of the Asian continent where resistant elms are found today. Unfortunately, these elms do not have the desired umbrella-like growth habit, but have provided genes for resistance to tree breeders.

In America, during the nineteen thirties, the disease appeared suddenly and almost simultaneously in widely scattered areas from Indiana to Connecticut and Maryland. It seems that entry of the fungus occurred in several seaports along the Atlantic and Gulf of Mexico in elm logs imported from Europe for veneer. From the ports, the logs moved on open flatcars along thousands of miles of railway rights-of-way to a score of destinations.

State and Federal authorities immediately entered upon a concentrated and expensive program of extermination. Many infested areas were found by tracing the shipment of logs from the ports to the veneer factories. These efforts were later abandoned.

Prior to introduction of the fungus on logs, nursery stock from overseas had been quarantined for years. New Federal quarantines were established to prohibit further importation of elm logs without permit. These efforts were abandoned in the early forties after the extensive program of early eradication failed.

Causal Organism. The sexual stage (Fig. 25-5 F, G), *Ceratocystis ulmi,* occurs infrequently and is of little importance in America but does occur in Europe. The asexual stage is *Graphium ulmi.* Very small, hyaline (colorless), one-celled conidia are borne in a sticky drop at the apex of a dark-stalked synnema (Fig. 25-3, 25-5 B, C). When mounted in water, the conidia disperse like a cloud since the matrix holding the droplet together is water soluble. The fungus overwinters as mycelia and conidia in dead trees and logs (Fig. 25-7 A).

The Vector. The fungus is spread primarily by the smaller European elm bark beetle (*Scolytus multistriatus*) (Fig. 24-4 E). The beetle overwinters as a grub in the bark of unhealthy trees, recently cut logs, or firewood (Fig. 25-6 A). During May or early June, adult beetles emerge (Fig. 25-4 C), fly to healthy trees (Fig. 25-6 B), and feed in the crotches of the twigs (Figs. 25-6 C, 25-4 D). If they have emerged from infected trees or wood, they may carry spores of the fungus on their bodies or mouth parts and deposit them in feeding wounds. The spores constitute the primary inoculum.

Fig. 25-1. (A) Destruction of a grove of elms by *Ceratocystis ulmi.* (B) Early symptoms of Dutch elm disease, and (C) a natural root graft between two roots (oak).

The native elm bark beetle (*Hyluropinus rufipes*) also transmits the disease. It overwinters as an adult and has only one generation a year. The European beetle overwinters as larvae and may have a small number of second generation individuals in a year. The American beetle makes its gallery across the grain whereas the European beetle constructs its gallery with the grain (Fig. 25-4 A, B).

The Disease. If the tree is infected near the top, the foliage on a single limb turns yellow, wilts, and collapses (Fig. 25-1 B). It may take several weeks or even months for the entire tree to become diseased. However, if lower branches of the tree, especially small twigs issuing from the main trunk, are infected first, the progress is very rapid and the entire tree may die in a few weeks. The fungus grows in the xylem elements and the spores are rapidly carried to higher parts of the tree until they lodge. They germinate, grow through the obstruction, and sporulate. Conidia of this new crop are then carried rapidly upward until they too strike an obstruction.

Infections also take place between adjacent trees through natural **root grafts** (Figs. 25-1 C, 25-7 G). These infections occur throughout the growing season. They are primary infections when the fungus overwinters in the donor tree prior to infection of the recipient tree by means of the root graft. Infections through root grafts are secondary infections when the fungus passes through a root graft from the donor tree during same growing season that the donor tree first became infected.

The wood in branches of infected twigs is streaked with brown (Fig. 25-2). However since certain other fungi induce similar symptoms, positive diagnosis cannot be made on the basis of the wilting and streaking symptoms. This may be accomplished by culturing the fungus on wood chips (Fig. 25-3).

The disease cycle (Fig. 25-7) involves the life cycle of the beetle as well as the life cycle of the fungus (Fig. 25-5).

Fig. 25-2. Characteristic streaking in the wood of twigs infected with *Ceratocystis ulmi*.

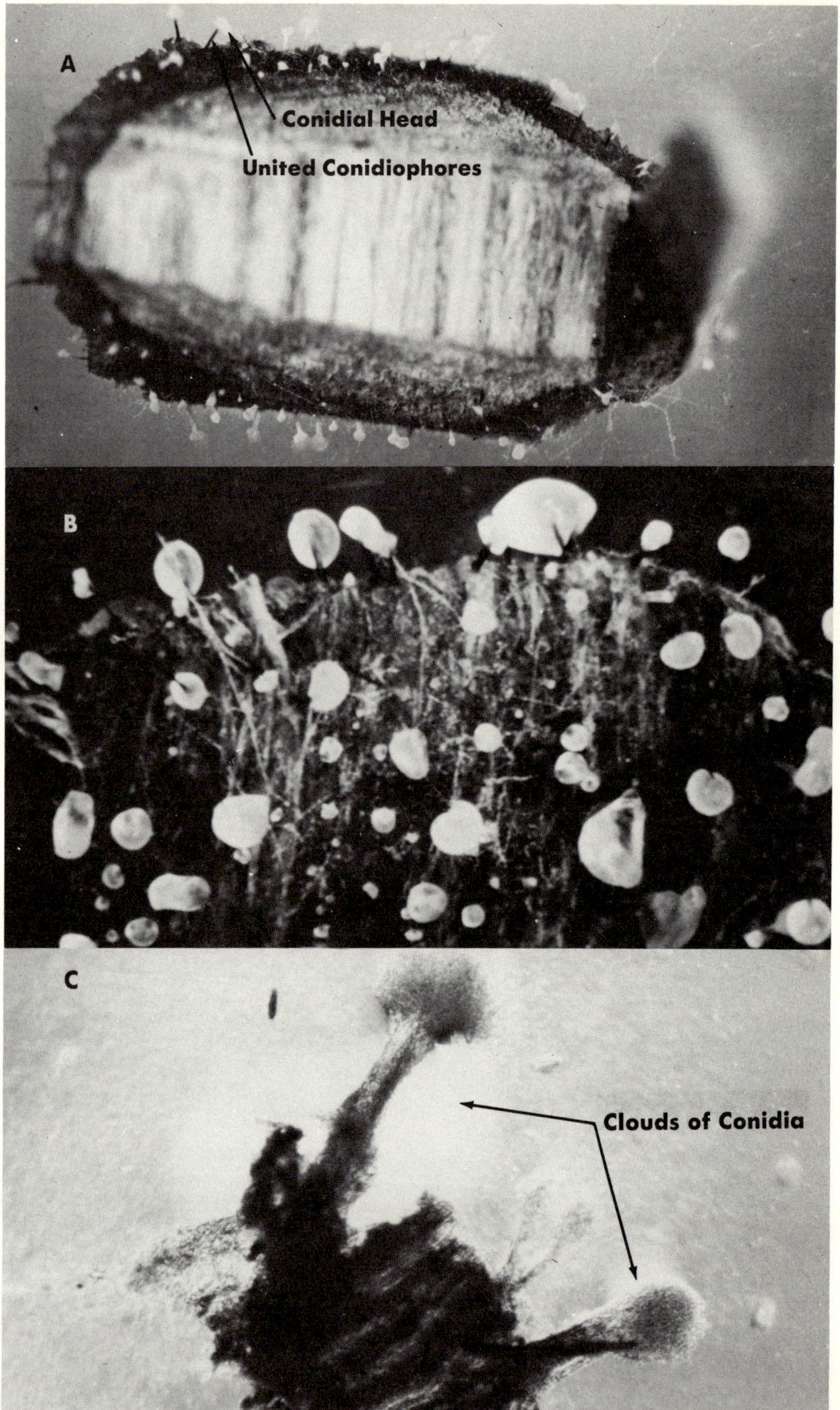

Fig. 25-3. (A) Synnemata of *Ceratocystis ulmi* on an elm chip in culture. (B) Enlarged detail of synnemata. (C) Cloud of spores dispersing from the conidial head of a synnema mounted in water on a microscope slide.

Fig. 25-4. (A, B) Galleries of the European elm bark beetle. (C) Emergence holes (arrows) of the beetles in the bark. (D) Feeding wounds (arrows) of elm bark beetles. (E) The European elm bark beetle.

Control: based on the reduction of sources of inoculum and vectors to move it. Control measures include: (1) All infected trees, logs, and firewood should be burned, stripped of bark, or treated with chemicals to eliminate egg-laying sites for the beetle. (2) Healthy trees should be sprayed with appropriate insecticides to kill the beetles after emergence and before they have much time to feed and spread the fungus. (3) All elms should be pruned so as to remove dead, dying, and poor vigor limbs. These latter ones are the favorite feeding sites of the beetles. (4) Chemical fumigation of the soil should be used between trees to kill root grafts and prohibit the spread of the fungus through them. (5) If elms are replanted, they should be spaced so as to avoid future root grafts.

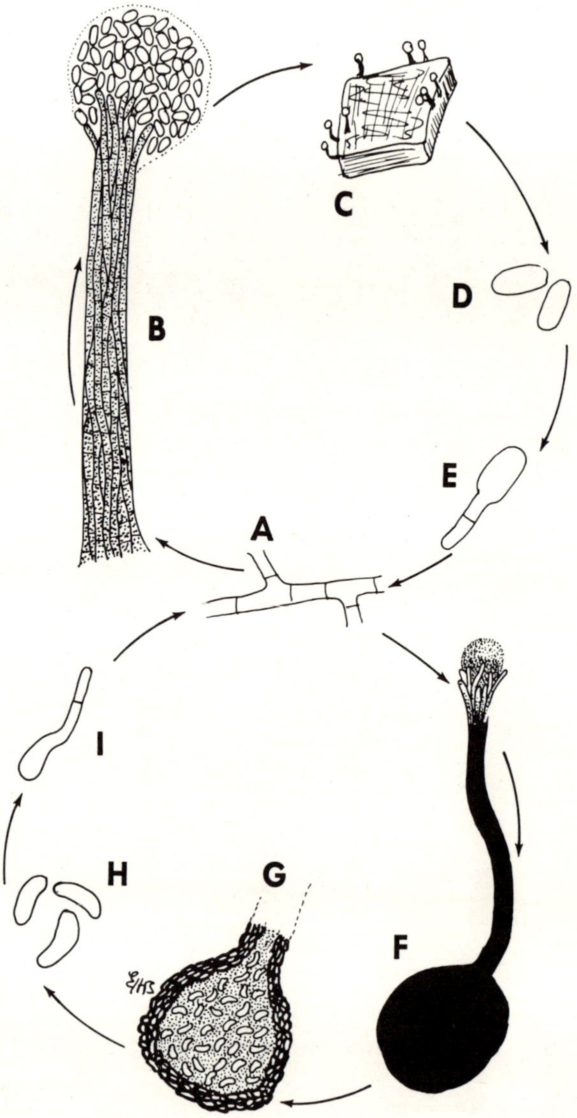

Fig. 25-5. Life cycle of *Ceratocystis ulmi*.

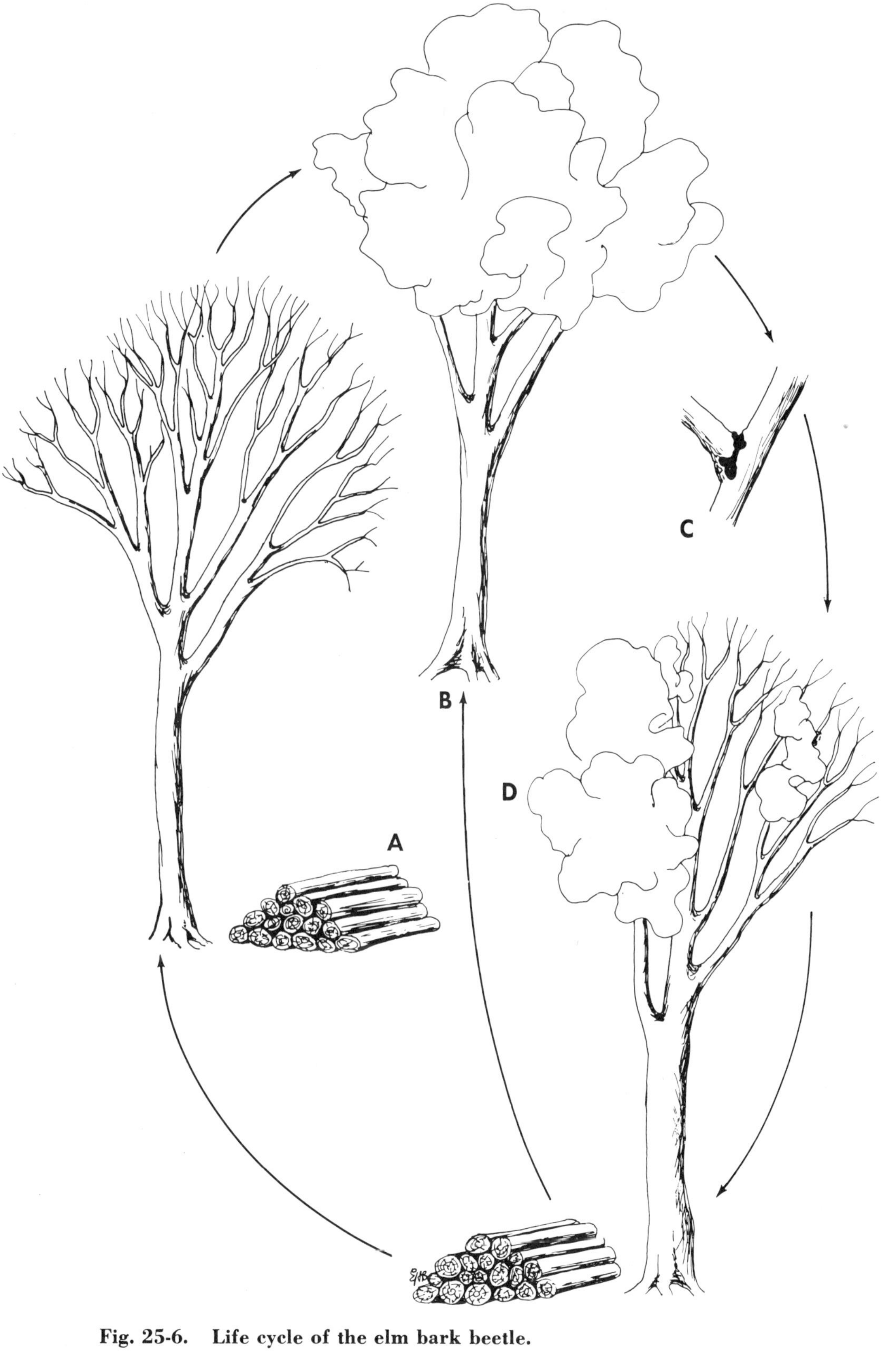

Fig. 25-6. Life cycle of the elm bark beetle.

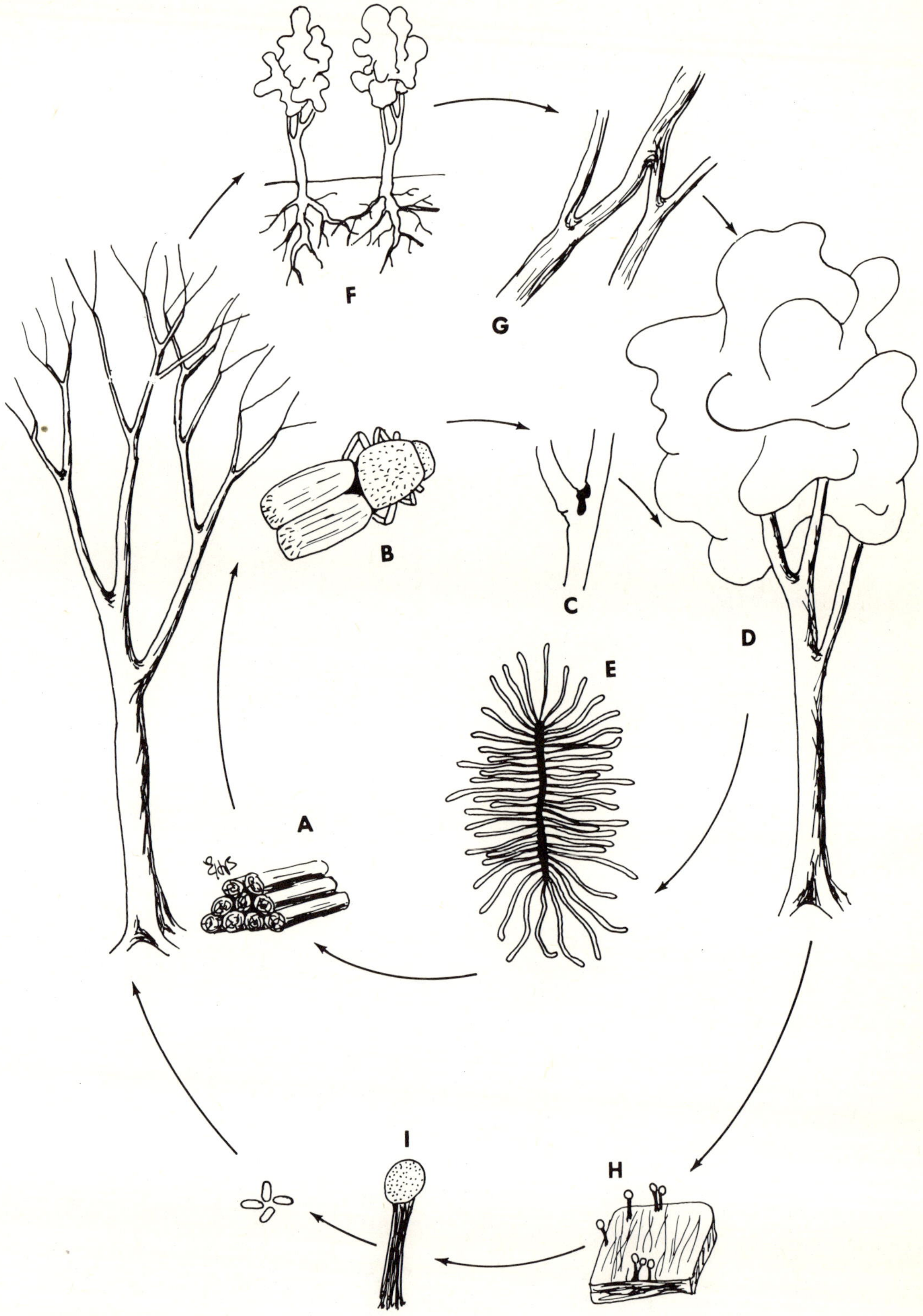

Fig. 25-7. Disease cycle of Dutch elm disease (A combination of the life cycles of the causal fungus and beetle vector).

26

The Powdery Mildews

Throughout the temperate zones of the world, on many different kinds of plants, there occurs a whitish fungus growth of the powdery mildew fungi (Figs. 26-1,-3,-6). These fungi, which comprise the family Erysiphaceae, are all obligate parasites and all but a few are ectoparasites. Ectoparasites grow on the surface of the host and invade only with haustoria (Fig. 26-2). Even these structures are limited usually to the epidermal cells. Some members of the Erysiphaceae also occur in tropical areas.

In early stages, scattered and diffuse spots of white appear on aerial parts of plants. These enlarge, coalesce, and may involve most of the plant surface. Entire new shoots may be deformed. The diseases are called powdery mildew diseases because the white areas are powdery due to the masses of conidia which are readily dislodged. They are one-celled, ovoid, hyaline, and borne in chains on a conidiophore with cells not unlike the conidia (Fig. 26-1). The conidia are cut-off successively at the base. The majority of powdery mildew fungi have asexual stages which belong to the form-genus *Oidium*.

On annual hosts, cleistothecia begin to appear as the hosts mature (Fig. 26-1 A). They can be observed with unaided eye as pepper-like spots (Fig. 26-3 B). On perennial plants, they usually appear as the days grow colder and the dormant season approaches. On lilac, they can be observed first on the north, or cooler side.

The fungus overwinters in different ways, depending on the species and location. On crucifers, and occasionally beans, the fungi overwinter as cleistothecia. Thus, the ascospores constitute primary inoculum. The conidia of the fungus on peas are seed-borne and thus are the overwintering and primary inocula. On many perennial hosts the dormant mycelium overwinters in living hosts, especially the buds. For cucurbits, and occasionally beans, conidia are blown north with the wind from southern areas, and in the south, conidia commonly overwinter on many hosts. On grasses and grains, mycelial mats are often the overwintering structures.

Six genera of powdery mildew fungi occur in North America. They are readily recognized by two characteristics: the kind of cleistothecial appendages (Fig. 26-7) and the number of asci in each cleistothecium. The cleistothecia of *Sphaerotheca* and *Podosphaera* contain a single, large ascus, whereas the cleistothecia of the other genera contain several asci.

Control: consists of dormant and protective sprays, use of resistant varieties, and seed treatment. The method of control depends primarily on the mechanism of overwintering for the particular causal organism. With many plants, the economic significance is believed to be slight and the disease is ignored.

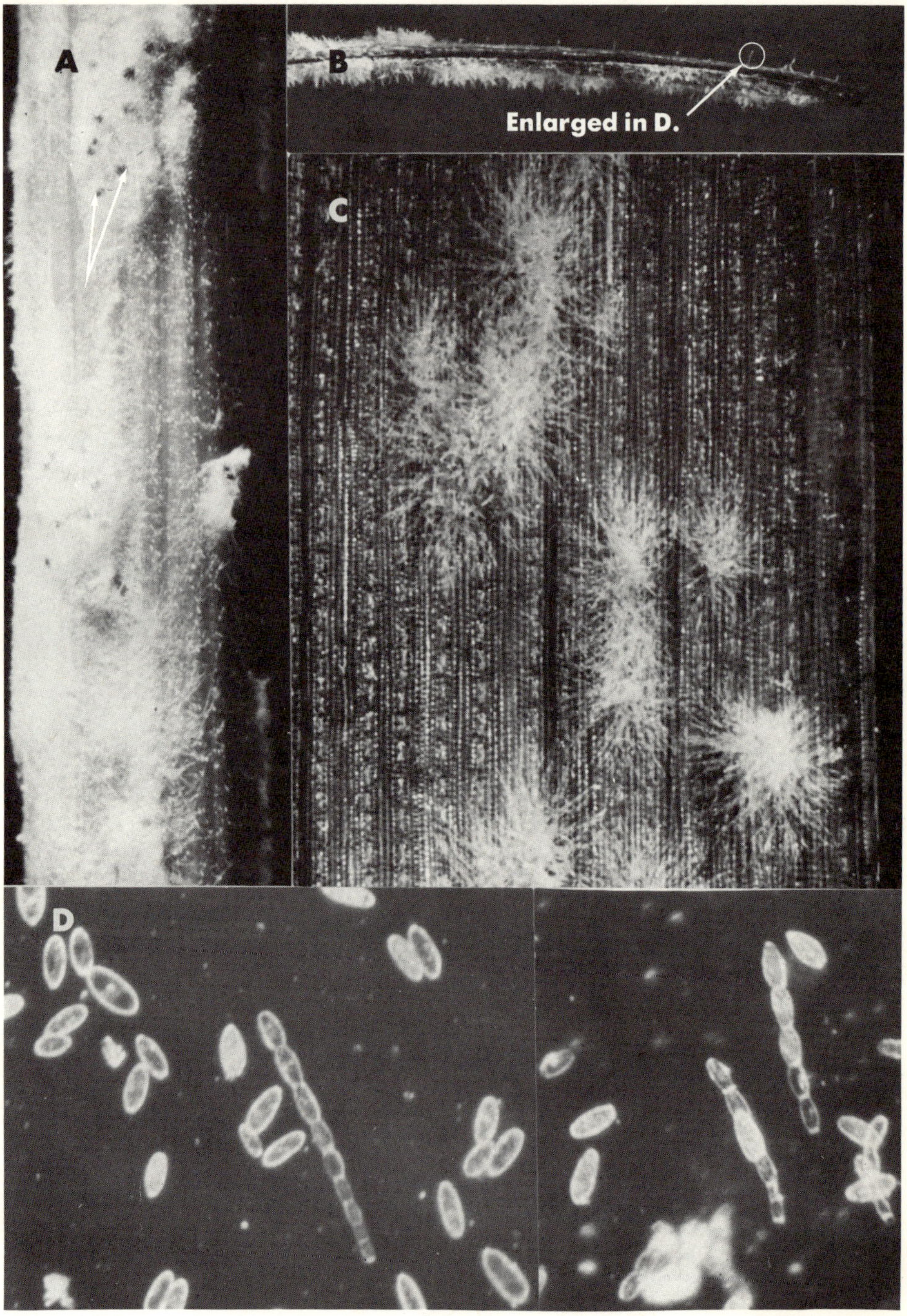

Fig. 26-1. **(A) Mycelia and young cleistothecia (arrows) of *Erysiphe graminis* on barley; (B) mycelia and conidia in side view; (C) top view of colonies of mycelia and conidia on leaf; (D) photomicrograph of conidia.**

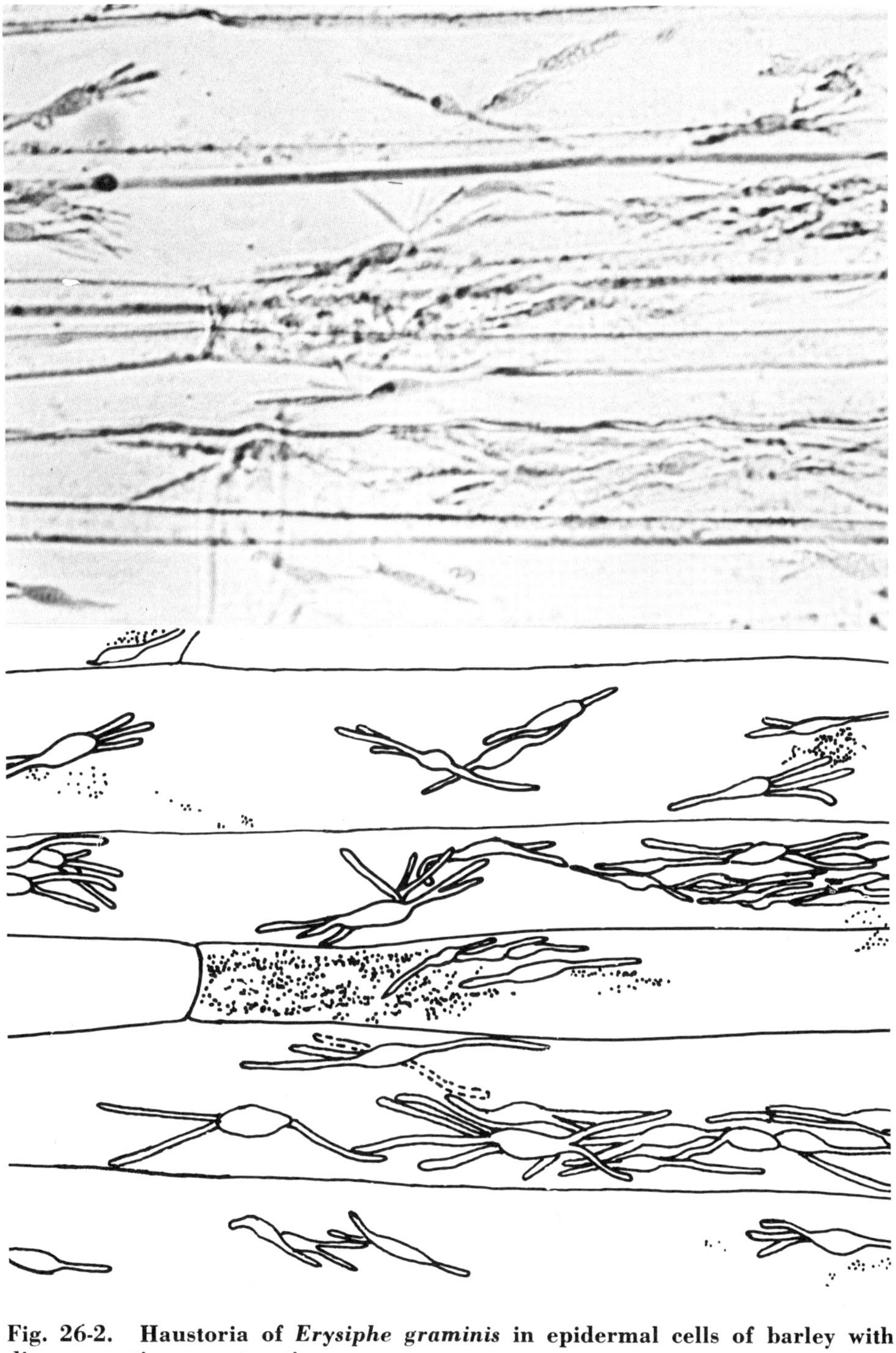

Fig. 26-2. Haustoria of *Erysiphe graminis* in epidermal cells of barley with diagrammatic reconstruction.

Fig. 26-3. (A) Powdery mildew of lilac; (B) mycelia and cleistothecia of *Microsphaera alni* on lilac leaf; (C) enlargement to show cleistothecia (arrows) on the leaf; (D) photomicrograph ($\times$ *215*) of a cleistothecium displaying dicotomously branched appendages.

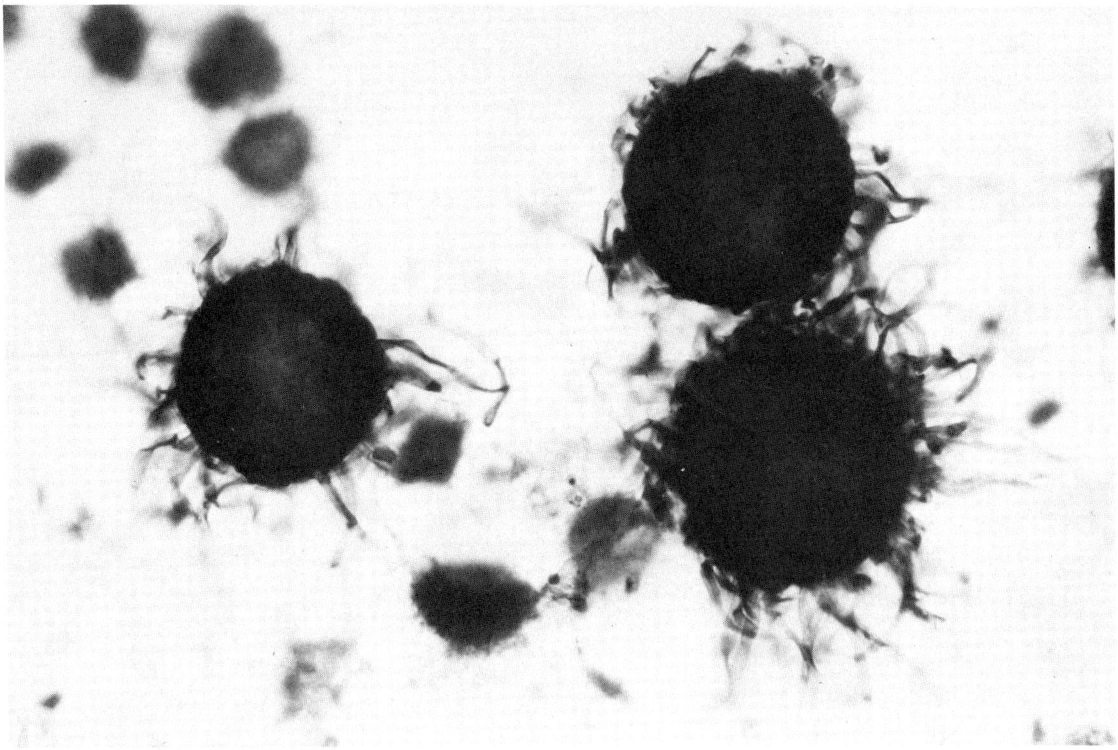

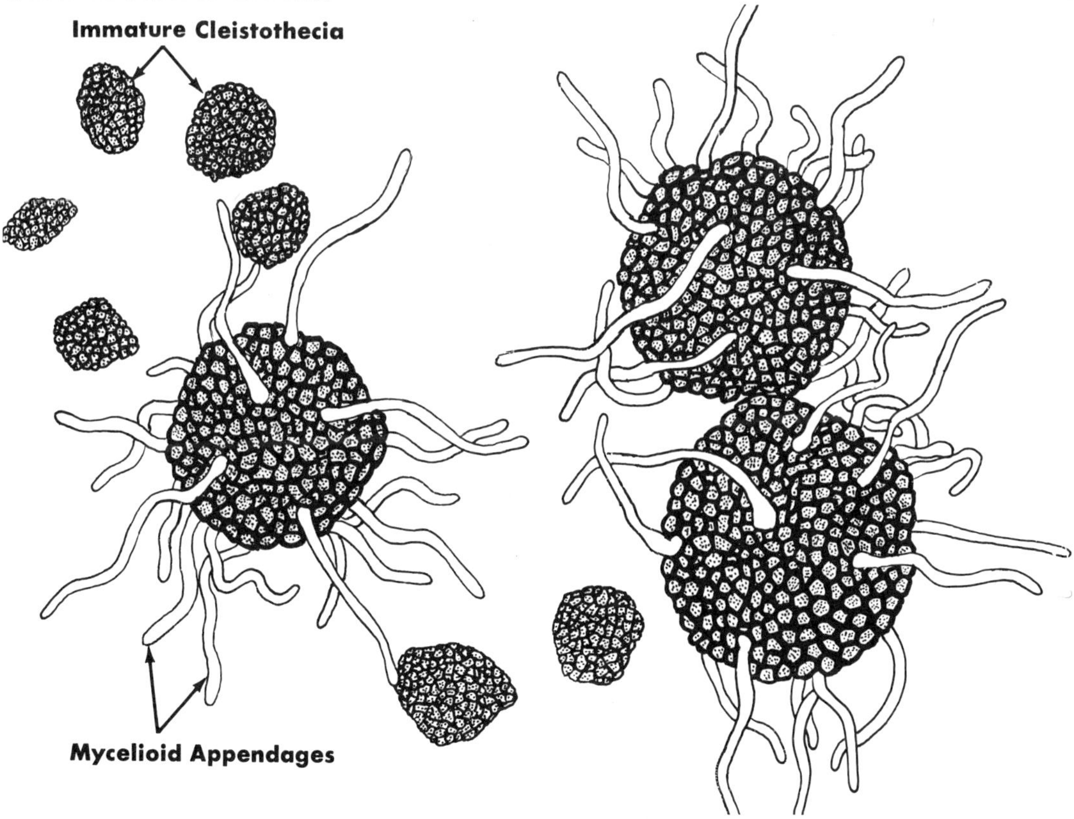

Fig. 26-4. Photomicrograph of a cleistothecium of *Erysiphe graminis* and diagrammatic reconstruction (× *240*).

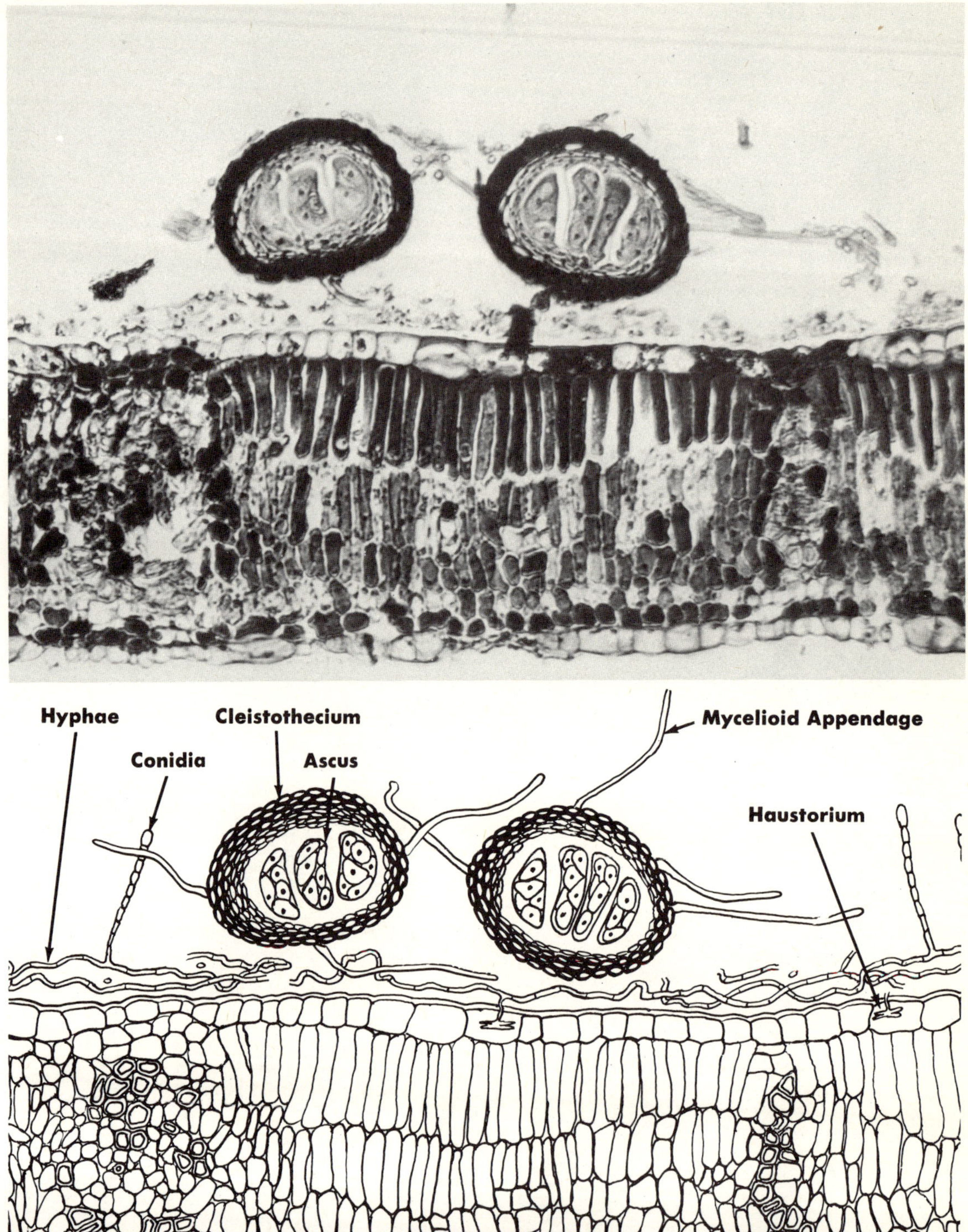

Fig. 26-5. Cross section of cleistothecia of *Erysiphe graminis* and diagrammatic reconstruction (× *240*).

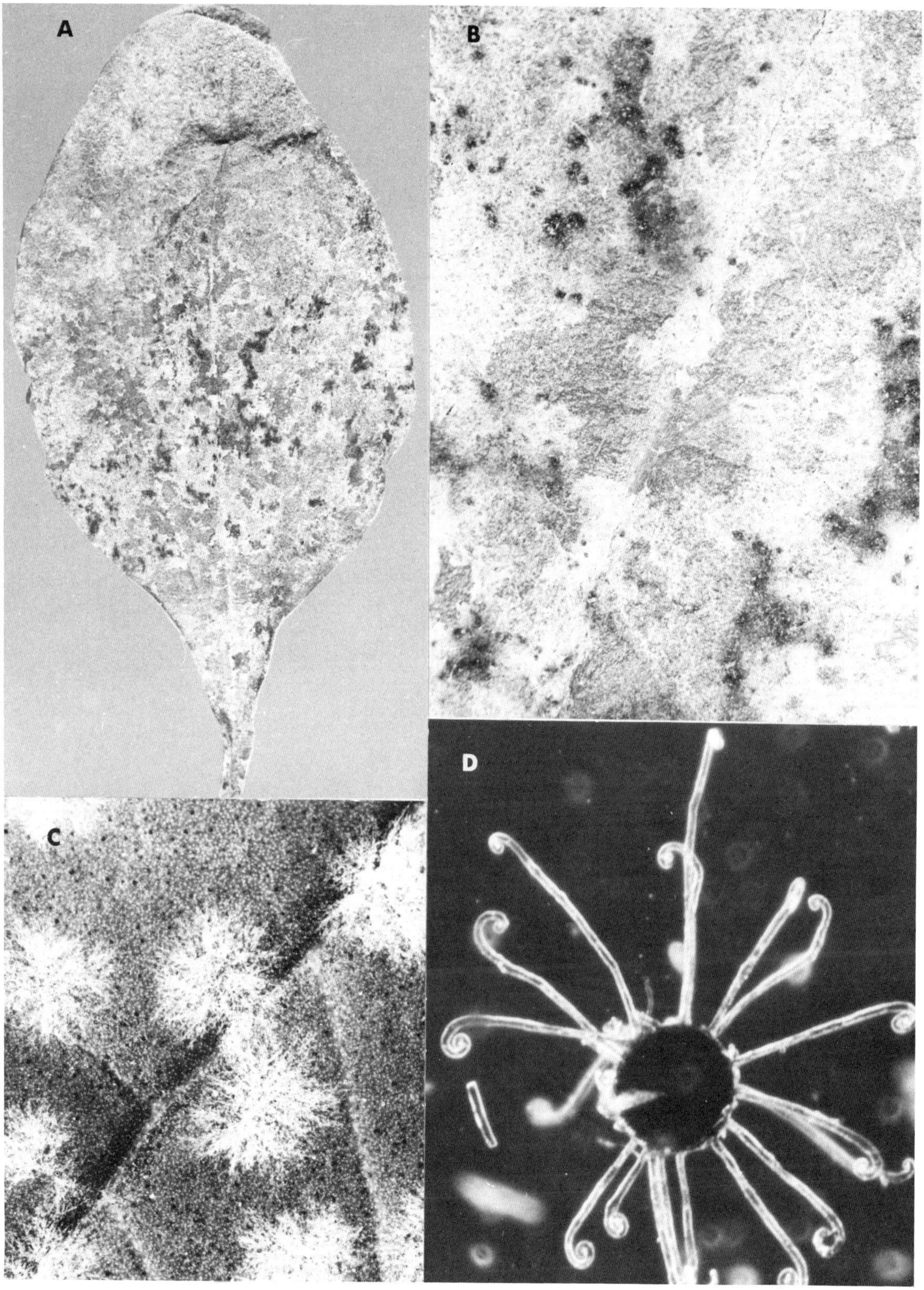

Fig. 26-6. Powdery mildew of plantain caused by *Erysiphe cichoracearum:* (A) symptoms, (B) mycelium and cleistothecia, (C) young colonies of mycelia, and (D) photomicrograph of a cleistothecium of *Uncinula necator*, the causal fungus of the powdery mildew disease of grape.

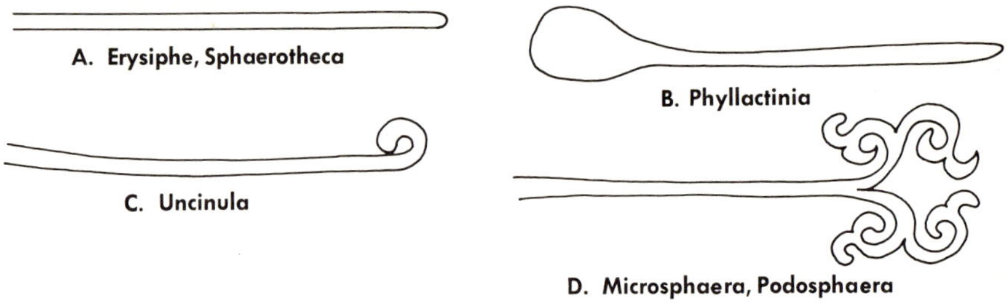

Fig. 26-7. Types of cleistothecial appendages found on some common genera of powdery mildew fungi.

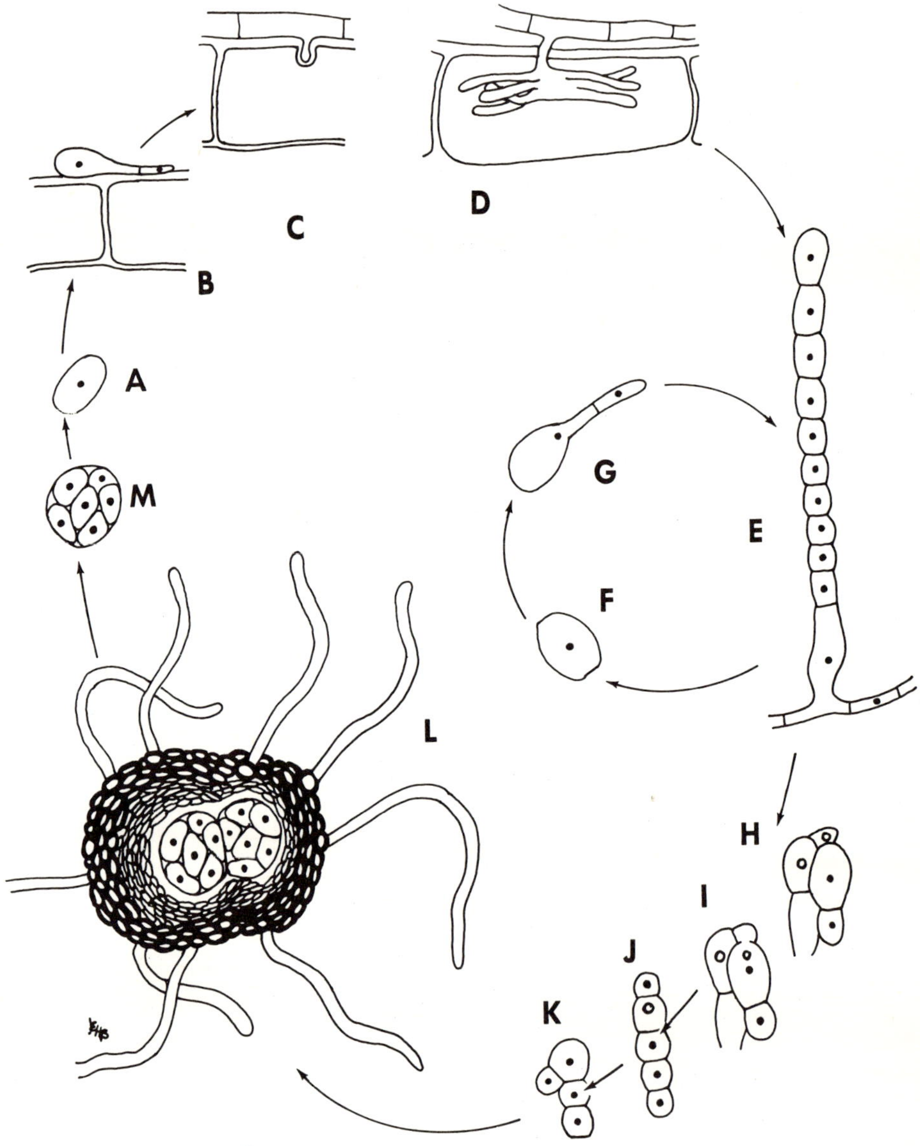

Fig. 26-8. Life cycle of *Erysiphe graminis*.

SUGGESTED READING

Schnathorst, W. C., "Spread and Life Cycle of the Lettuce Powdery Mildew Fungus." *Phytopathology*, Vol. 49 (August, 1959), pp. 464–468; 562–571.

27

Hypoxylon Canker

Hypoxylon occurs on aspen, alder, birch, maple, oak, and willow. Although some authorities separate the different forms of the fungus into three species, we combine all the forms into one species. *Hypoxylon pruinatum.* The disease ranges across southern Canada and northern United States and most probably throughout the range of the aspen. It is easily spotted in more advanced stages because the trees are girdled, killed, and broken at the position of the canker (Fig. 27-1). Small, yellow leaves are borne on seriously affected trees and they persist until spring if the tree is killed during the summer. Dead branches bearing persistent leaves are called **flags**.

The fungus is a **wound parasite** of the bark tissues and is a low-grade pathogen. It enters insect punctures, branch nodes, and injuries caused by wind breakage and can be found in the cortex, phloem, and cambium. Although it is also found in newly formed elements of the xylem with little lignification, it does not attack the wood to any appreciable extent.

Fig. 27-1. Aspen tree broken at the point of a canker caused by *Hypoxylon pruninatum.*

Fig. 27-2. (A) Pruininose perithecia and (B) conidial pillars of ***Hypoxylon pruinatum.*** (C) View of conidial blisters.

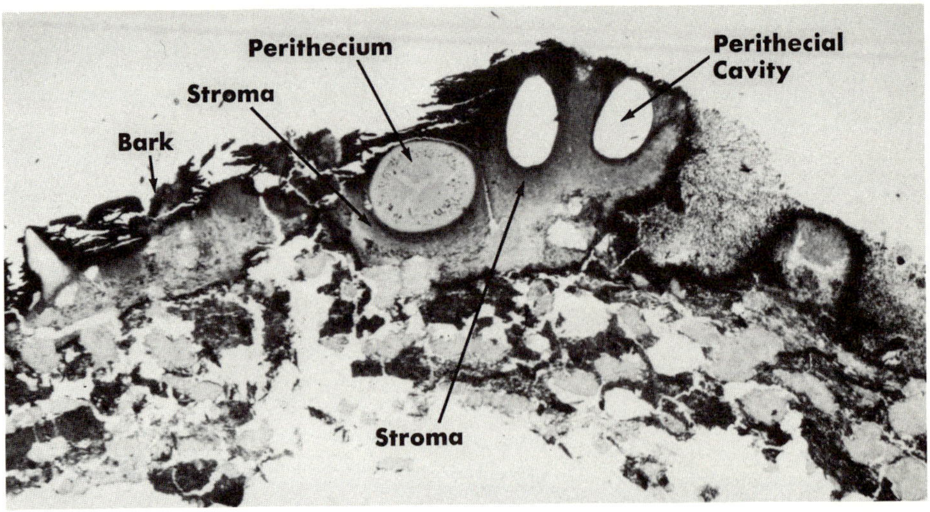

Fig. 27-3. Cross section through bark and stroma of *Hypoxylon pruinatum* showing perithecial cavities, some of which have lost their contents during preparation (× *10*).

Life Cycle. Young cankers appear as irregular to reddish-brown areas around a wound. As the cankers enlarge, the older portions become mottled (Fig. 27-2C). At the end of the first year and during the second, the periderm is raised in spots by the formation of **conidial pillars** (Fig. 27-6 B, I, J). The periderm ruptures during May and June forming **conidial blisters** and exposing the grayish conidial layer over the entire surface of the cortex and conidial pillars (Fig. 27-6J). The conidia are oblong to ovoid, one-celled, hyaline or brownish in mass, and borne on brownish, branched, septate conidophores. Conidia are first borne apically but become lateral in position on a shoulder due to continued growth of the tip of the conidiophore which then produces another conidium (Fig. 27-6 A, C). The ruptured blisters dry-out and the conidial layer disappears, leaving shrunken, dried pillars (Fig. 27-2B).

By the second or third year, when the cankers are two to three feet long, immature perithecial stromata appear as grayish to black areas (Fig. 27-2A). They become abundant on old cankers. The outer surface of the stroma tends to be flat with the sides of the perithecia projecting slightly and covered with a whitish, pruinose coat which disappears, leaving the stroma black. The perithecia occur in a single layer, seated directly in the diseased bark and covered by the dark stroma material (Figs. 27-3, 27-6 D). The base and walls of the perithecia are lined with filiform paraphyses and cylindrical asci. The asci typically contain eight, one-celled, brown, oblong to elliptical ascospores (Figs. 27-4, 27-5, 27-6F). They contain one or two large oil globules and an elongate, spiral, hyaline slit which is a germination pore (Fig. 27-6 G). A **germination pore** is a thin area in a heavy spore wall from which a germ tube usually issues (Fig. 27-6 H).

Ascospore discharge can be observed with the naked eye when a strong beam of light is placed at a level immediately above the ostioles (necks) of moist, viable perithecia in a darkened room. Ascospores can be discharged for two consecutive summers from the same stroma.

Control: not yet devised. Pulp wood should be harvested when the amount of new wood being formed is greater than the amount of wood being destroyed.

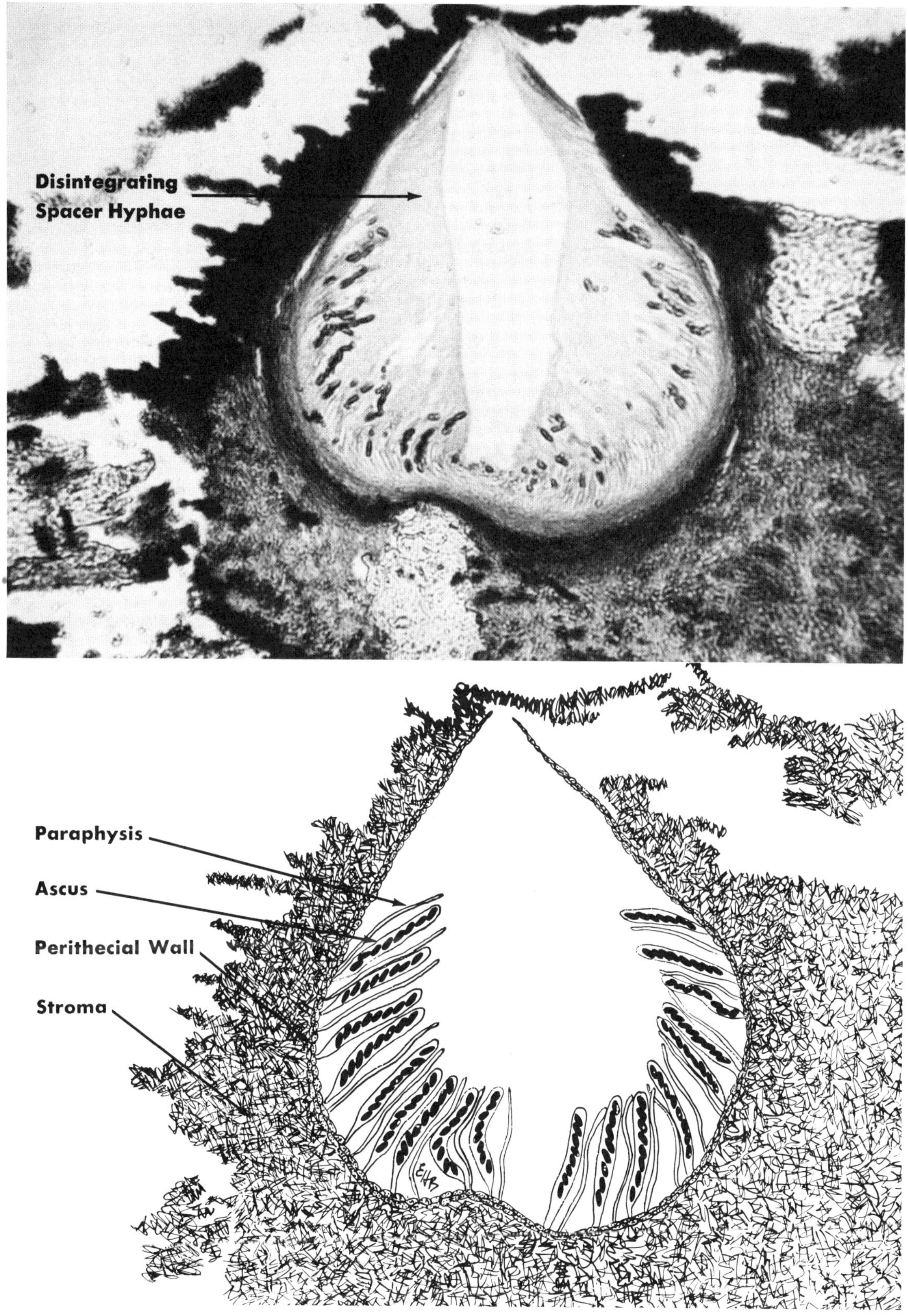

Fig. 27-4. Photomicrograph and diagrammatic representation of a longitudinal section through a perithecium of *Hypoxylon pruinatum.* Dotted line represents the plane of the section in Fig. 27-5 ($\times$ *80*).

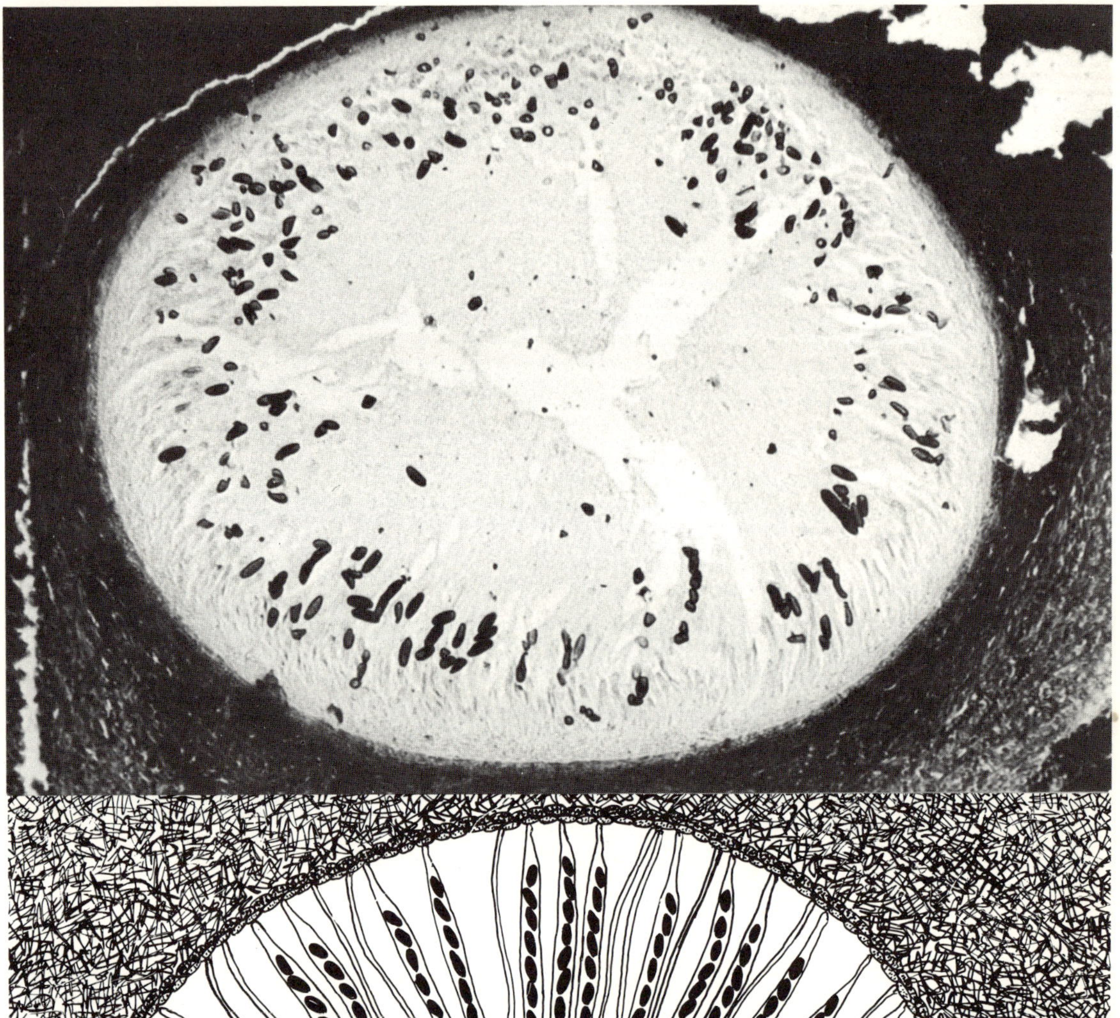

Fig. 27-5. Photomicrograph and diagrammatic representation of a cross section through a perithecium of *Hypoxylon pruinatum* made at right angles to the section in Fig. 27-4 (× 96).

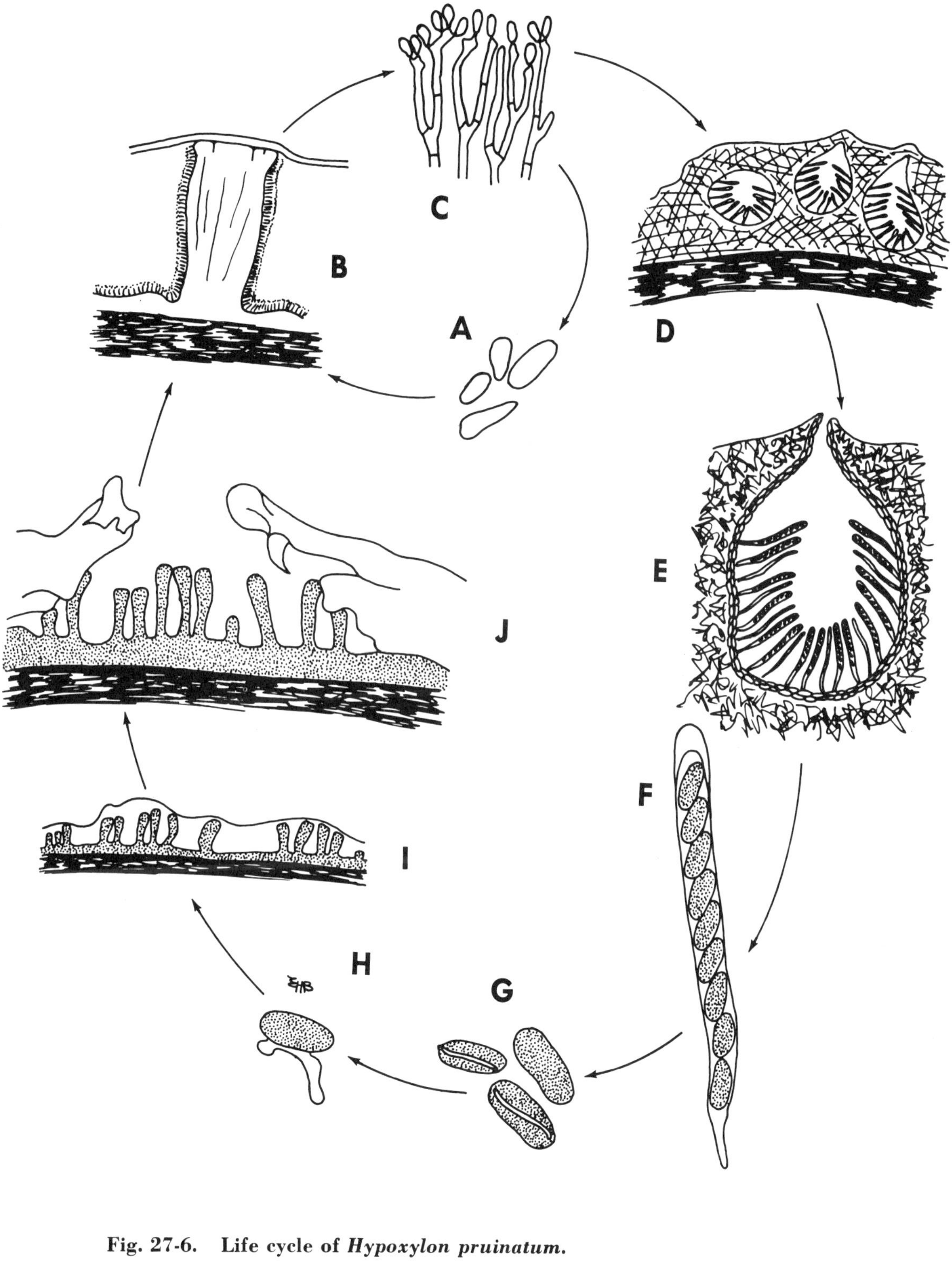

Fig. 27-6. Life cycle of *Hypoxylon pruinatum*.

SUGGESTED READING

Bier, J. E., *Hypoxylon Canker of Popular,* Studies in Forest Pathology. III. Tech. Bul. 27 (Publ. 691) Dom. Can. Dept. Agric., 1940.

28

Ergot of Grain

Ergot is a disease of wild and cultivated species of grass throughout the world. Incidence is always very low and no threat is posed to production of grain or survival of natural grasslands. Ergot is a disease only of the flowers. When a grass blooms, the ovaries are attacked, and once the head matures, the seeds, or kernels, are replaced by larger, hard, purplish bodies of the pathogen called **sclerotia** (Fig. 28-1). The sclerotia are harvested along with the grain and contaminate animal feedstuffs and flour with poisonous alkaloids which cause sickness, abortion, and death. This is called ergotism. The extracted alkaloids are used as drugs to induce abortion and to control hemorrhage.

There are some two dozen species of *Claviceps* of which *Claviceps purpurea* is the most well-known and occurs commonly on rye. Recently, a new species, *Claviceps gigantea,* was reported from Mexico causing horse's tooth, DIENTE DE CABALLO, on corn. Some of the kernels are replaced by very large, long sclerotia which resemble other familiar objects on the farm.

Disease Cycle. The fungus overwinters as sclerotia on the ground or mixed with seed. In the spring, these germinate to produce several stalked stroma with heads (Figs. 28-3, 28-5, I, J). The heads contain a large number of embedded perithecia (Figs. 28-4, 28-5 K). The ostiolar canal of the perithecium is lined with periphyses. The asci contain eight, long filamentous, thread-like, septate ascospores (Fig. 28-5 L).

The primary inoculum is the ascospores which are forcibly ejected and carried by the wind to flowers of susceptible plants. Ejection of ascospores is timed to coincide with the blooming period of susceptible hosts. Certainly this indicates a long evolutionary history of interaction. The ovary is at first covered and then replaced by a white cottony growth of hyphae (Fig. 28-5 C). This mass of hyphae bears short, simple conidiophores which produce large numbers of small, one-celled, hyaline conidia (Fig. 28-2, 28-5 D). A sticky, sugary fluid called honeydew is produced which drips from the aborted flower. The honeydew is populated with multitudes of conidia.

The conidia cause secondary infection of the same host species, or more frequently of another susceptible species which blooms slightly later. Wild grasses may be important sources of inoculum for nearby cultivated fields. Insects are important as vectors of the secondary inoculum.

Control: provided through sclerotia-free grain and the mowing of headlands before the honeydew stage occurs on early blooming wild species of grasses.

Fig. 28-1. Ergot of grain, comparing diseased and healthy florets and normal seed and sclerotium.

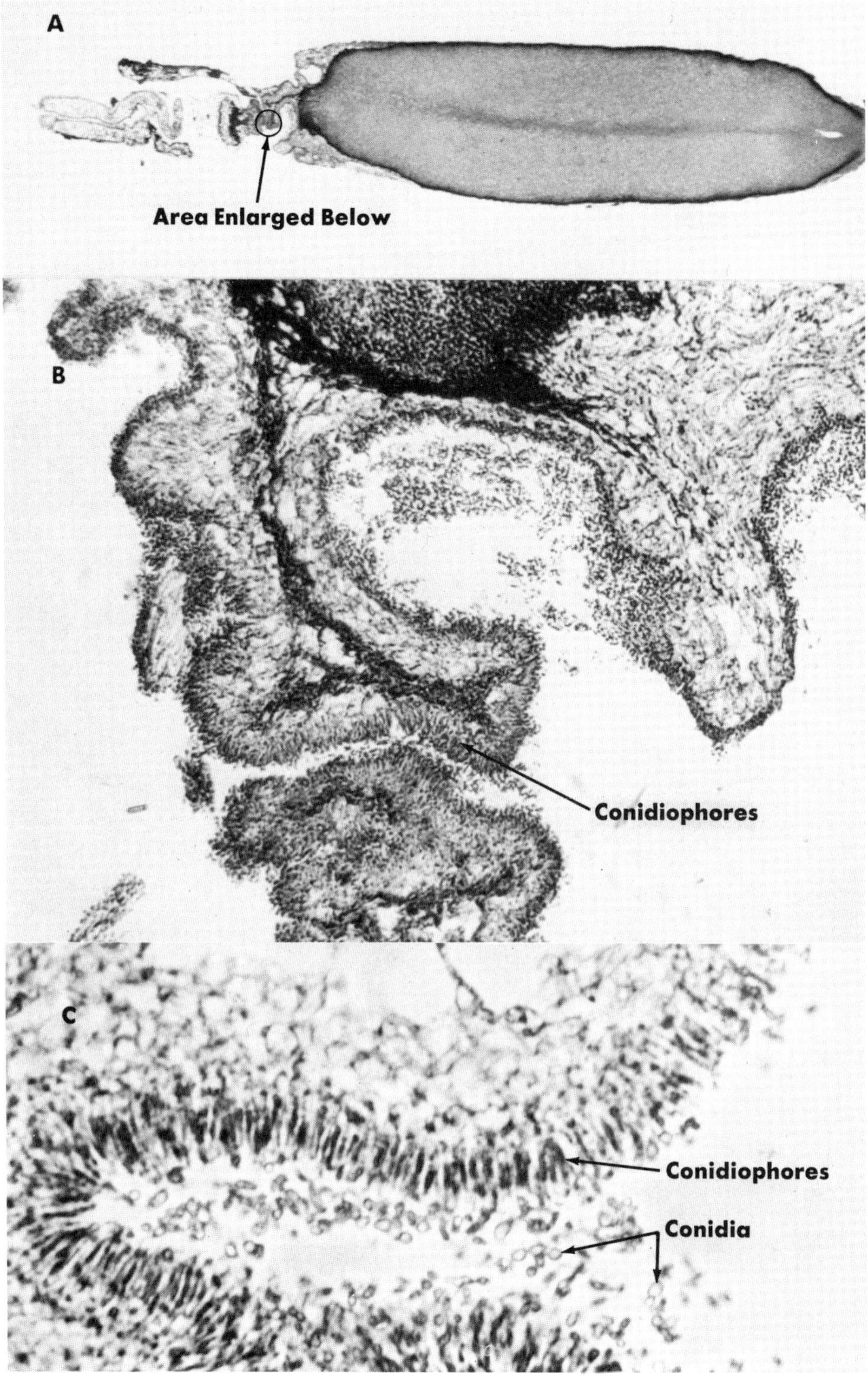

Fig. 28-2. (A) Longitudinal section through an aborted ovule showing the immature sclerotium (× *16.6*) and (B) conidiophores (× *415*) and (C) conidia (× *830*).

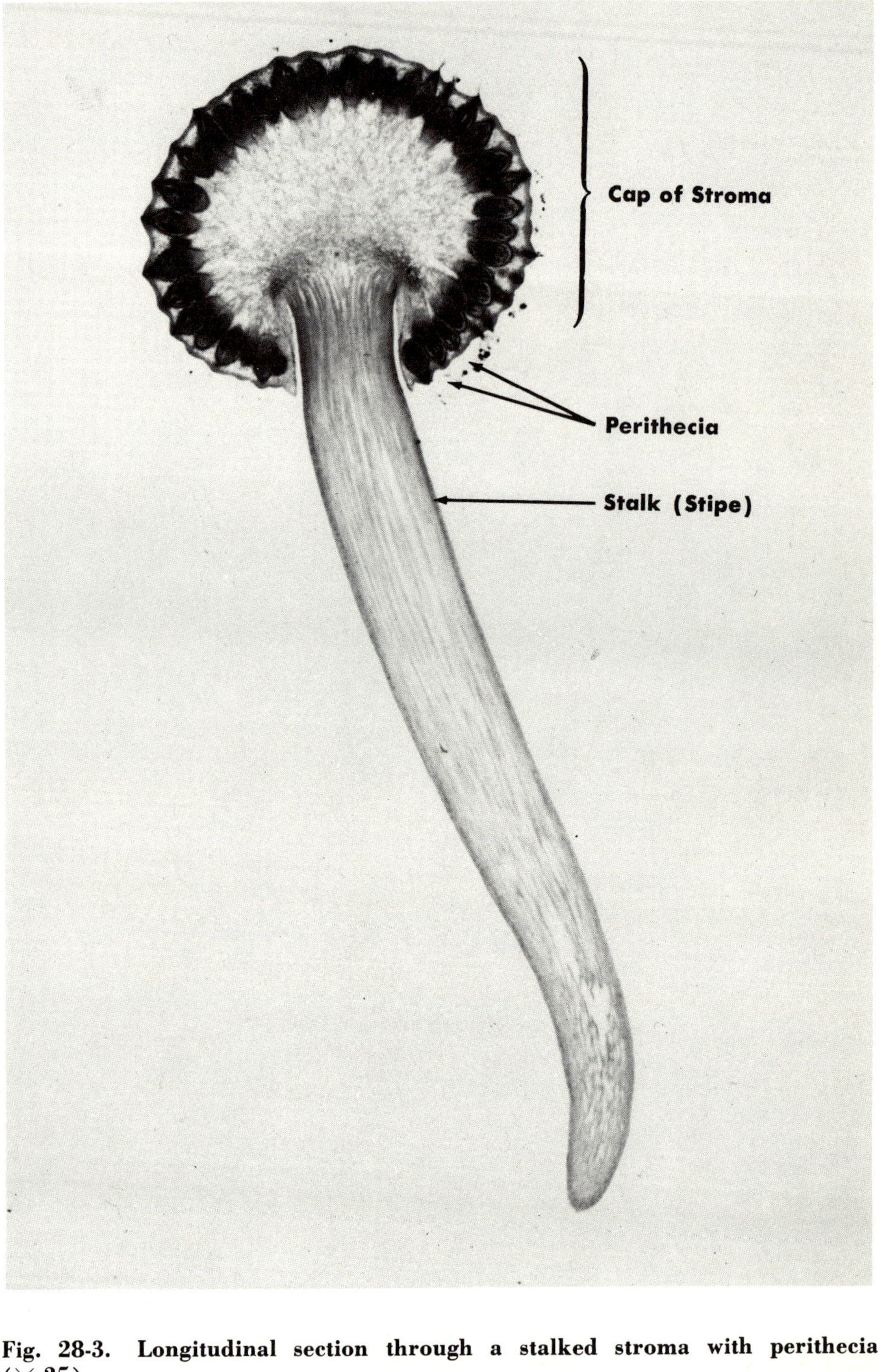

Fig. 28-3. Longitudinal section through a stalked stroma with perithecia (× 25).

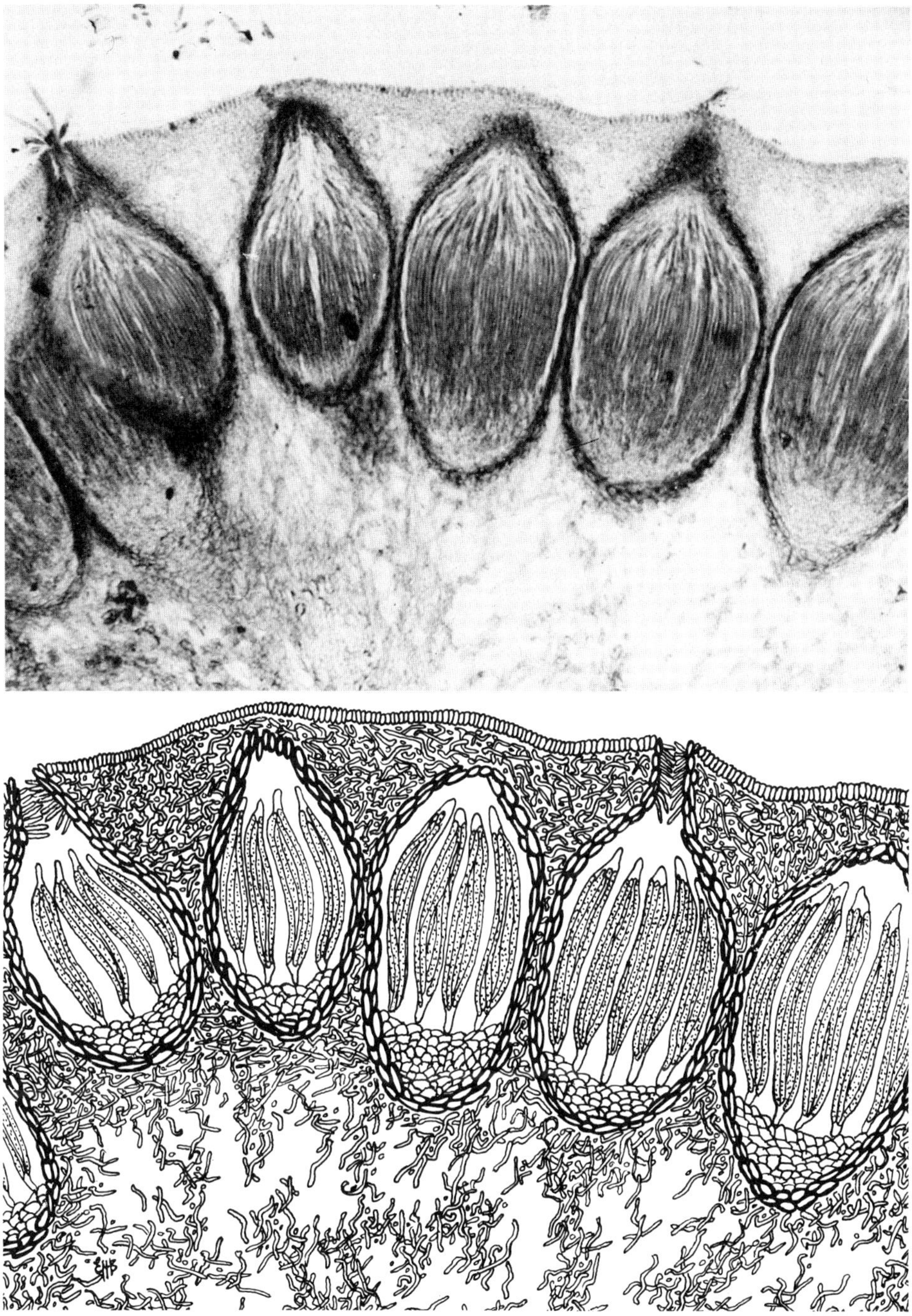

Fig. 28-4. Perithecia of *Claviceps purpurea* with diagrammatic reconstruction (× 250).

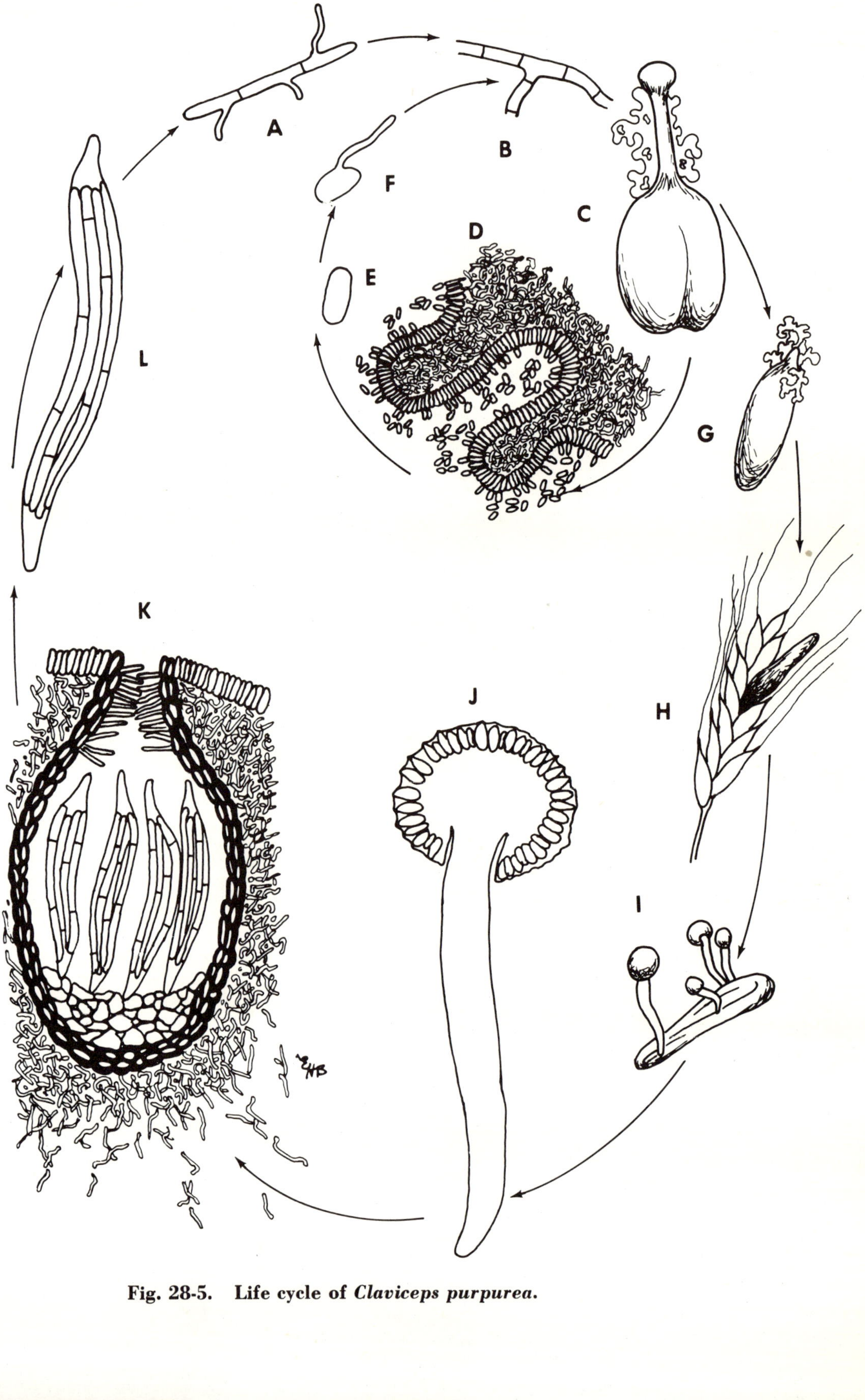

Fig. 28-5. Life cycle of *Claviceps purpurea*.

29

Black Leaf Spot of Elm

The most prevalent leaf-spotting disease of the elms is that caused by *Gnomonia ulmea*. Although it appears early in the spring, it is most noticeable in the autumn when the fallen brown leaves on the ground exhibit a plethora of little black spots (Fig. 29-1). Usually no harm is done but, on occasion, severe attacks in early spring bring about considerable defoliation.

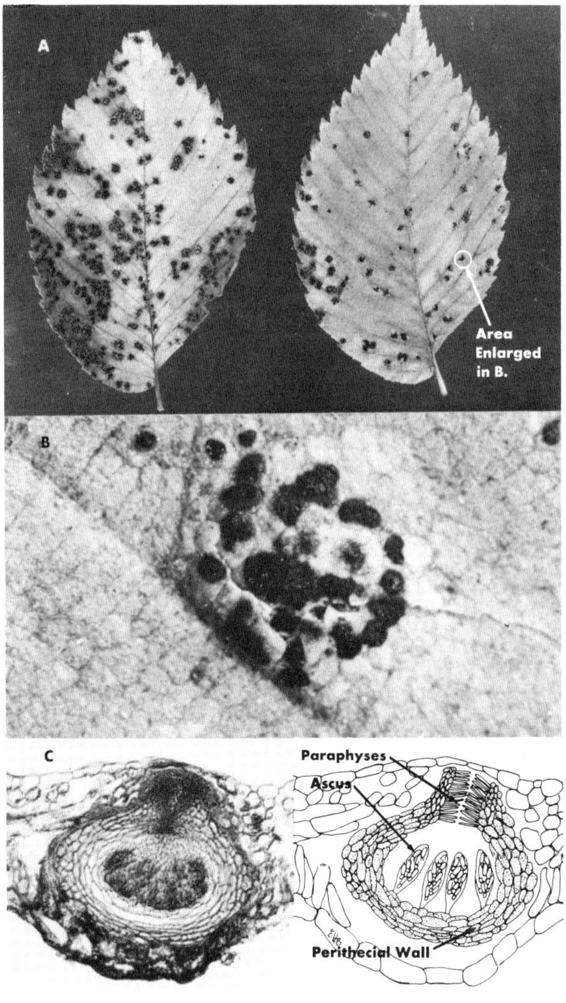

Fig. 29-1. (A) Black leaf spot of elm. (B) Macrophotograph of a cluster of perithecia. (C) Photomicrograph and diagrammatic representation of a perithecium of *Gnomonia ulmea* (× 208).

30

Sycamore Anthracnose

Gnomonia veneta causes a common leaf and twig blight of sycamore and brings about persistent and perennial defoliation (Fig. 30-1) when sycamore is grown under stress such as at the extremes of its range. Lesions on leaves are very pronounced (Fig. 30-2 A). They are dark-bordered, irregular, spreading, and tend to follow the veins. The imperfect stage is a *Gleosporium*. Cankers on branches are deep, perennial, and destructive (Fig. 30-2 B, C). Removal of cankers and fallen leaves reduces the inoculum and fungicidal sprays sometimes seem to offer a little relief. Satisfactory control, however, is wanting.

Anthracnose of hickory, caused by *Gnomonia caryae,* also exhibits dramatic symptoms (Fig. 30-3).

Fig. 30-1. Sycamore anthracnose.

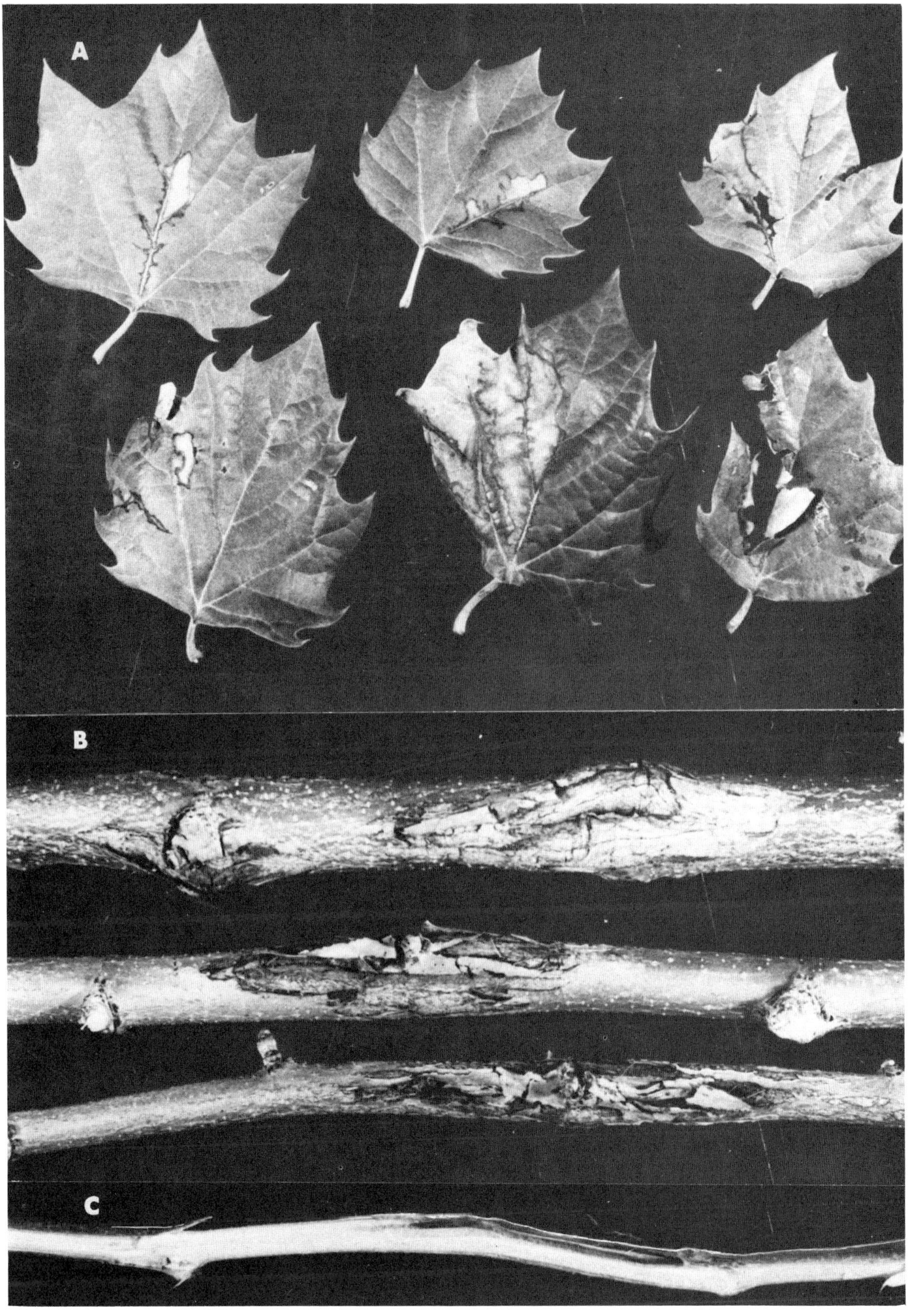

Fig. 30-2. (A) Leaf symptoms and (B) branch cankers of sycamore anthracnose. (C) Longitudinal section through a branch to show depth of a canker.

Fig. 30-3. Hickory anthracnose.

31

Beech Bark Disease Complex

This is a story of ecological succession—of interrelationships between a fungus and an insect, and the secondary roles played by other fungi and other insects. The disease, or more properly the disease complex, occurs on the American beech in the Northeastern United States and the Maritime provinces of Canada. The beech scale, *Cryptococcus fagi*, feeds on the bark (Fig. 31-1B) and the fungus, *Nectria coccinea* var.

Fig. 31-1. (A) Perennial stem canker of *Nectria cinnabarina* and the two major participants in the beech bark disease complex: (B) the insect scale and (C) the fungus.

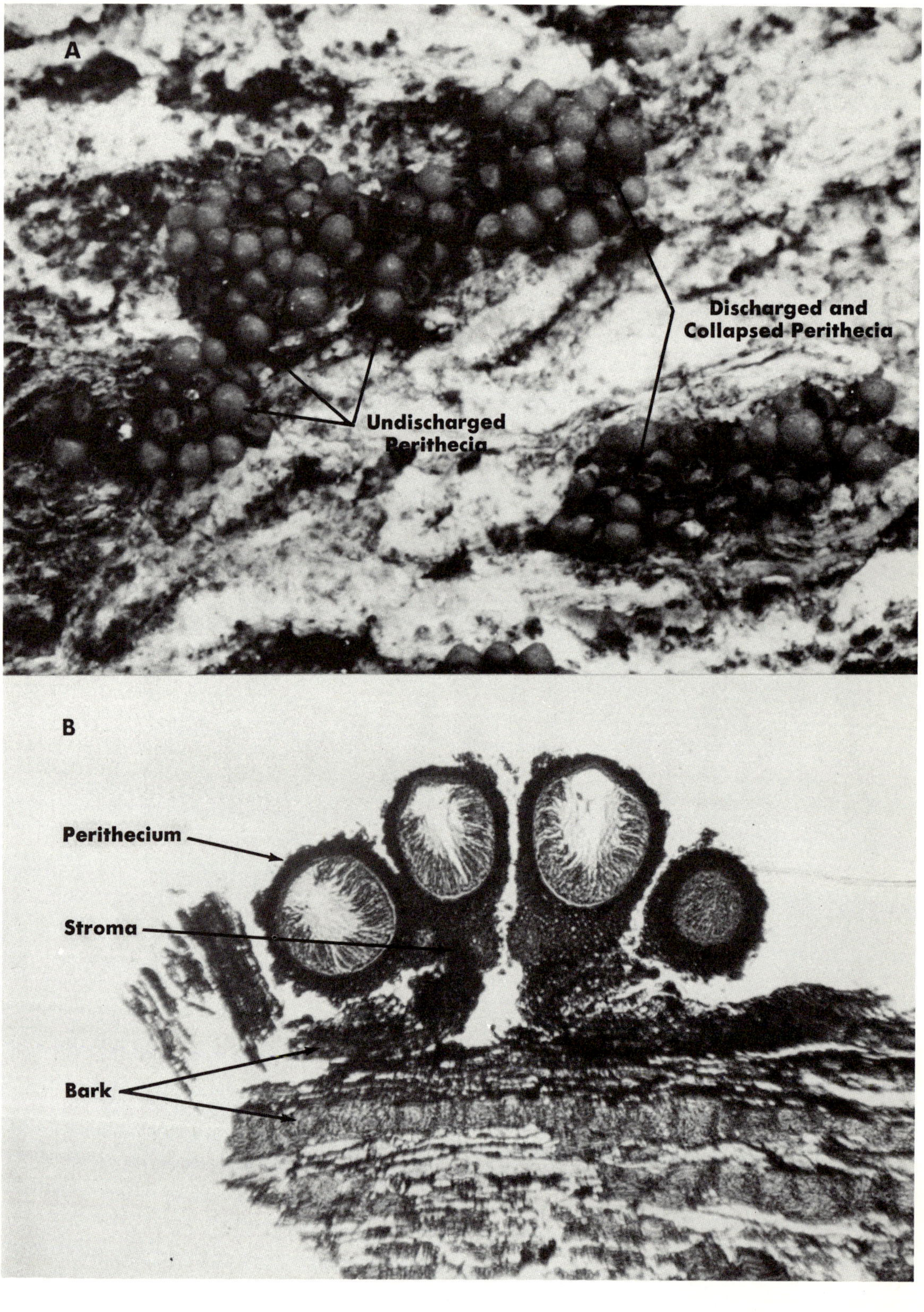

Fig. 31-2. **(A) Close-up of the bright red perithecia of *Nectria coccinea* var. *faginata* and (B) photomicrograph of a cross section through stromata and perithecia (× *80*).**

faginata, enters the wounds (Fig. 31-1C). *Nectria galligena* can also enter the wounds but is secondary in importance. In Europe, a similar disease complex occurs on the European beech with the same scale insect and *Nectria galligena* as the major fungus partner. Little damage is done by either the insect or the fungus alone, even though the insect can attack the tree alone. Some species of *Nectria*, such as *Nectria cinnabarina* (Fig. 31-1 A), cause serious cankers without the aid of other agents.

The scale insect, *Cryptococcus fagi*, is yellow, soft-bodied, and elliptical, with red eyes and a secretion of white wooly wax over its entire body. The fungus, *Nectria coccinea* var. *faginata*, is an ascomycete which produces conidia in the summer on white cushion-shaped masses of hyphae with conidiophores over the surface (Fig. 31-3 A-D). This is called a **sporodochium.** Bright red perithecia are produced in the fall on the same stromata (Figs. 31-2, 31-3 G).

The *Nectria* kills the bark but barely penetrates into the wood, then other, more

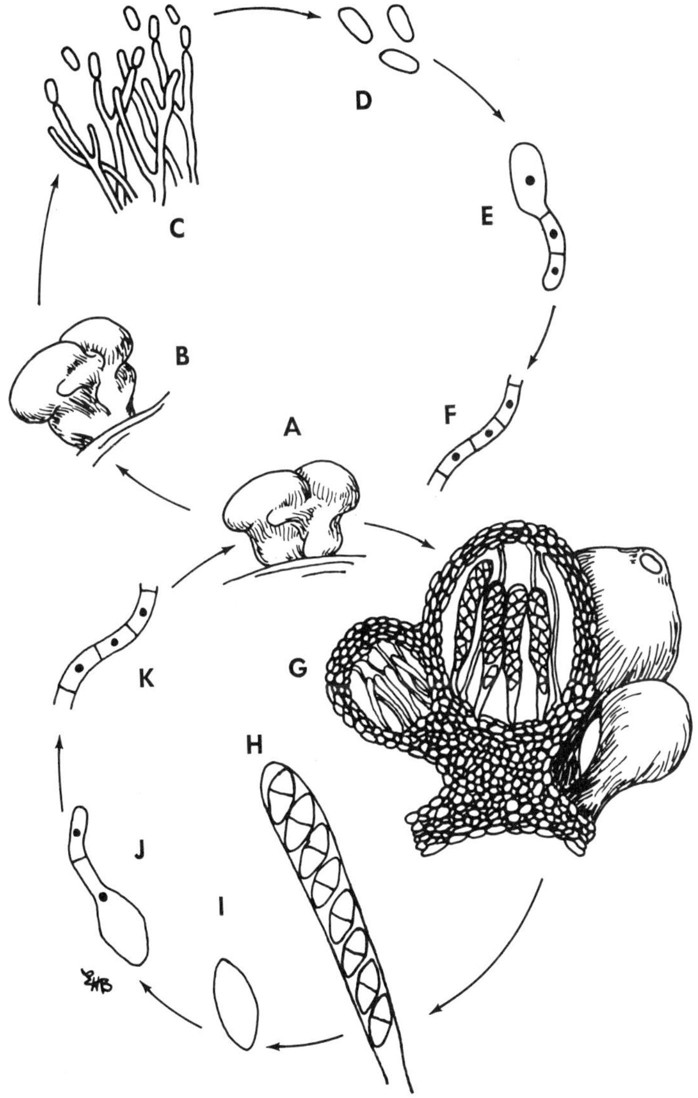

Fig. 31-3. Life cycle of *Nectria coccinea* var. *faginata*.

aggressive fungi enter. Species of *Hypoxylon* grow rapidly into the diseased area and the *Nectria* is retarded. Inoculum is reduced since *Nectria* lives for several years on trees not invaded by the species of *Hypoxylon*. Following the *Hypoxylon* are the decay fungi, especially species of *Fomes* and *Polyporus*. After the trees die, many other kinds of insects move in.

Both the larvae and the adults of the twice-stabbed ladybird beetle, *Chilococcus stigma*, feed on the scale insects. The ladybird beetles have little effect on the populations of the scale insect but are strong fliers and probably carry the conidia of *Nectria* to other trees. There are still other insects which prey on the scale insects. A fungus, *Gonatorrhodiella highlei*, parasitizes the *Nectria*. That makes it a **mycoparasite**, or a parasite of a fungus.

Control: is inadequate and fear has been expressed that the beech bark disease complex might spread from the Northeastern area into the northern midwest beech forests. Prompt salvage of diseased trees should be made before the *Hypoxylon* and decay fungi destroy the usable timber.

32

Tar Spot of Maple

Tar spot diseases are bold-appearing leaf lesions (Fig. 32-1) which appear as stripes when mature. Tar spot of maple is caused by *Rhytisma acerinum.* Another species also attacks maple and still others willow, golden rod, and other plants.

Life Cycle. Slender, thread-like ascospores (Fig. 23-3E) mature in apothecia in the black stroma (Fig. 32-2, 32-3 A-C) of overwintered leaves and are forcibly ejected during May and June. Wind currents carry them to susceptible leaves where they germinate (F) and penetrate the stomata. Hyphae grow into the epidermal and mesophyll tissues and secrete a black gummy substance which cements fungus and plant material together in a shiny black stroma. The upper portion of the epidermal cells are pushed upward in the stroma as the fungus tissue continues to develop. During the summer, many small rod-shaped bodies are produced in acervulus-like structures. They are variously called spermatia and conidia. They have never been germinated nor demonstrated to cause infections and their function, if any, is unknown. After the leaves fall, immature apothecia form. The fungus overwinters in this stage and the apothecia mature in early spring.

Control. This is a relatively uncommon disease and control is not exercised. If incidence is a problem in nurseries, destruction of fallen leaves and application of Bordeaux mixture is helpful.

Fig. 32-1. (A) View of tar spot of maple from the upper surface and (B) lower surface of the leaf.

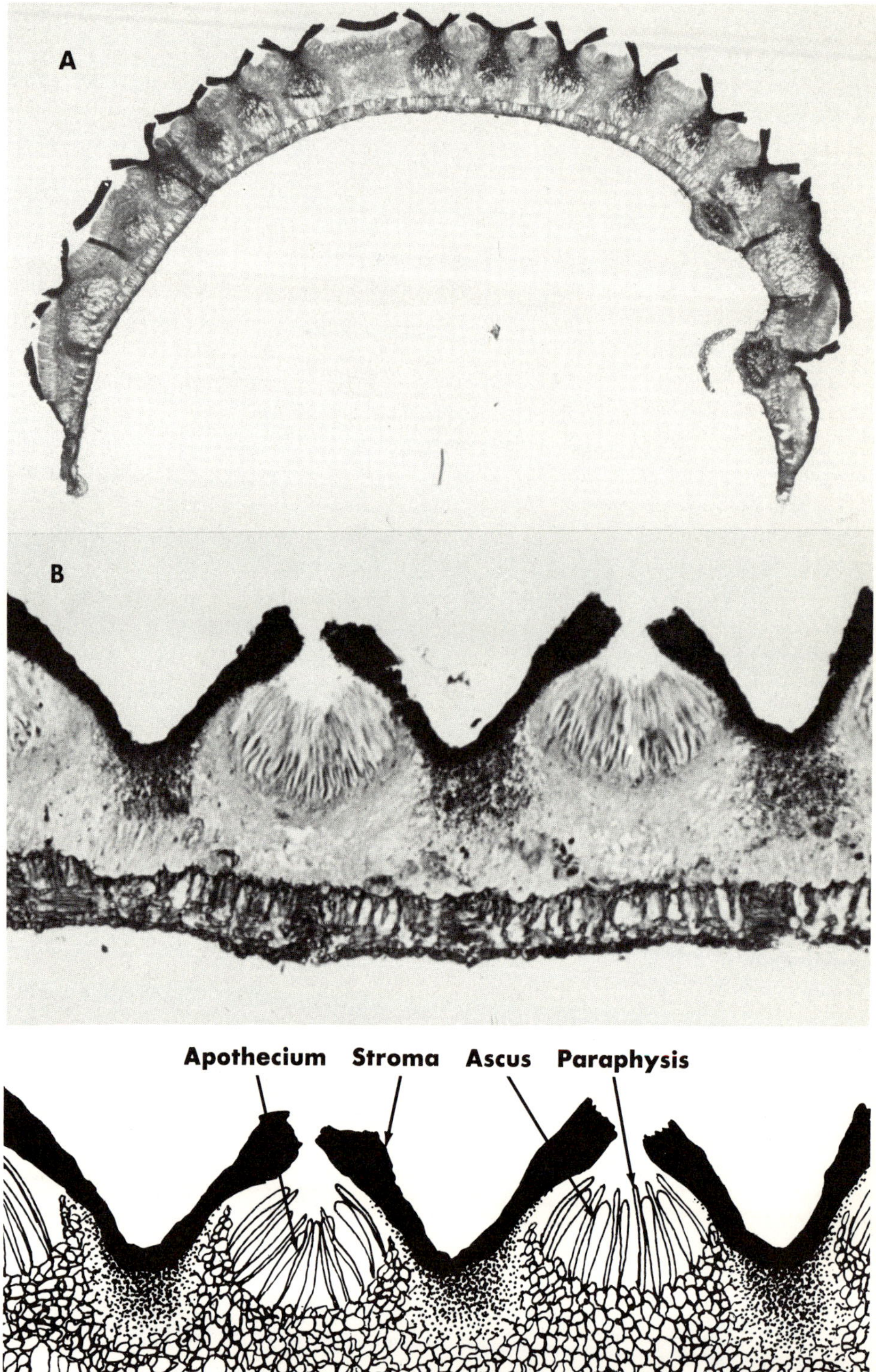

Fig. 32-2. **(A) Cross section through an entire infected leaf (× 7) and (B) enlargement and diagrammatic representation (× *14*).**

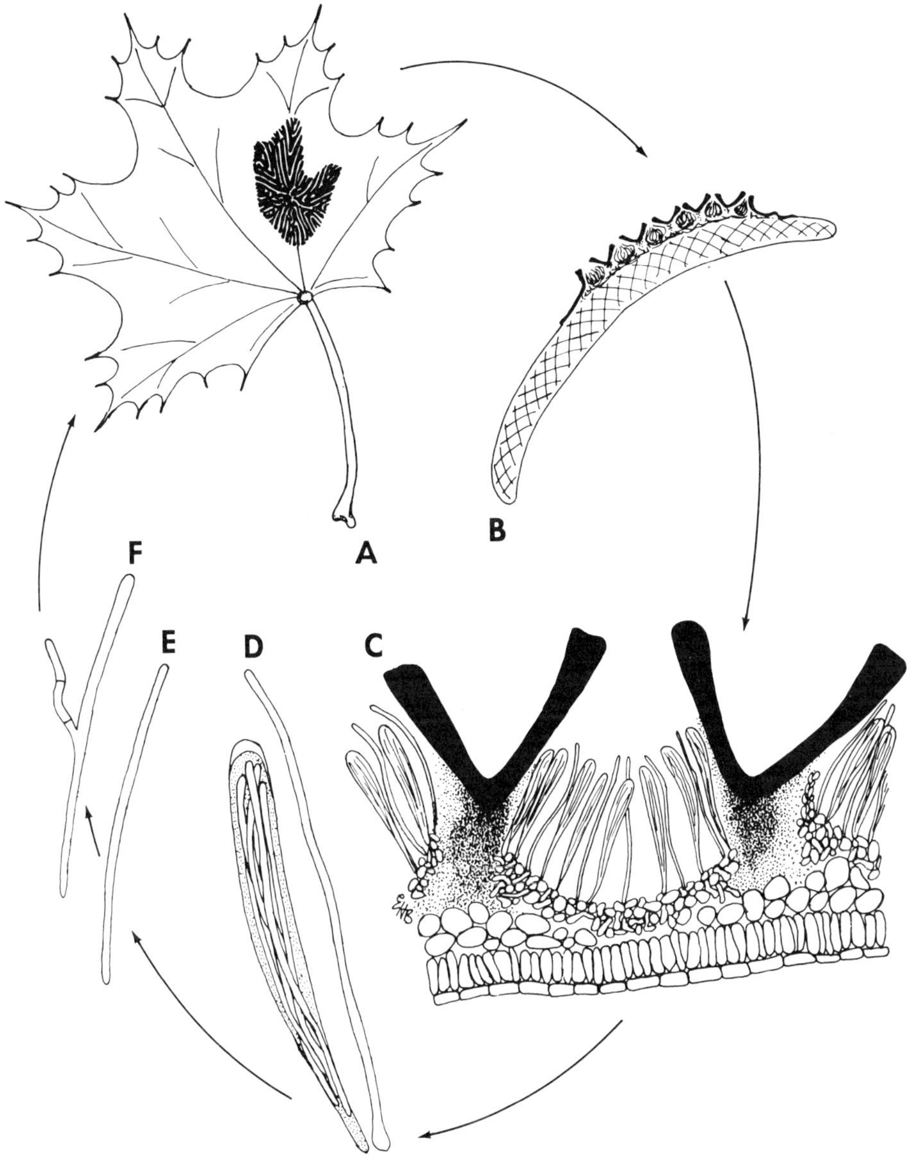

Fig. 32-3. Life cycle of *Rhytisma acerinum*.

33

Sclerotinia Diseases

SCLEROTINIA diseases occur as dry cankers, rots, blights, and damping-off on succulent plants throughout the world in the field, in transit, and in storage. They are especially significant on vegetables such as cabbage, lettuce, and beans. If dry conditions prevail after infection, which requires relative humidity in excess of 90%, then only dry brown cankers develop. If humid conditions continue after infection, the entire plant may be covered with a weft of white mycelium in which large black sclerotia are formed (Fig. 33-1). The fungus thrives under cool, moist conditions.

No conidia are produced by *Sclerotinia sclerotiorum* but do occur in *Sclerotinia porri*, a pathogen of onion. The imperfect stage is *Botryotinia*. *Monilia* is the imperfect stage of species in the genus *Monilinia*, which is very closely related to SCLEROTINIA.

Disease Cycle. The fungus overwinters as mycelium but mostly as sclerotia. These germinate in the spring with the production of apothecia (Fig. 33-2). The ascospores are violently discharged in clouds and are wind-blown to susceptible plants. The fungus is widely distributed with infected produce and seed contaminated with sclerotia.

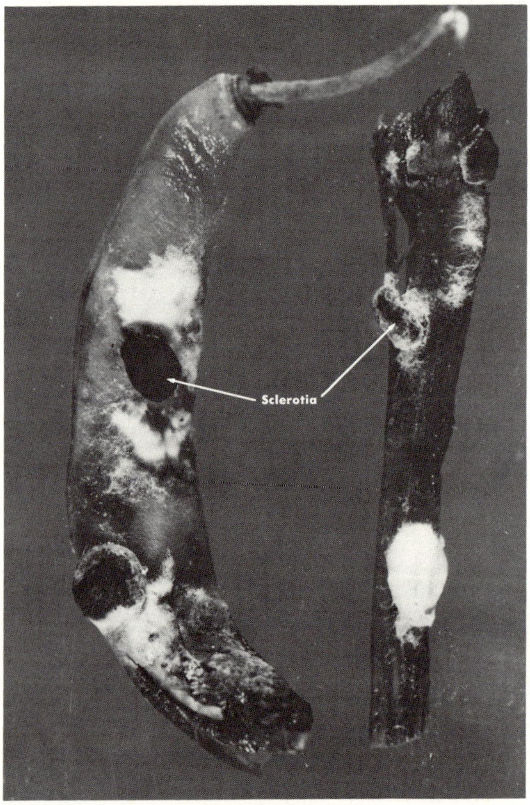

Fig. 33-1. Mycelial wefts and sclerotia of *Sclerotinia sclerotiorum* on the pod and stem of bean.

Control: unsatisfactory but can be approached with such measures as soil drainage, long rotations, and sanitation to remove infested debris. Some interesting control measures include the use of sheep in Arizona to eat the crop refuse and infested debris with a subsequent quarantine to prohibit distribution of manure in fields where susceptible plants will be grown. In Florida, some areas of lowland are planted with rice in alternate years to destroy the sclerotia by flooding.

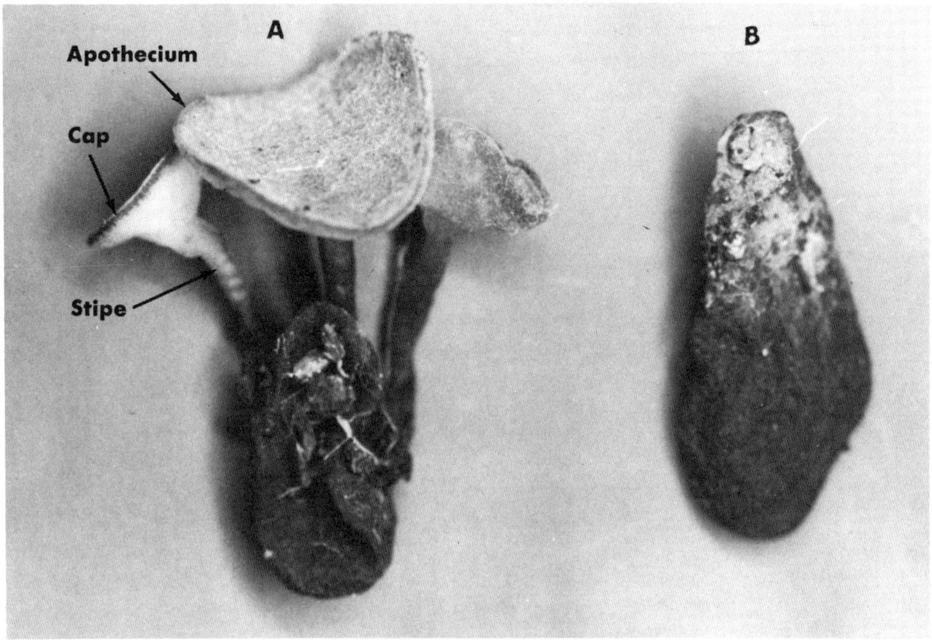

Fig. 33-2. Germinated (A) and ungerminated (B) sclerotia of *Sclerotinia sclerotiorum*.

34

Brown Rot of Stone Fruits

The stone fruit is botanically a drupe and contains a single, large pit or stone. The edible, fleshy portion is derived from the pistil. This fruit is in the genus *Prunus.*

Three species of *Monilinia* attack such stone fruits as peach, plum, cherry, apricot, and almond. *Monilinia fructicola* is found only in the United States and New Zealand, *Monilinia fructigena* only in Europe, and *Monilinia laxa* in both the United States and Europe. Another species attacks blueberry. Although these fungi sometimes attack other rosaceous plants with fleshy fruits, they are not important except for a rot of apple in England.

The fungus overwinters primarily as dormant mycelium in diseased mummies hanging in trees or on the ground (Figs. 43-3 B, 34-5 I). The mummies are fruit which were infected, rotted, and dried. The mummies hanging in the trees produce large crops of conidia in the spring. This is a strategic position for production of primary inoculum. The mummies on the ground produce apothecia (Fig. 34-3, 34-4, 34-5 J, K) which look like golf tees. Ascospores (A) are forcibly ejected and are carried by wind currents. Primary inoculum may be either the conidia produced on hanging mummies or ascospores ejected from the apothecia produced on the ground mummies.

Primary inoculum is produced during the blossom stage of the host and the blossoms and tender new shoots are blighted and turn brown. Secondary inoculum

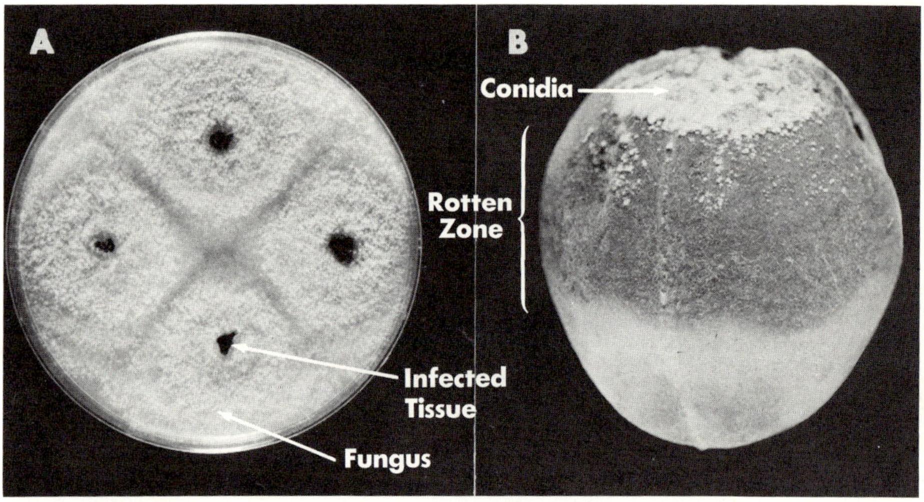

Fig. 34-1. (A) *Sclerotinia fructicola* sporulating in culture and (B) on a rotted peach.

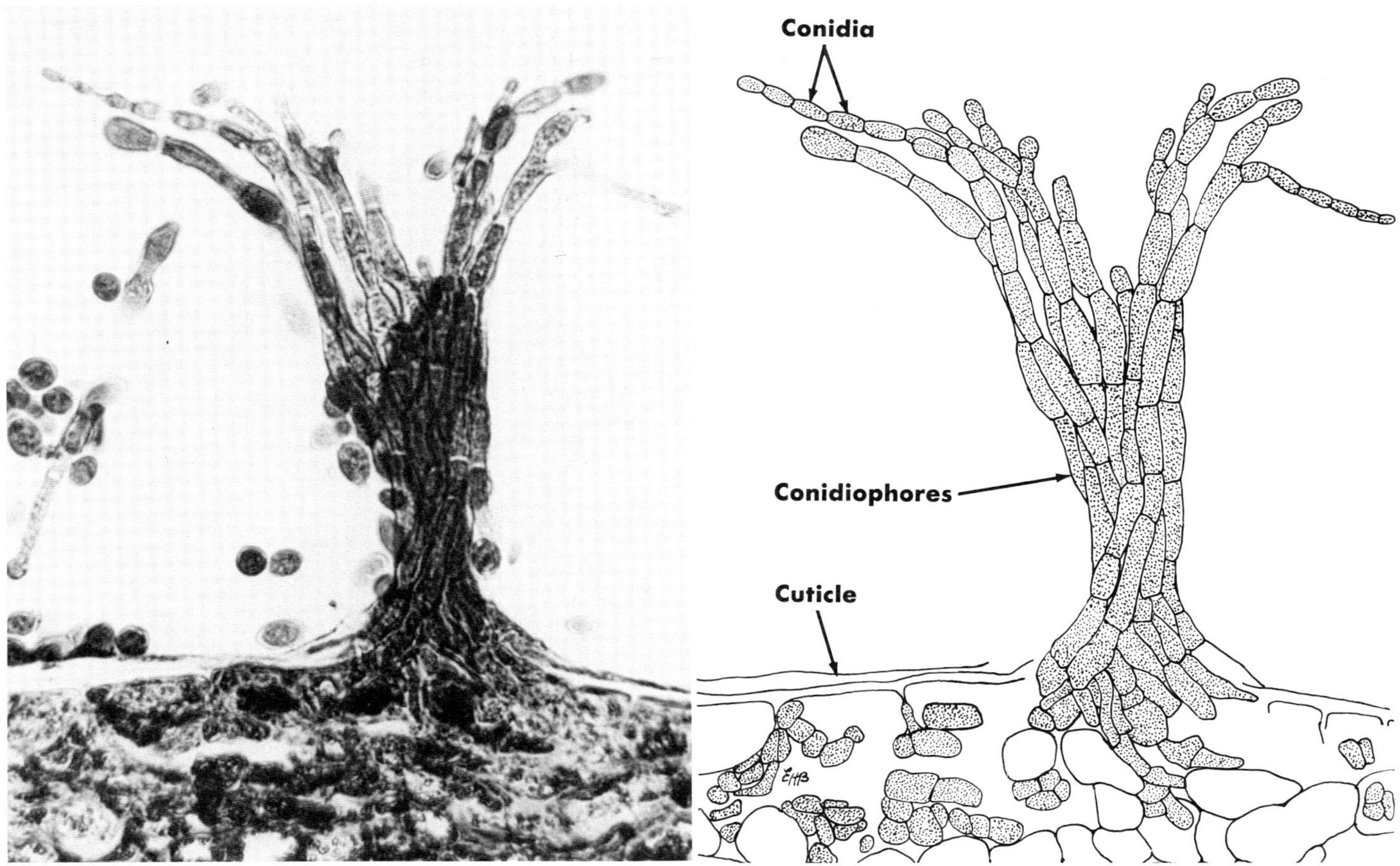

Fig. 34-2. Conidiophores and conidia of *Sclerotinia fructicola* and diagrammatic reconstruction (× 500).

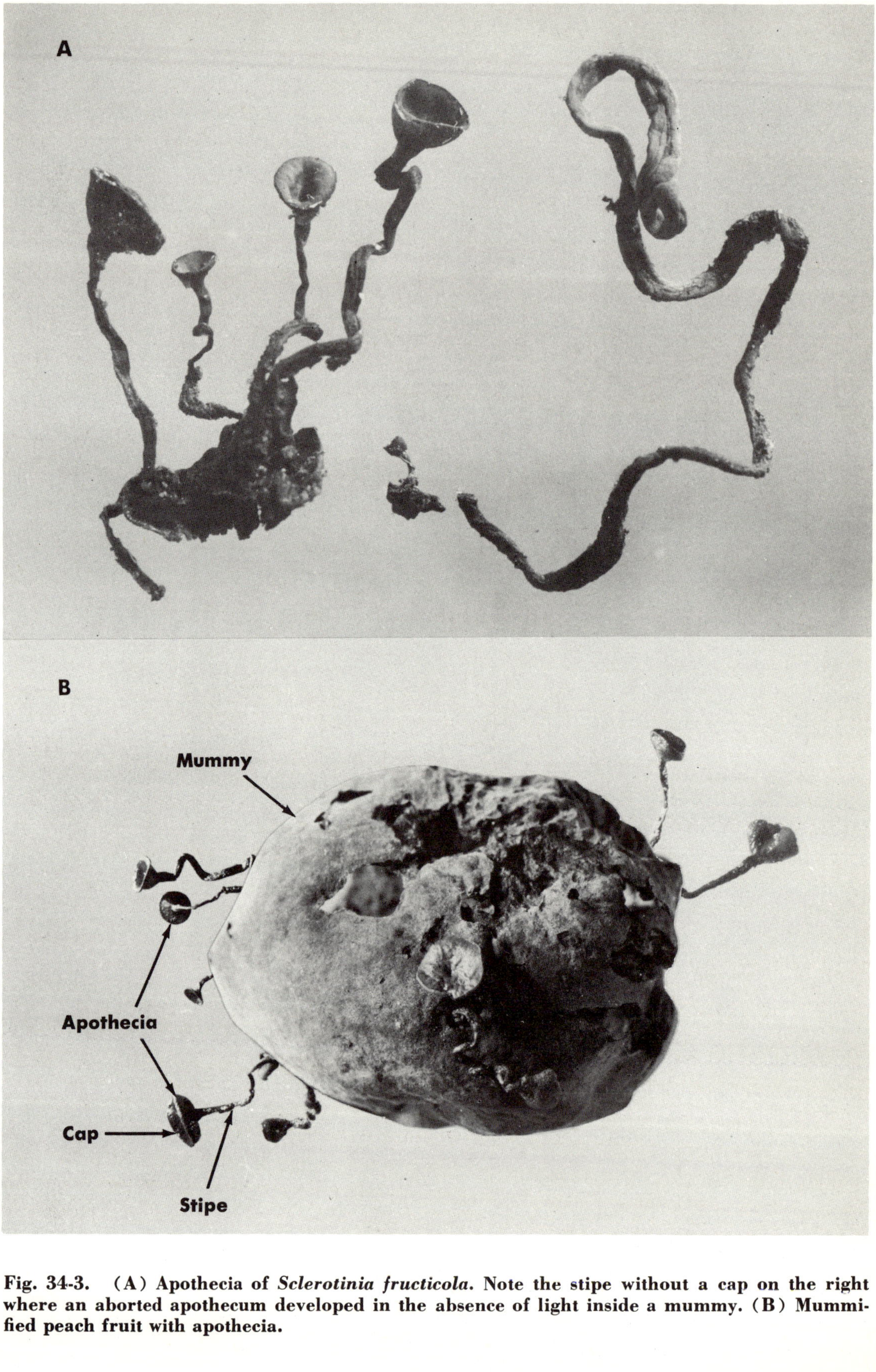

Fig. 34-3. (A) Apothecia of *Sclerotinia fructicola*. Note the stipe without a cap on the right where an aborted apothecum developed in the absence of light inside a mummy. (B) Mummified peach fruit with apothecia.

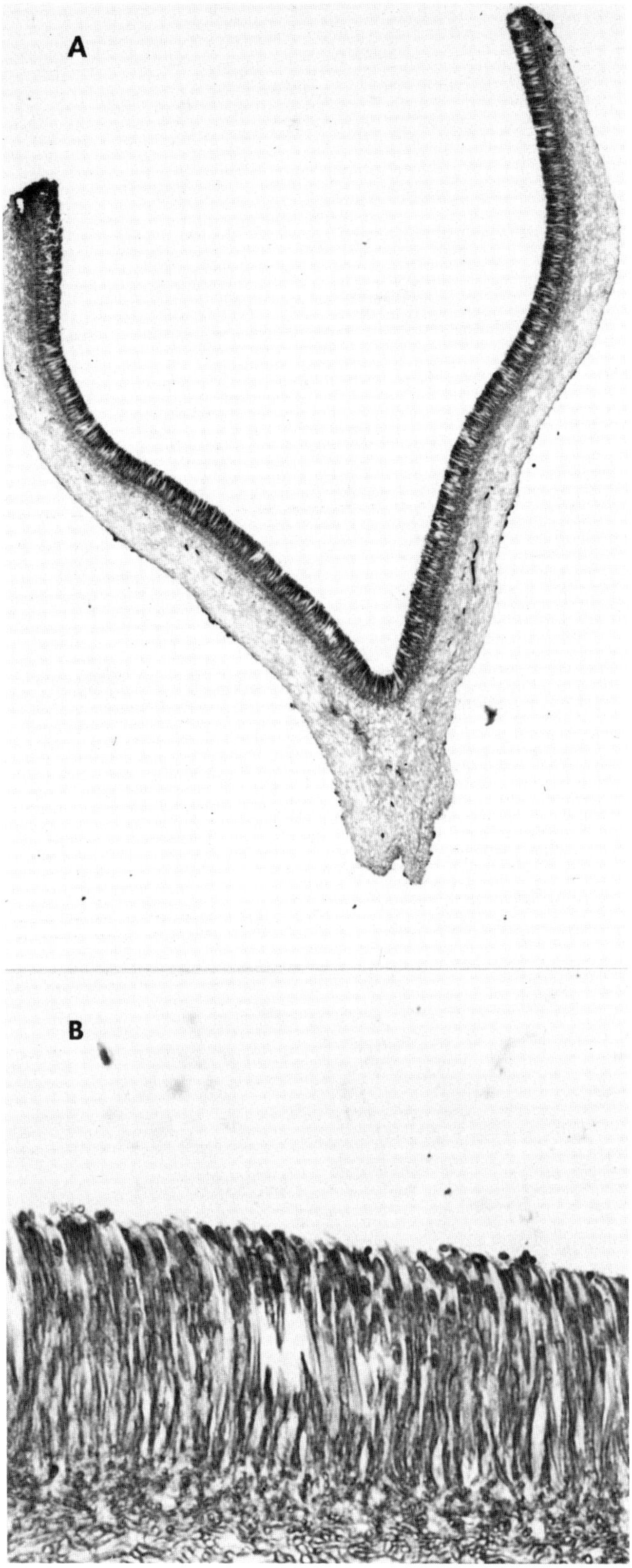

Fig. 34-4. (A) Longitudinal section through the cap of an apothecium (× *24.2*) and (B) enlargement (× *242*).

is abundant since conidia are produced profusely on infected tissues during humid periods. Green fruit is not attacked unless wounded. Ripe, or nearly ripe, fruit is penetrated through stomates, wounds, or directly through the cuticle. A mushy brown rot develops rapidly, over which a layer of buff-colored conidia are produced in profusion (Fig. 34-1). The infection of adjacent fruit in the area of contact is very common.

Control: achieved by the removal of hanging mummies and dormant eradicant spray applicants to inactivate the overwintering inoculum.

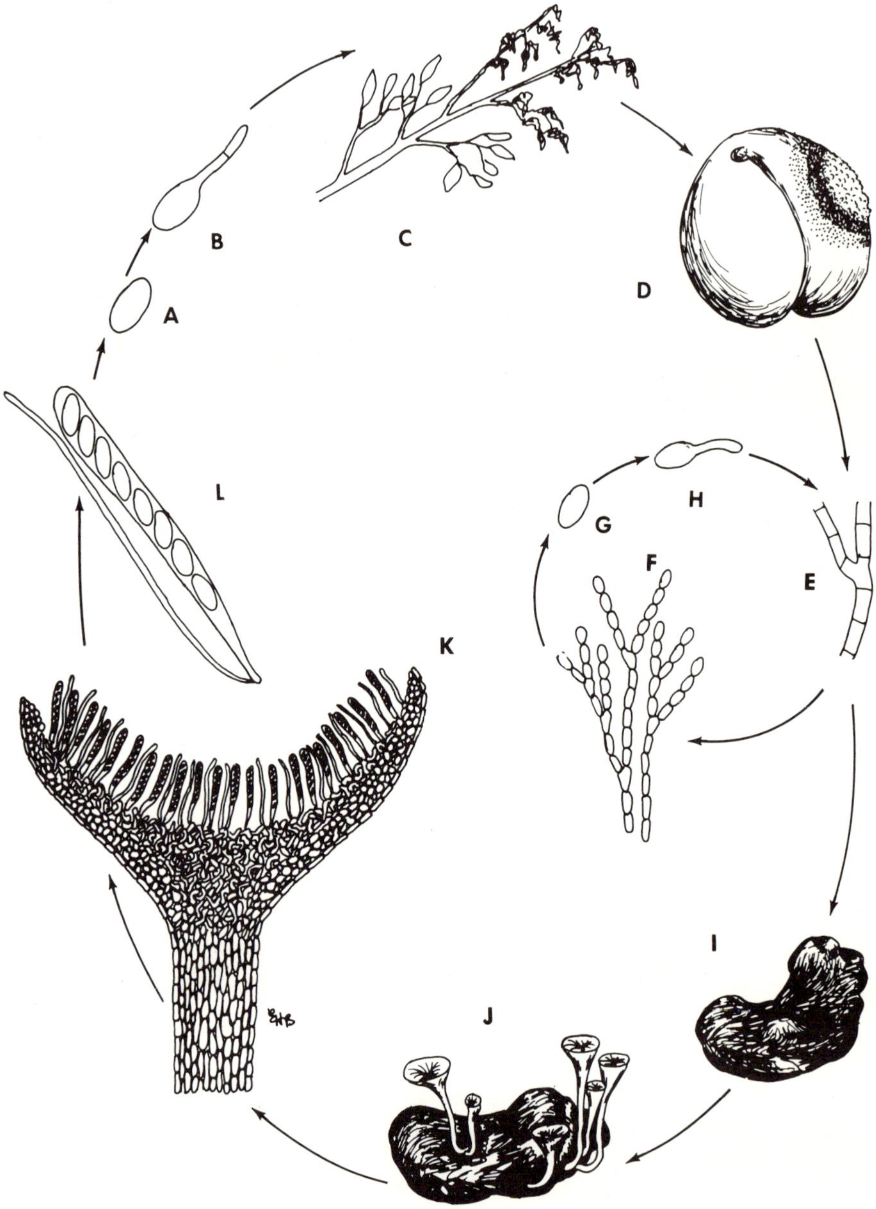

Fig. 34-5. Life cycle of *Sclerotinia fructicola*.

Experiment XV

Brown Rot of Stone Fruits: Inoculation and Isolation

Procedures used for isolation of organisms from diseased tissues vary with the type of causal organism, type of tissue, and the condition of the tissues at the time of isolation. The refinements are of three basic techniques:

(1) dilution of materials containing propagules
(2) tissue transplants from infected tissues containing propagules
(3) isolation of spores from diseased tissues.

Tissue transplants may be taken from the center of a lesion, or from the advancing margins, where often the contaminants are less frequent and sometimes the pathogen more readily recovered. The success of the technique depends on the elimination of surface contaminants by sterilization, commonly with 10%—20% commercial bleach (active ingredient: sodium hypochlorite) for 10 to 20 min., or by flaming after dipping in alcohol. The technique is especially useful for the isolation of pathogens from fleshy tissues since a thorough surface sterilization for elimination of surface contaminants can be achieved without elimination of the pathogen. Whole sections of fleshy parts can be surface sterilized and then implanted in agar. In the case of apples, potatoes, melons, and large woody parts, the entire organ, or a section of it, can be surface sterilized, the outer parts removed with a sterile instrument, and the inner, infected tissues implanted directly in agar without additional sterilization. Contaminants often can be avoided by working from the healthy side of the plant structure towards the infected side. Tissue transplants are also good for isolation of organisms causing systemic diseases.

The spore transfer technique is effected by just touching the surface of a newly sporulating lesion with a cool, sterilized needle and then gently stroking the surface of agar to transfer the spores adhering to the needle. Often the spores are more readily picked up if the needle is dipped in sterile water immediately before touching the sporulating surface.

EXPERIMENT

1. Inoculate healthy peaches by inserting bits of rotten peach flesh, or agar with mycelia of the causal organism, under a ¼ inch square flap of skin. Seal with vaseline. Observe the progress of the disease *daily*.

2. Isolate the causal organism when symptoms are well developed by two methods: (1) tissue transplant (Fig. 34-1 A) and (2) spore transfer.

Materials

vaseline
rotten peach or culture of the causal organism
1 healthy peach per group (for inoculation)
2 plates of PDA per group (for isolation)

DISCUSSION QUESTIONS

1. What are the similarities between soft rot of potato, caused by *Pectobacterium carotovorum*, and brown rot of peach, caused by *Monilinia fructicola?* What are the differences?

2. Which phase of the disease, twig and blossom blight or fruit rot, is most important to a grower, shipper, grocer, and consumer?

3. If you can control the twig and blossom blight phase of the disease, is the fruit rot phase also controlled?

4. What role do insects and injuries play in the occurrence and severity of the disease?

SUGGESTED READING

Cole, H. W., "The Pathogenicity of *Botrytis cinerea, Sclerotinia fructigena,* and *Sclerotinia laxa* with Special Reference to the Part Played by Pectolytic Enzymes." Studies on the Physiology of Parastism, *Annals of Botany,* Vol. 20, 1956, pp. 15–38.

35

Black Rot of Grape

Black rot and downy mildew of grape are both apparently of American origin on wild species of grape. Importation of *Phylloxera*-resistant American rootstocks into France spread the causal fungi of these diseases into Europe. Black rot is now found in most all countries where grapes are grown and where the climate is cool and wet. Black rot is of little importance in California because of the arid and warm climate.

Guignardia bidwellii, the causal agent of black rot, also occurs on species of *Parthenocissus*, the Boston ivy and Virginia creeper (Fig. 35-1, 35-2). Although the fungi are morphologically alike on all hosts, the strain from grape will not attack the species of *Parthenocissus* and vice versa. This biological specialization indicates a long evolutionary history of the fungus on these hosts in America.

Black rot on all hosts is characterized by definitely bordered lesions with tan centers. They develop small black bodies, pycnidia, which are easily seen by the unaided eye and typically occur in a ring about the periphery of the tan center (Figs. 35-1, 35-2). On fruit, a rapid, black-colored rot sets in which often exhibits concentric zonations. The pycnidia are also readily seen on the rotted fruit. The fruit dries, shrivels, and becomes mummified (Figs. 35-3, 35-4, 35-5).

Disease Cycle. The fungus overwinters primarily as immature perithecia on fallen leaves and mummified fruit. The asci mature in the spring and violently discharge ascospores at intervals throughout the growing season. Long periods of wetting are required for ejection and germination of ascospores and for infection of susceptible host tissues. Penetration is presumed to be directly through the cuticle.

Pycnidia are produced abundantly and rapidly. Conidia are borne in a sticky mass which dissolves readily in water and splashing rain is important in their dispersal. Long wetting periods are also required for germination and infection.

During the latter part of the summer, pycnosclerotia form. These are similar to the pycnidia in appearance but have a slightly thicker wall and contain only oil droplets. After a long incubation period these structures produce conidia. Spermagonia are also produced but the spermatia have not been germinated and their role in the life cycle of the fungus has not been demonstrated. Ascocarps eventually form and it is not clear whether or not the pycnidia and pycnoscelerotia represent stages in the development of the ascocarps.

Control: achieved by the repetitive application of protectant fungicides from prebloom stages of growth of the host onward. Often, these can be limited to rainy periods, reducing the number of sprays required considerably. Early mechanical cultivation of the soil, especially where the ground beneath the vines is hilled-up, buries much of the overwintered inoculum. The increasing use of herbicides for chemical control of weeds might cause increasing incidence of the disease by permitting a greater amount of the primary inoculum to be effective.

Fig. 35-1. (A) Black rot of Boston Ivy and (B) close-up showing pycnidia.

Fig. 35-2. (A) Black rot of grape and (B) close-up showing pycnidia.

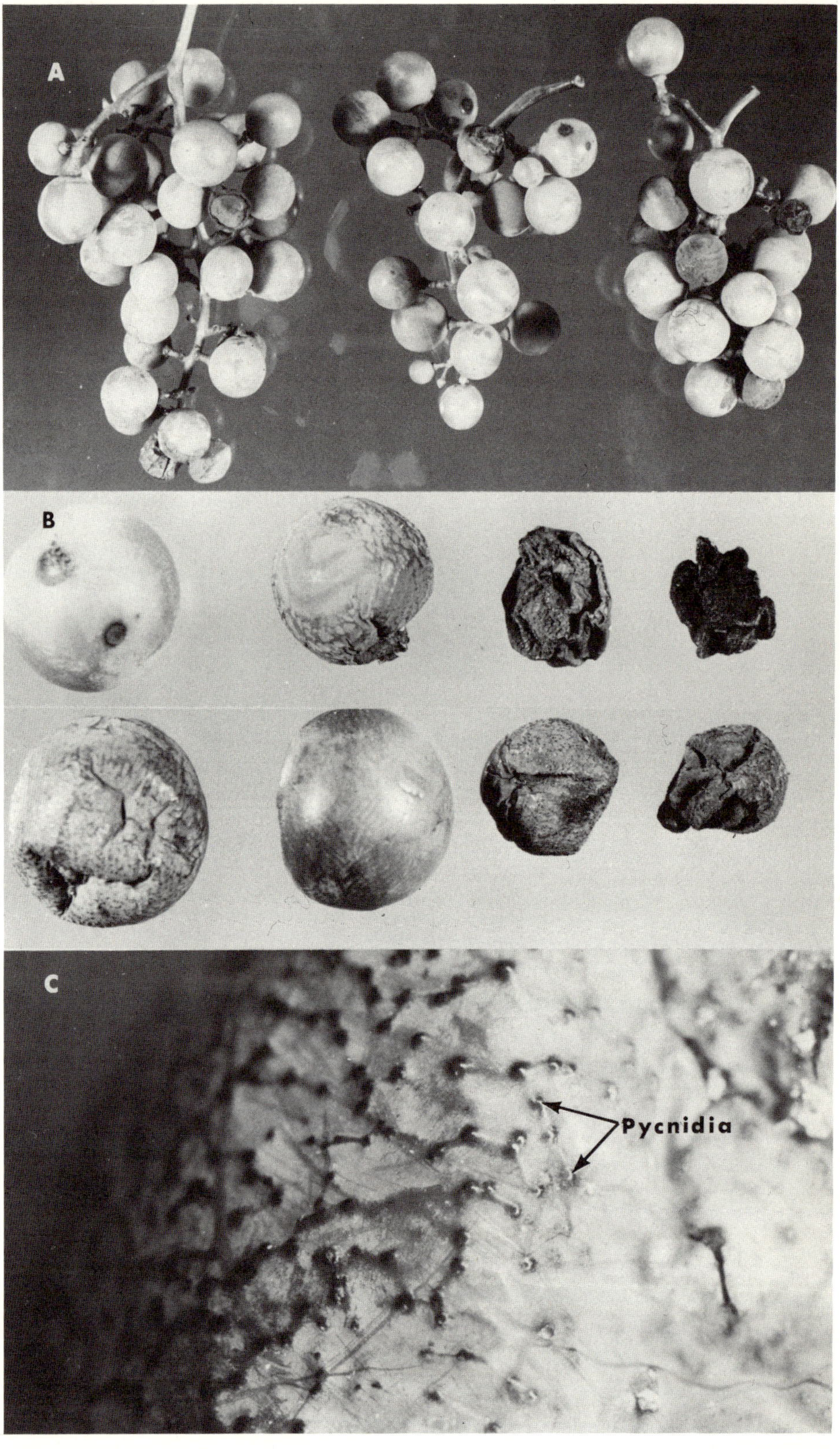

Fig. 35-3. (A) Black rot of grape on fruit and (B) close-up showing stages in mummification. Upper left is not black rot. (C) Close-up of pycnidia on mummifying fruit. Concentric patterns visible in (A) and (B) are typical.

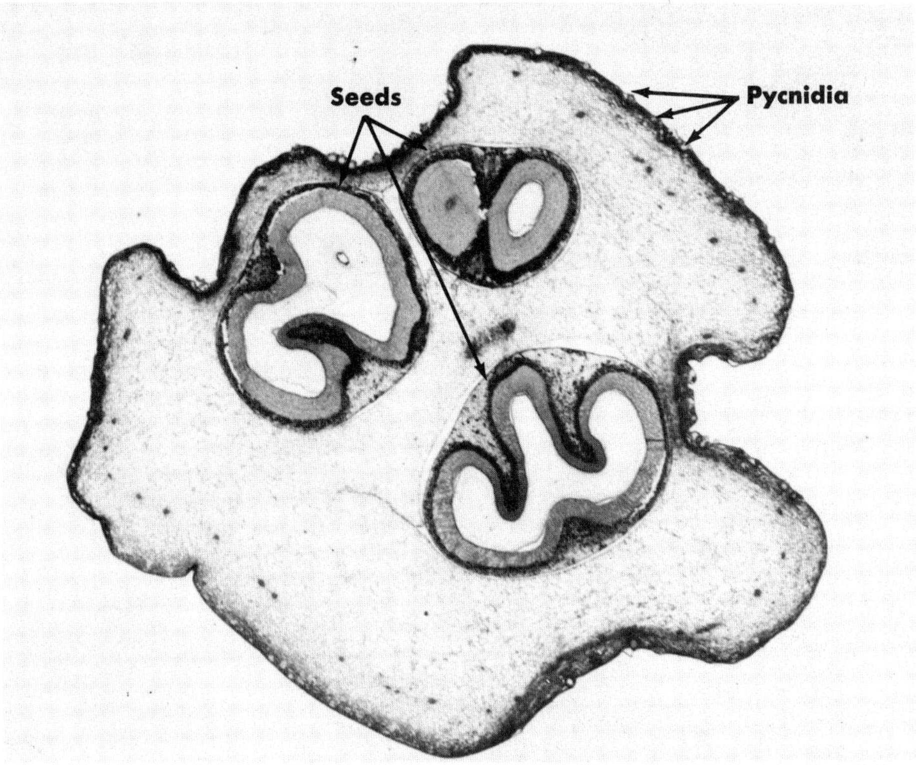

Fig. 35-4. Photomicrograph of a cross section through a mummified grape berry showing pycnidia embedded around the periphery (× *20*).

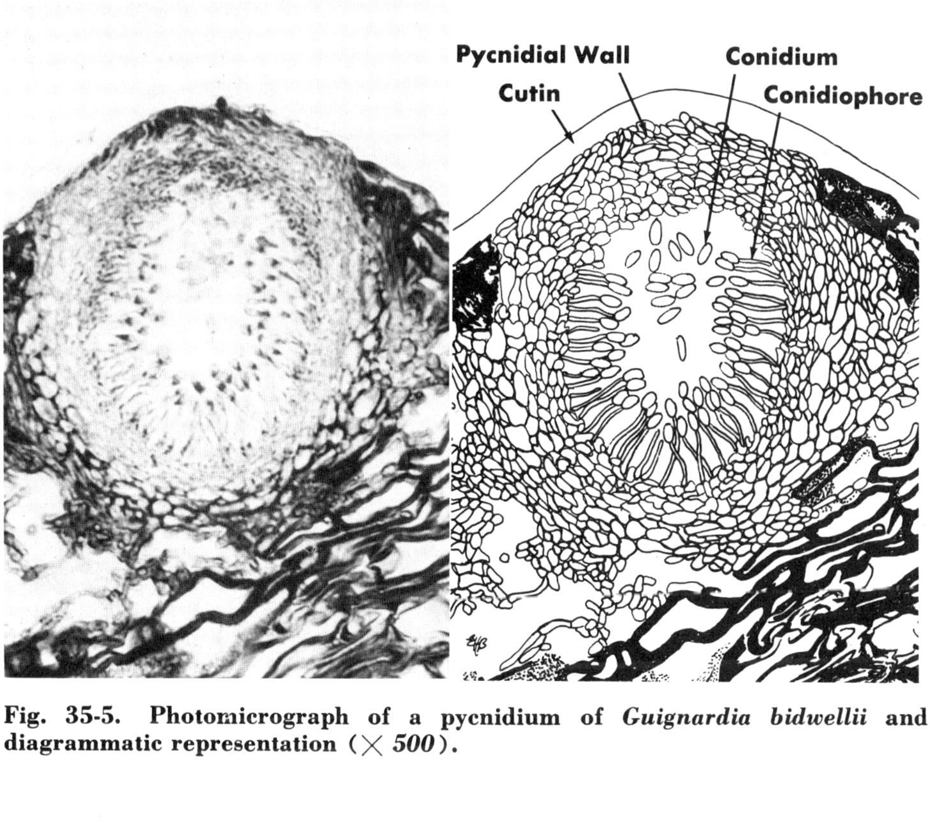

Fig. 35-5. Photomicrograph of a pycnidium of *Guignardia bidwellii* and diagrammatic representation (× *500*).

36

Black Knot of Plum and Cherry

Black knot of plum, caused by *Dibotryon morbosum,* occurs on cultivated plum and wild cherry (Fig. 36-1A). It is a common disease in northeastern and north-central United States. One cannot wander about the north woods without encountering it on wild cherry. The disease is of little economic importance and is easily controlled by removal of cankers in orchards and nearby wild cherry trees.

The knots are composed of overgrowths of host tissue and hyphae of the causal fungus. Primary infections occur mostly on the wood of the current season's growth. Ascospores are unequally two-celled and are generally produced on two-year old knots. Penetration is through unwounded surfaces.

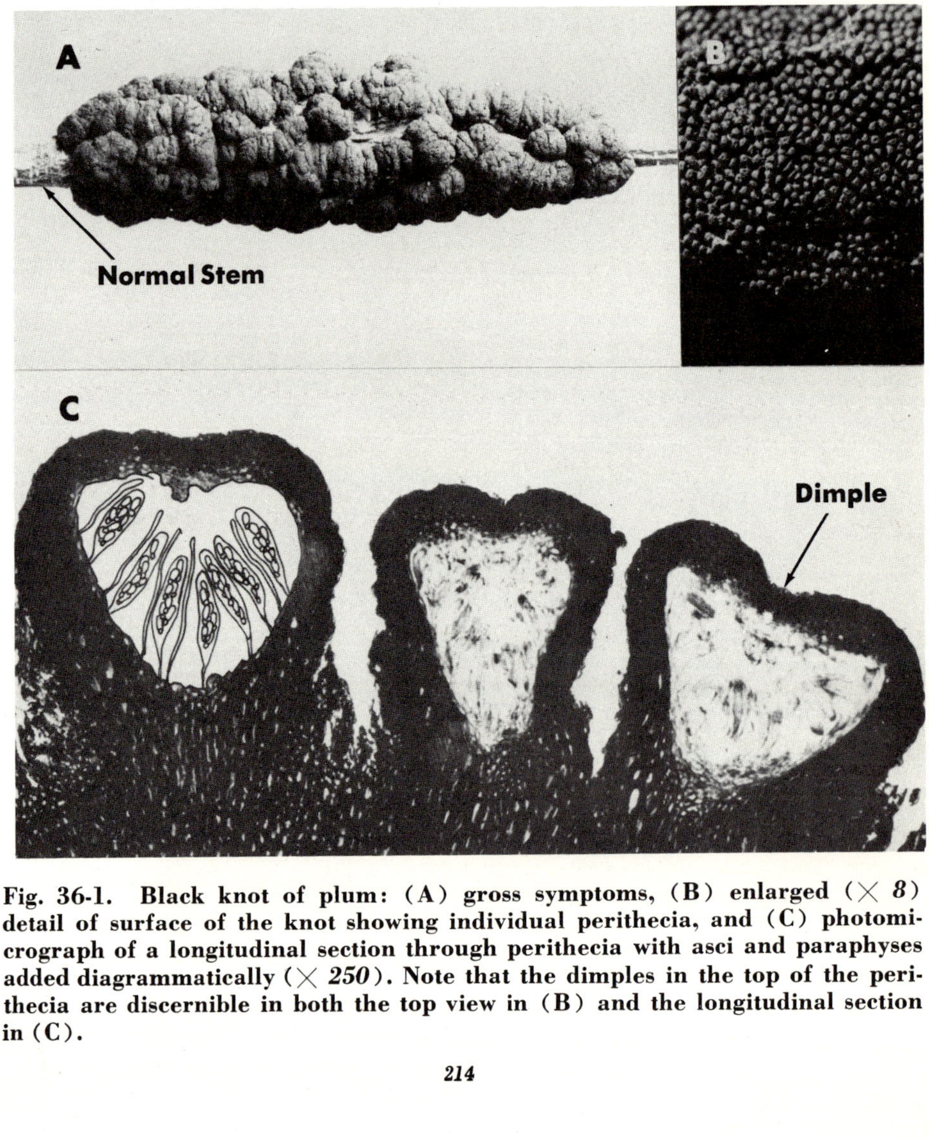

Fig. 36-1. Black knot of plum: (A) gross symptoms, (B) enlarged (× *8*) detail of surface of the knot showing individual perithecia, and (C) photomicrograph of a longitudinal section through perithecia with asci and paraphyses added diagrammatically (× *250*). Note that the dimples in the top of the perithecia are discernible in both the top view in (B) and the longitudinal section in (C).

37

Apple Scab

Apple scab occurs around the world in all countries where the apple is grown. It occurs primarily on apple, crabapple, and other species of the genus *Malus*. Apple scab is the most important disease of apple whether or not it is controlled. If the disease is not controlled, large losses in quality and quantity of fruit result, and if the disease is controlled, great expenses in materials, equipment, and labor are incurred. The most severe infestations occur where cool, moist spring weather prevails, whereas light infections occur in irrigated areas with dry, warm summers.

Disease Cycle. The causal organism, *Venturia inaequalis*, is a typical ascomycete fungus. We begin our observations with the ascospores in the spring (Fig. 37-5). They are divided by a septum into two cells of unequal size, and hence the name of the species. Ascospores are produced from overwintered inoculum, and therefore constitute primary inoculum. Since the ascospores are primary inoculum, infections resulting from them are primary infections.

The ascospore issues a germ tube, the tip of which becomes swelled and closely appressed to the host surface (Fig. 37-6A). Such a structure is called an **appressorium.** A fine, thread-like, hyphal peg grows from the appressorium through the cuticle. This is direct penetration. A mass of closely packed hyphal cells forms between the cuticle and the epidermis (B). Here, the fungus secures its nutrients and engages in physiological combat with the host. It does not penetrate the cells of the host until the leaf falls.

Several days after infection, short, erect, simple conidiophores develop on the mass of hyphal cells. They bear single flame-shaped conidia at the tip (C), (Fig. 37-3). Conidia are produced in succession and a series of ridges develop on the conidiophores, marking the positions of conidia produced earlier. The pressure of developing conidia ruptures the cuticle. A mass of underlying hyphal cells with short, erect conidiophores, such as now are exposed, is called an **acervulus.**

Conidia are carried by wind and splashing rain to susceptible tissues. They germinate, produce appressoria (E), and infect the host in a manner similar to the ascospores. These are secondary infections since they result from inoculum produced by primary infections during the current season. The secondary infections soon produce conidia and the cycle of conidia-infection-conidia-infection is underway. This series of infections is called the **repeating cycle** and is caused by repetitive, asexual reproduction.

As a result of primary and secondary infections in the spring and early summer, leaves, green flower parts, and young fruit become infected (Figs. 37-1, 37-2). The disease gets its name from the scabby appearance of the dark, hard, and cracked lesions on the fruit. The lesions are mostly superficial but enlargement of the young fruit is restricted in the infected area and the resulting fruit is knobby and malformed when mature.

As the rains become infrequent with the passage of spring and temperatures rise with the onset of summer, infections decrease until they cease. Then, late in the sum-

Fig. 37-1. Apple scab on (A) the ornamental *Malus purpurea* and (B) close-up on fruit.

Fig. 37-2. Apple scab on (A) foliage and young fruit and (B) typical scab lesions on mature fruit.

mer, with a drop in temperature and the commencement of early fall rains, secondary infections occur anew. Fruit, which are infected late in the season, are well-developed in size and malformation is nearly absent. Frequently, small, pinpoint lesions occur which enlarge later in storage.

When the leaves fall, the fungus ramifies through the tissues and produces gametangia and perithecial initials. These are partially formed perithecia. Growth ceases during the cold winter months but renews with the early spring warmth and rains. The perithecia then mature and appear as small pinpoint bumps on the overwintered leaves (Fig. 37-4). The short, fat necks of the perithecia protrude through the surface. Upon wetting, the ascospores are violently discharged from the asci. Thus the overwintering inoculum is the immature perfect stage and not the ascospores themselves.

Control: is achieved at great expense by repeated application of protectant fungicidal sprays, beginning in the spring when ascospores first mature and weather conditions are favorable. Perithecia can be crushed and an evaluation made of the stage of maturity (Fig. 37-5). Once the ascospores begin to mature, sprays need to be applied whenever favorable weather for ejection of ascospores and infection occurs.

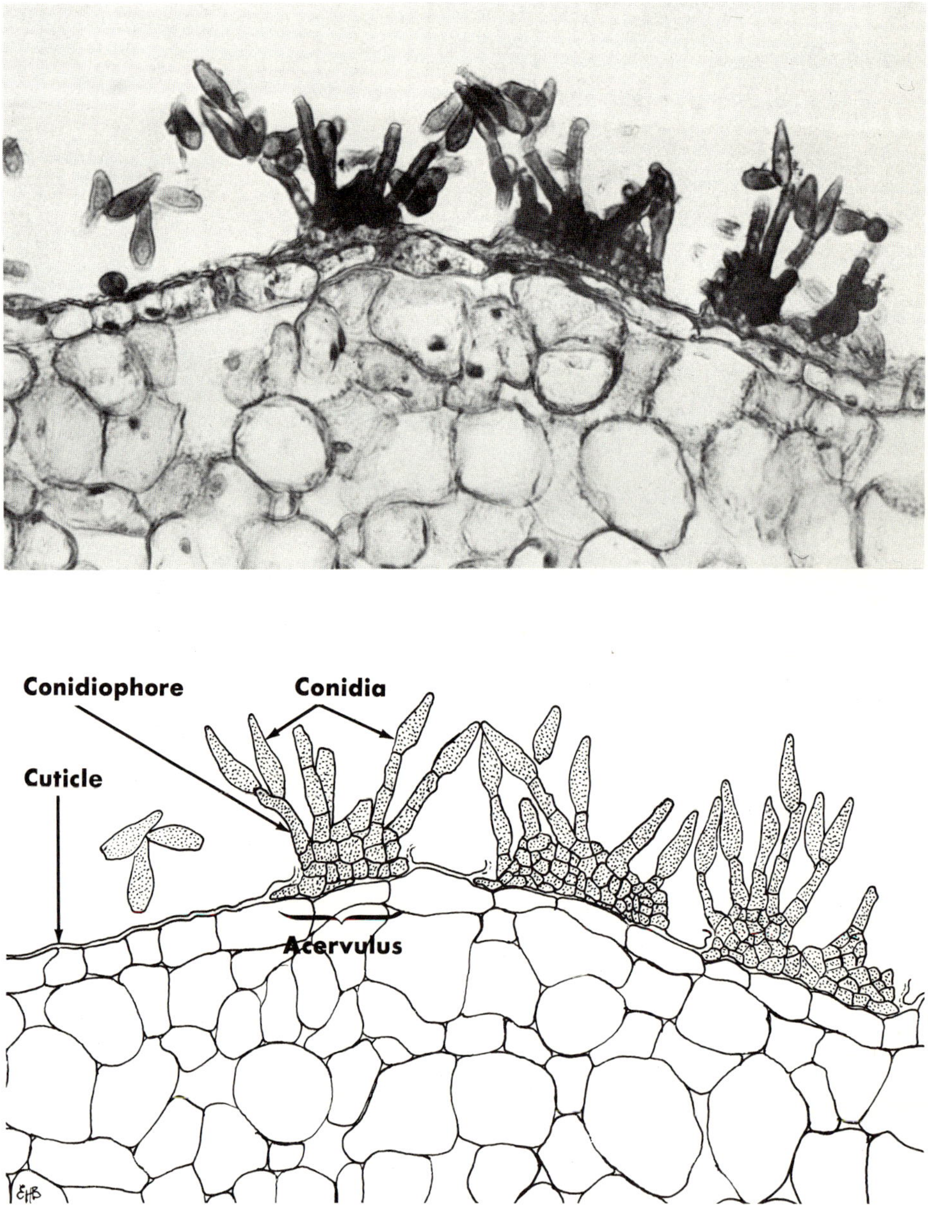

Fig. 37-3. Photomicrograph of acervuli of *Venturia inaequalis* showing conidiophores and conidia and diagrammatic representation (× *208*).

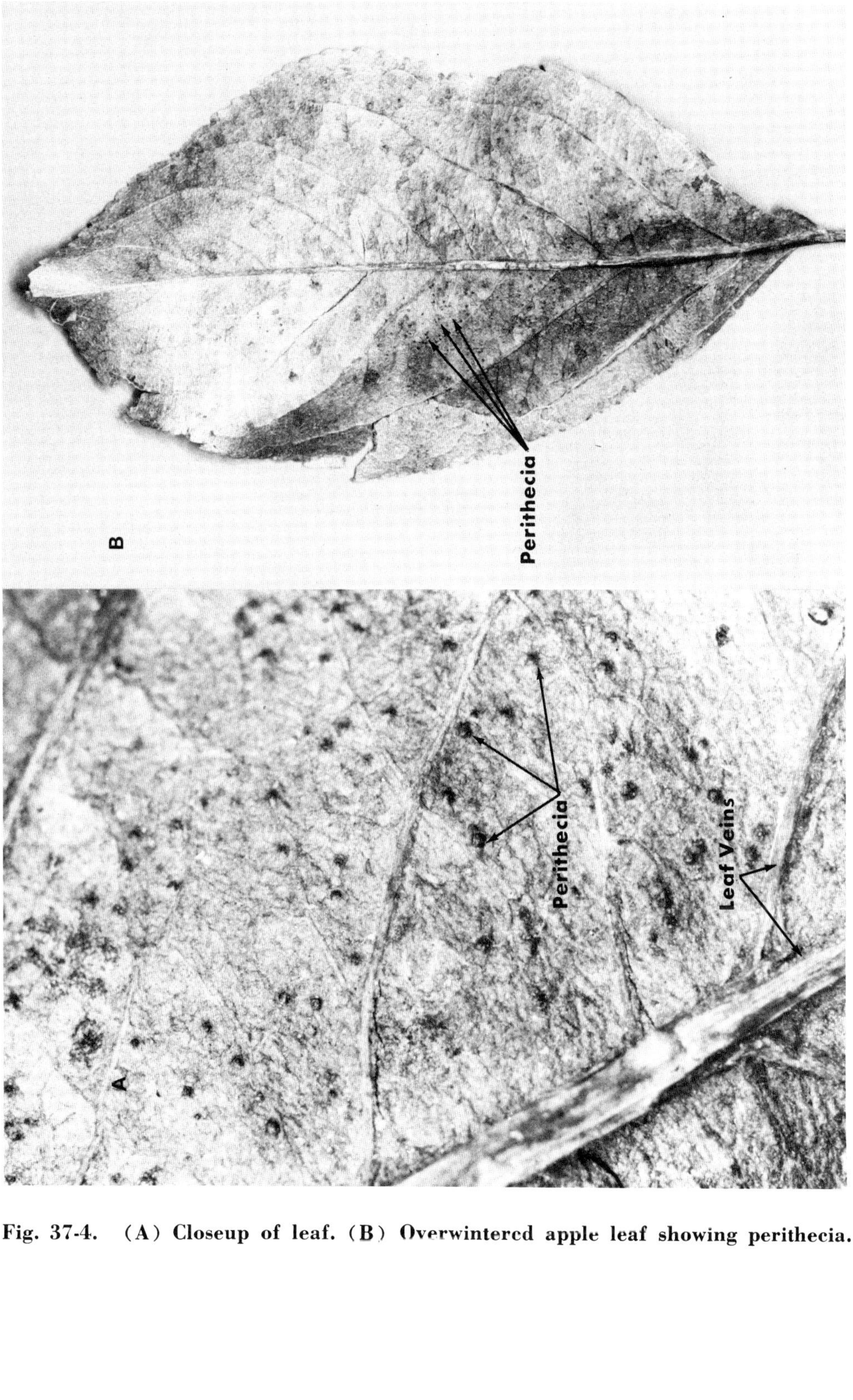

Fig. 37-4. (A) Closeup of leaf. (B) Overwintered apple leaf showing perithecia.

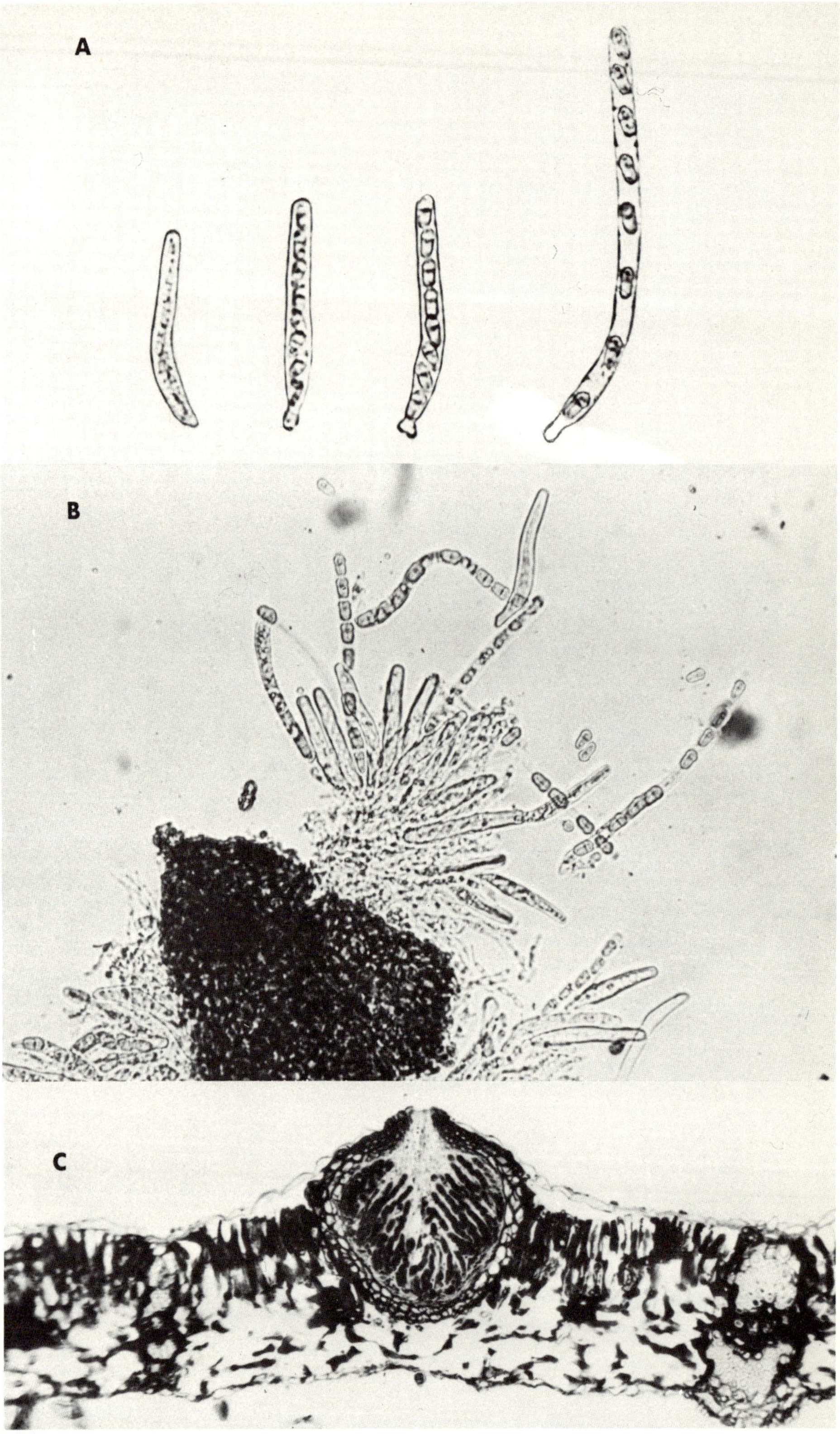

Fig. 37-5. (A) Stages in the maturation of asci of *Venturia inaequalis* from (B) a squashed perithecium. (C) Photomicrograph (× *125*) of a longitudinal section through an embedded perithecium with mature asci.

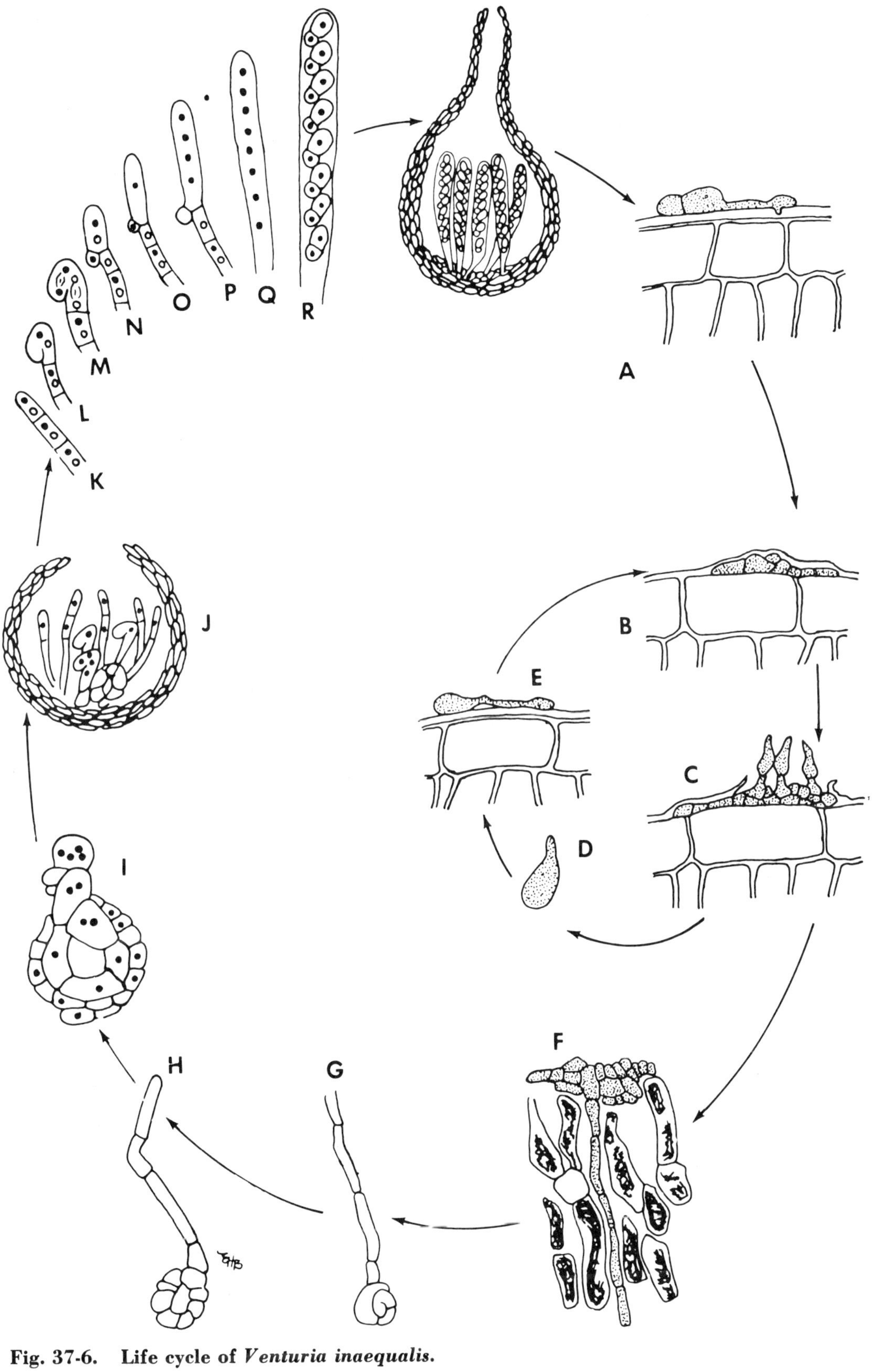

Fig. 37-6. Life cycle of *Venturia inaequalis*.

38

The Deuteromycetes: (The Fungi Imperfecti)

There are thousands of septate fungi which reproduce only by means of conidia and are not known to have any sexual stages. They cannot be classified along with the other fungi which are grouped according to sexual characteristics. Consequently, an artificial class, the Deuteromycetes, was organized. This group and all taxonomic subdivisions of it are called forms. Thus, there is the form-class Deuteromycetes with form-orders, form-families, and so forth (see also Chapter 15).

The conidia of some Deuteromycetes are very similar to the conidia of Ascomycetes. A few, such as *Rhizoctonia,* have a Basidiomycete for a perfect stage. For the most part, many of the Fungi Imperfecti are probably Ascomycetes. For convenience, the imperfect, or conidial stages of Ascomycetes are classified often in the Deuteromycetes.

Four form-orders are recognized and separated on the basis of the type of fructification (Fig. 23-1): Sphaeropsidales (pycnidia), Melanconiales (acervuli), Moniliales (free conidiophores, sporodochia, synnemata), and Mycelia Sterilia in which no spores of any kind are formed.

The type of conidiophore and the color, shape, and septation of the conidia are the characteristics used to separate form-genera. Form-species are separated almost entirely on the basis of conidial size and host. These may be convenient characteristics but are not indicative of natural relationships or real differences at times. Diagrams of several genera of the Fungi Imperfecti are presented to give some idea of the variation in architecture. Moniliales (Fig. 38-1; 38-2, *Fusarium, Fusicladium,* and *Graphium*), Melanconilaes (Fig. 38-2, middle row), and Mycelia Sterilia (Fig. 38-2, *Rhizoctonia*).

Many Fungi Imperfecti, both those with known perfect stages and no known perfect stages, are important pathogens of plants, animals, and man.

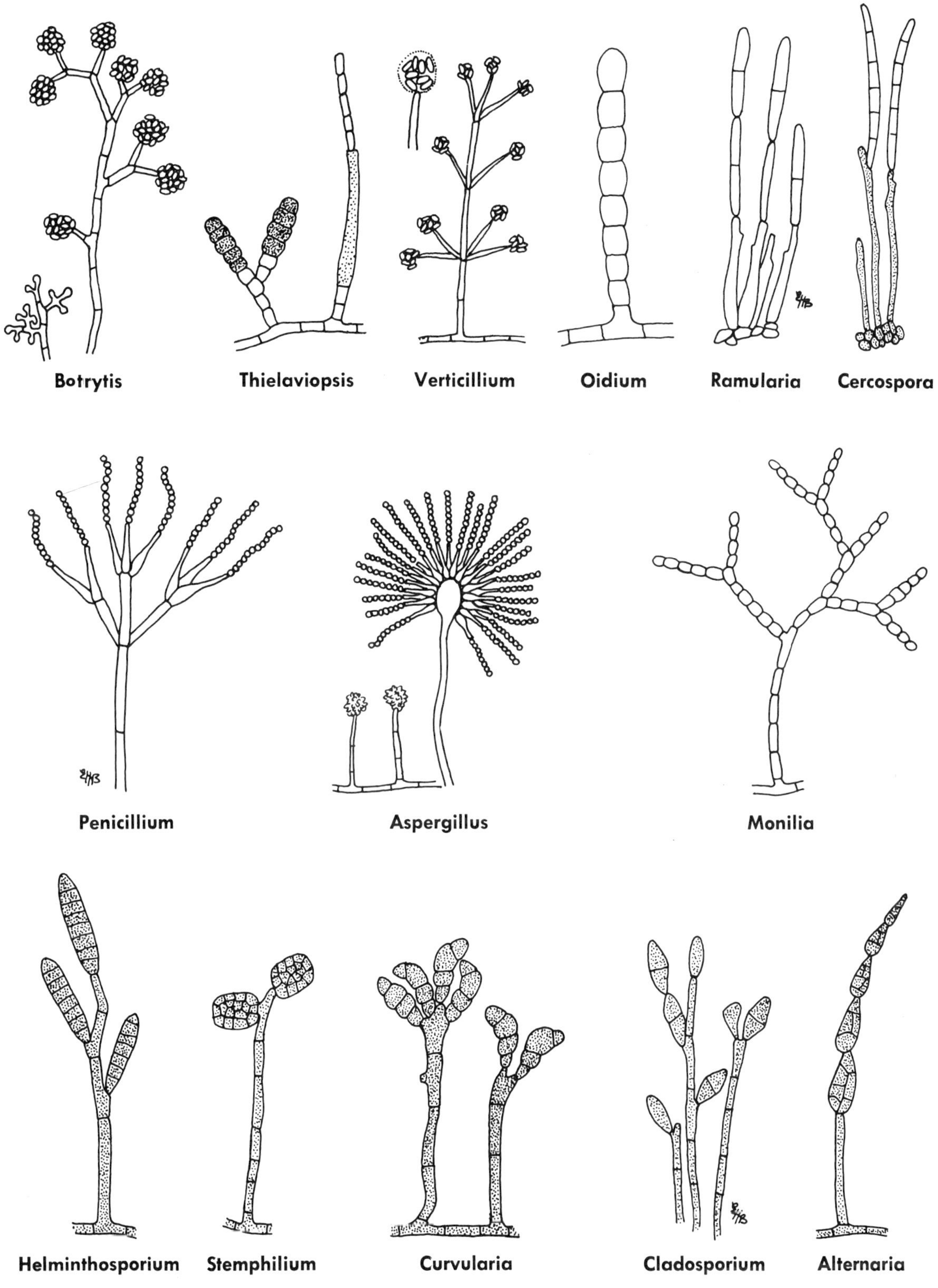

Fig. 38-1. Some genera of the Fungi Imperfecti.

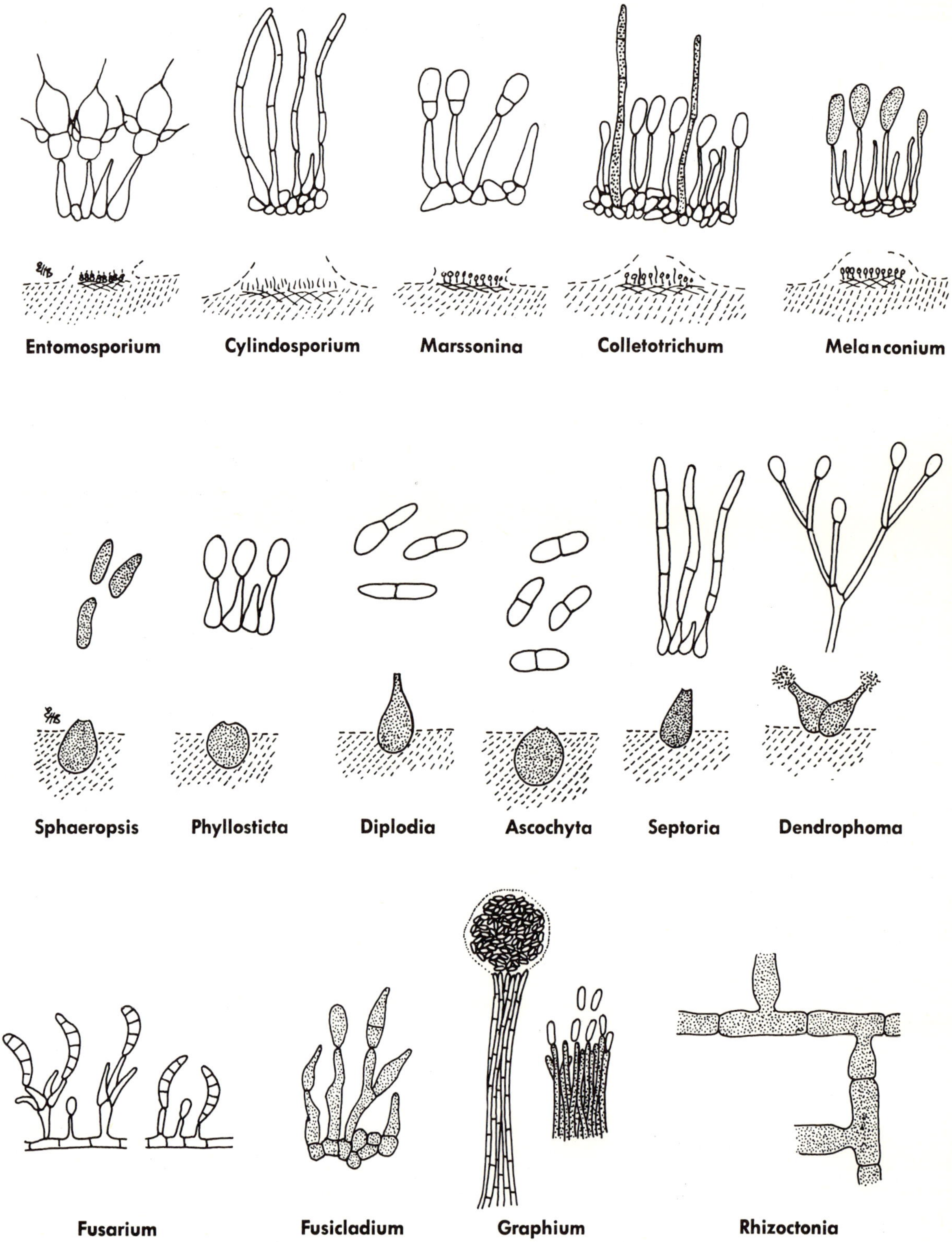

Fig. 38-2. Some genera of the Fungi Imperfecti.

39

Alternaria Diseases

Alternaria is a universal fungus and occurs most frequently as a saprobe on dead and decaying organic materials. Spores of *Alternaria* are common in house dust and the fungus is probably more significant as the chief fungal cause of hay fever than as a plant pathogen.

Alternaria is significant in microbiological work because it is a common contaminant in the laboratory. *Alternaria* is of further interest to the plant pathologist as a weak pathogen and a secondary parasite.

Causal Organism. *Alternaria* is dark in color. Conidia are rather large, pear-shaped, and multicellular with both transverse and longitudinal septations. Conidia are catenulate, or in chains. Conidiophores on diseased tissues are short and erect but in artificial media are often nearly indistinguishable from the hyphae. No chlamydospores or sclerotia are formed and no perfect stages are known (Fig. 39-1,-2,-3).

Pathogenicity. *Alternaria* can be grouped into (1) those that are saprobes and secondary parasites only and (2) those which are not only saprobes but also primary pathogens that can attack certain plants and cause similar diseases with similar disease cycles. Four species of primary pathogens and their hosts are:

Alternaria brassicae (Cruciferae) cabbage, radish, cauliflower
Alternaria cucumerina (Cucurbitaceae) cucumber, watermelon, squash
Alternaria dauci (Umbelliferae) carrot, celery, parsley
Alternaria solani (Solanaceae) potato, tomato, nightshade

These four fungi cause post-emergence damping-off, collar rots, stem lesions, leaf spots, and tuber and fruit rots of their respective hosts.

Disease Cycle. The pathogenic *Alternaria* typically overwinter as mycelia in diseased plant refuse of susceptible weeds and crops. They may be carried on and in the seeds. Primary infection results from seed transmission, which frequently causes post-emergence damping-off, and from wind-blown spores from other plantings and weed hosts. Dispersion within a field is by means of wind, splashing rain, and implements.

These diseases, as most diseases caused by weak pathogens, are more prevalent on plants of low vigor and poor nutrition. They occur most frequently in relation to injuries, especially insect injuries, and unfavorable environmental conditions for plant growth.

Control. Resistance is of little value in the control of *Alternaria* diseases—another characteristic of disease caused by weak pathogens. Chemical spray application, seed treatments, crop rotation, and weed host eradication are the most important and effective control measures. The treatment of seeds in hot water or mercury phosphate eliminates the seed-carried inoculum. Rotation permits the dying-out of overwintered inoculum in diseased plant refuse.

Secondary Parasitism. Species of *Alternaria,* particularly *Alternaria tenuis,* are found on senescent or old tissues such as petals, old leaves, and ripe fruit. They are especially prevalent in the necrotic, or dead, tissues of lesions caused by primary pathogens. These fungi represent classical **secondary parasites,** subsisting through someone else's hard work—the primary pathogen. They are not insignificant to the plant pathologist in this capacity; however he must learn to distinguish them from the real culprits. When he examines lesions from the field or a moist chamber, he discovers *Alternaria.* When he isolates organisms from diseased tissues, he finds that he has been carefully raising some more *Alternaria.* Is the *Alternaria* a pathogen and the cause of the lesion, or is it a tramp? If the pathologist cannot distinguish these alternatives from experience, he must utilize the principles of Koch's Postulates.

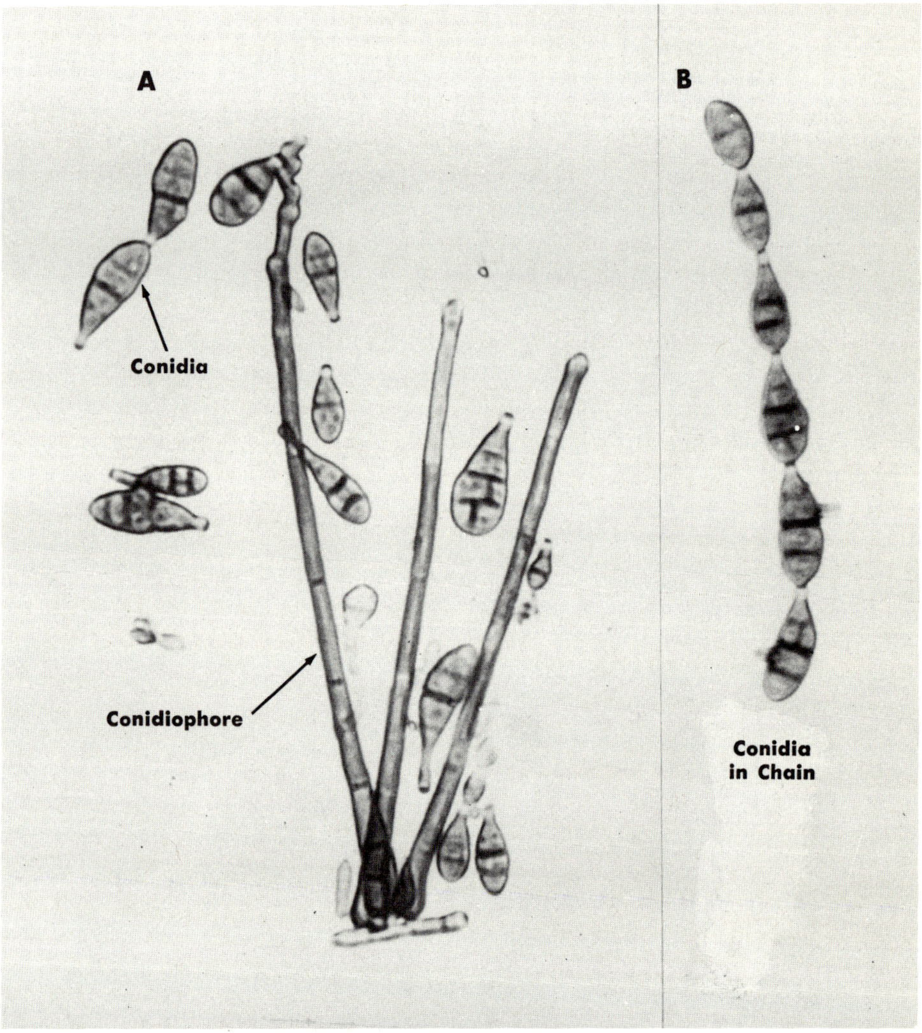

Fig. 39-1. Conidia and conidiophores of *Alternaria*.

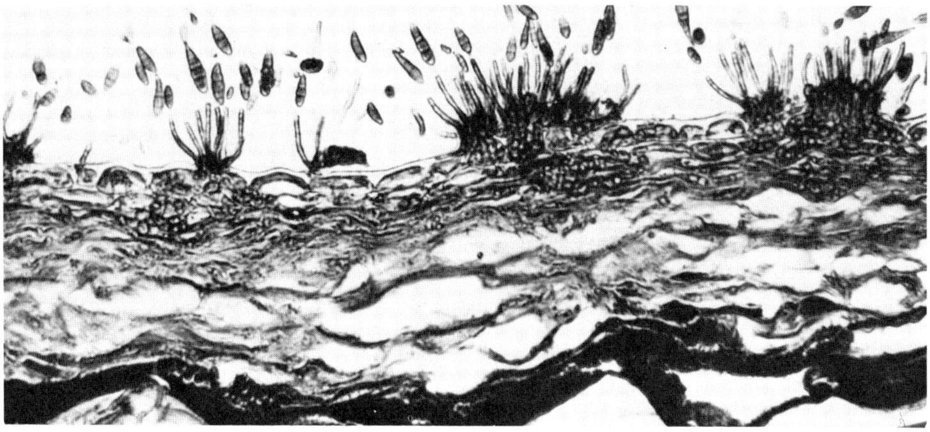

Fig. 39-2. Conidia and conidiophores of *Alternaria* on tomato fruit.

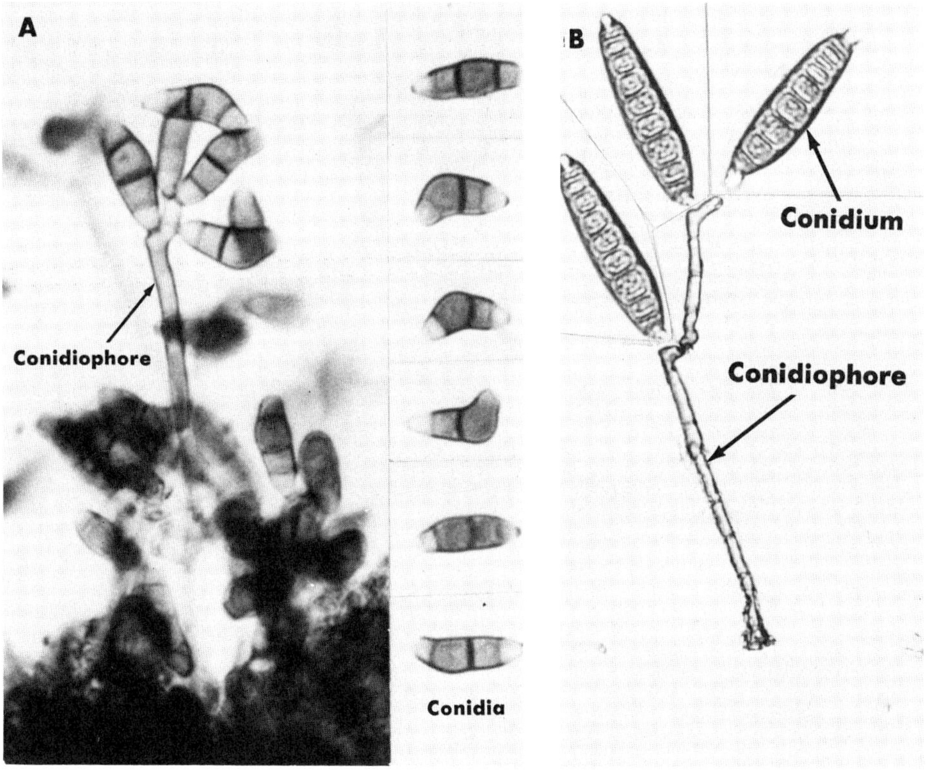

Fig. 39-3. (A) Conidia and conidiophore of *Curvularia* and (B) Conidia and conidiophore of *Helminthosporium*. Note the similarity of the three genera.

Botrytis Diseases

If *Alternaria* is not the most widely distributed fungus, then *Botrytis* is. This was one of the first fungi to be described. It is similar to *Alternaria* in that it is an efficient saprobe, a weak pathogen, and a secondary parasite. Since its development is favored by cool, moist conditions, it occurs typically in temperate zones and on winter crops in sub-tropical areas. *Botrytis* can cause damping-off, stem cankers, leaf spots, flecks, and tuber rots. However, it seldom rots potato tubers and is more commonly known to cause fruit rots and blossom blights (Fig. 40-1).

Causal Organism. Conidia are hyaline or brightly colored and are borne in grape-like clusters (Figs. 38-1, 40-2). They are attached singly to sterigmata on simple or branched conidiophores. Sclerotia are produced on diseased plant parts and in old cultures (Fig. 40-3). Some form-species of *Botrytis* have Discomycete perfect stages.

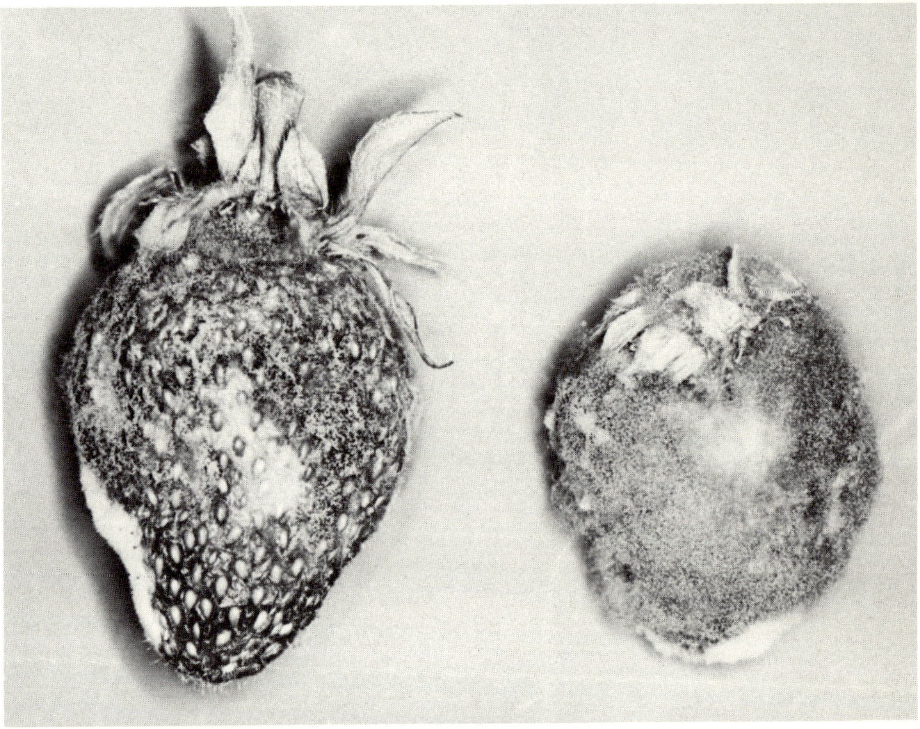

Fig. 40-1. Gray mold of strawberry caused by *Botrytis cinerea*.

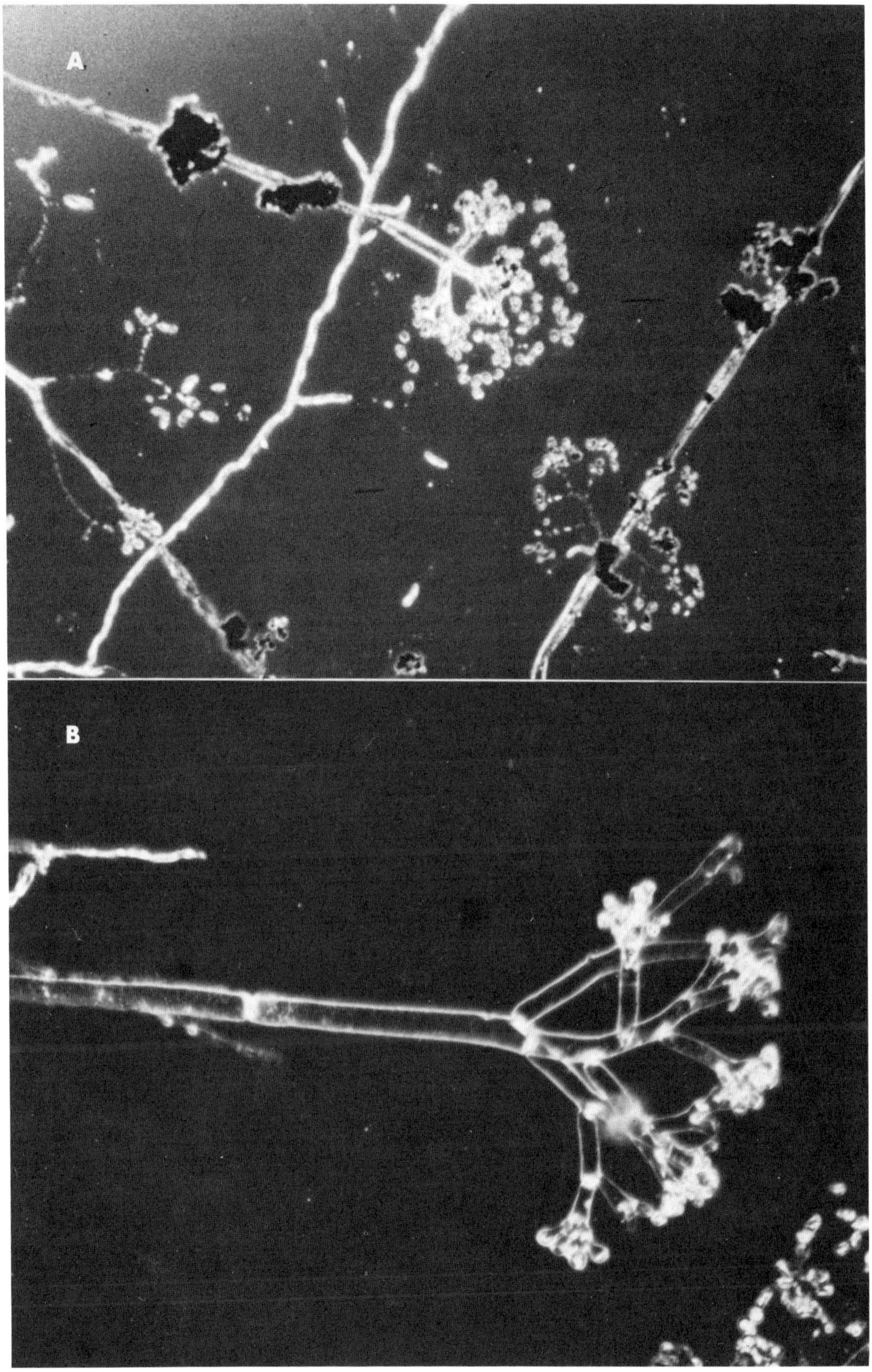

Fig. 40-2. Conidia and conidiophores of *Botrytis cinerea*.

Botrytis ricinia on castor bean has a *Botryotinia* perfect stage. *Botryotinia* is similar to *Monolinia* which has *Monilia* for an imperfect stage. Some of the pathogenic species and their hosts are:

Botrytis allii	onion
Botrytis elliptica	lily
Botrytis gladiolorum	gladiolus
Botrytis paeoniae	peony
Botrytis ricini	castor bean
Botrytis tulipae	tulip

Botrytis cinerea causes gray mold of strawberry (Fig. 40-1) and is frequently a secondary parasite on just about any plant material from aster to zinnia. It also occurs on the hosts listed above. *Botrytis byssoidea* and *Botrytis squamosa* are also pathogens of onion as well as *Botrytis alli:* The diseases of lily, gladiolus, and peony caused by species of *Botrytis* are reported to be the most common diseases on these plants.

Botrytis species are not seed-borne but sometimes are carried with the seed as infected bits of stems or as seed-size sclerotia. Commonly, *Botrytis* survives the winter as sclerotia and as mycelium on infected plant material such as winter killed rose canes, decaying refuse from infected blossoms, and infected bits of tissue carried with the seed.

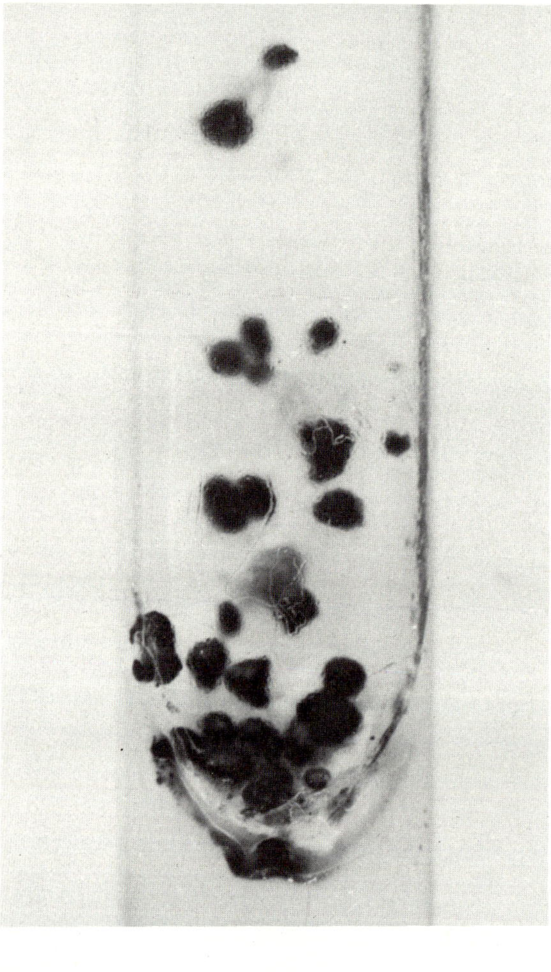

Fig. 40-3. Sclerotia of *Botrytis cinerea* in culture tube.

Control. Typically, for a weak pathogen and a secondary parasite, resistance is of little importance in the control of most BOTRYTIS diseases. The outer scales of colored onion varieties are resistant to penetration by *Botrytis* but the fleshy, white interior is not. Castor beans with long internodes in the inflorescence are not really resistant but have less disease incidence because of more rapid drying-out. Removal of infected and infested debris is helpful. Carbamate sprays are helpful with the leaf spotting and flecking diseases.

Control of BOTRYTIS fruit rots is greatly aided by control of earlier stages since inoculum is thereby considerably reduced. Frequently, infections of strawberry by *Botrytis cinerea* begin as blossom blight and calyx tip diseases. Berry rots then originate from the infected floral parts.

41

Fusarium Diseases

The *Fusaria* are distributed throughout the world in many forms which differ in morphology and pathogenicity. They are separated into various species which delineate morphologically uniform groups. Two of the most common and important species are *Fusarium oxysporum,* which typically causes vascular wilts (Fig. 41-3), and *Fusarium solani,* which commonly causes root rots (Fig. 41-1).

Some of the morphological species are further divided on the basis of host specificity into pathogenically uniform groups called **forms.** The form designation follows the species name of the latin binomial to give a trinomial, and is preceded by the initial "f." to indicate form. Thus, the form of *Fusarium oxysporum* which causes wilt of tomato, but not other plants, is designated *Fusarium oxysporum* f. *lycopersici.* Often the form designation is a derivative of the genus of the host. In this case, the host genus is *Lycopersicon.* A few other forms and their specific hosts are:

Causal Organism	Host
Fusarium oxysporum f. *callistephi*	aster
Fusarium oxysporum f. *conglutinans*	cabbage
Fusarium oxysporum f. *cubense*	banana
Fusarium oxysporum f. *pisi*	pea
Fusarium oxysporum f. *vasinfectum*	cotton

Likewise, *Fusarium solani* is subdivided into forms which are specific to different plant hosts, such as *Fusarium solani* f. *phseoli* on bean and *Fusarium solani* f. *cucurbitae* on pumpkin and squash. These two species and others cause damping-off, root rots, dry rots (Fig. 41-2), stem cankers, and vascular and nonvascular wilts. *Fusarium oxysporum* typically causes vascular wilts and *Fusarium solani* causes root rots, dry rots, and stem cankers. All species of *Fusarium* have a saprobic stage.

Although *Fusarium* appears colorless microscopically, it produces a pink, cottony growth in culture and on diseased plants parts. In addition, mycelium, macroconidia and microconidia are produced (Fig. 38-2). Sometimes one-celled, large round, thick-walled cells are formed from hyphal cells. These are **chlamydospores.** Sclerotia may also be produced.

FUSARIUM DRY ROOT ROT OF BEAN. Several species of *Fusarium* can be found in unhealthy roots of bean plants when growing conditions are poor—when it is too hot, too cold, or too wet. (Fig. 41-1). The most important root pathogen of bean is *Fusarium solani* f. *phaseoli* which can attack roots of healthy beans growing under fairly normal conditions. As the form *phaseoli* indicates, no other plant is probably infected naturally by the strain from bean.

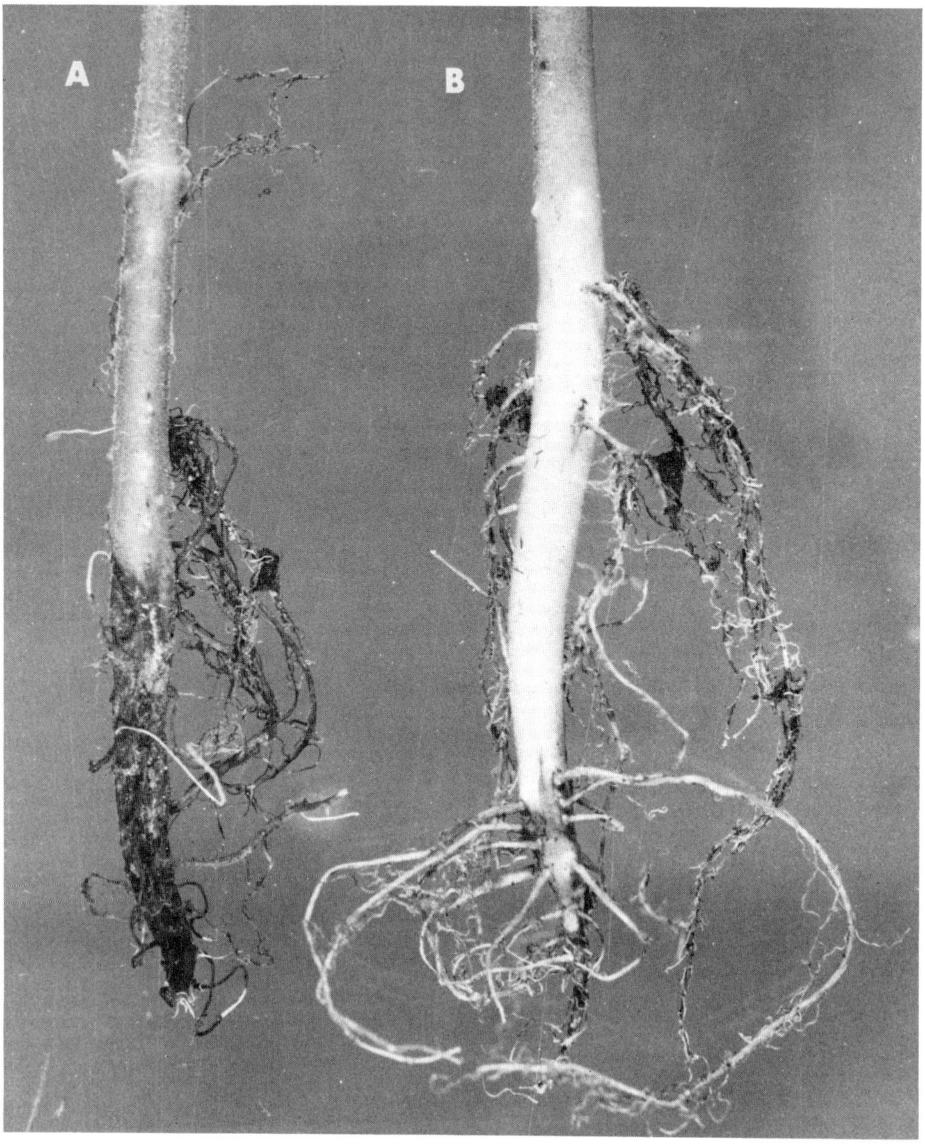

Fig. 41-1. Comparison of FUSARIUM root rot of bean (A) and healthy bean roots (B).

Striking lesions may appear on the lower stem, collar, and tap root. All roots may become involved. Red color, often in streaks, develops. The roots remain turgid. The shoot is not invaded as with *Fusarium oxysporum* and wilting is not an important symptom. The disease of the roots and cortical region of the lower stem is reflected by missing, dwarfed, and yellow plants in the row.

The fungus is a good saprobe in bean refuse and may survive for many years in the soil. Spores are not produced usually until the affected plant parts age and begin to decompose. Spread is through rain, contaminated implements, animals, and bean straw manure. Even so, the fungus spreads slowly.

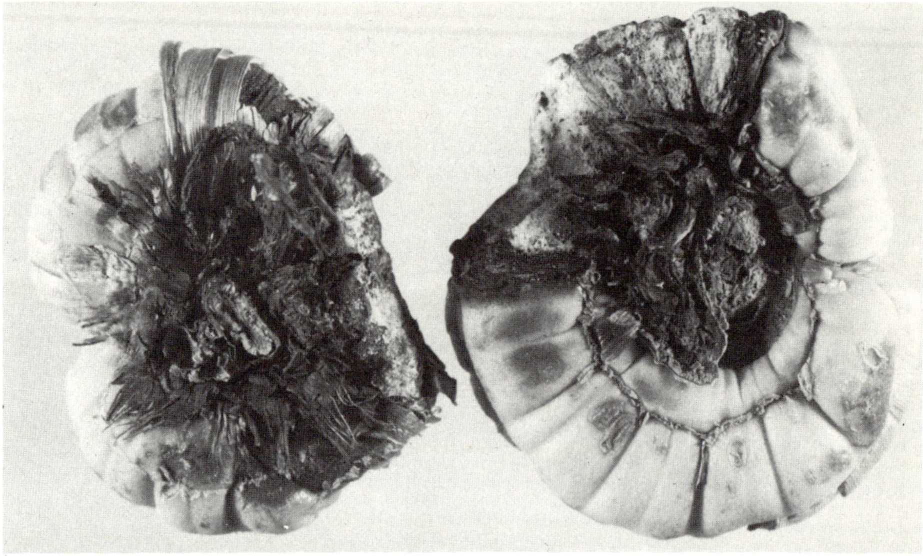

Fig. 41-2. FUSARIUM dry rot of gladiolus.

Control. Long rotation and restrictions on disposal of bean refuse where beans are to be grown provide some measure of control. Most any practice that favors good growth of adventitious roots reduces disease losses. Techniques of cultivation to hill-up more soil about the base of plants in irrigated areas encourages more adventitious root growth and curtails the potency of the disease.

FUSARIUM FOOT ROT OF SQUASH AND PUMPKIN is another example of a root rot caused by *Fusarium solani* f. *cucurbitae*. If young seedlings are attacked, a post-emergence damping-off occurs. The fungus can also infect through wounds in the fruit. Secondary organisms follow and a watery, smelly, quite disagreeable mass results.

This *Fusarium* is carried on and in the seed and seldom survives in soil or diseased refuse. Spread occurs through seeds and sometimes tools. A sexual stage of this species is known as *Hypomyces solani* f. *cucurbitae*, an Ascomycete. Use of disease-free seed gives complete control and practically eliminates the disease.

FUSARIUM DRY ROT OF GLADIOLUS. A typical dry rot caused by a *Fusarium* is the dry rot of gladiolus caused by *Fusarium oxysporum* f. *gladioli* (Fig. 41-2). In contrast to a wet rot, such a bacterial soft rot or brown rot of stone fruits, the diseased corms dessicate and have a sunken appearance. The organ remains fairly firm. Although this is the most important storage disease of gladiolus, reasonable control can be secured with proper drying and curing of corms.

FUSARIUM WILT OF TOMATO. Wilting is a symptom of a number of ills: (1) insufficient soil moisture, (2) inadequate uptake due to injury or disease of roots, (3) restricted transport due to disruption or blocking of vascular tissues from injury and general diseases of the stem, and (4) diseases of the vascular system. The true wilt diseases are diseases of the vascular system, primarily the xylem. These include bacterial wilts and fungus wilts. Examples are southern bacterial wilt caused by *Pseudomonas solanacearum*, bacterial wilt of cucumber caused by *Erwina tracheiphila*, and the fungus wilts caused by *Fusarium oxysporum*, VERTICILLIUM wilt caused by *Verticillium albo-atrum*, and Dutch elm disease caused by *Ceratocystis ulmi*.

Fig. 41-3. (A) FUSARIUM wilt of tomato. Free hand (B) cross section and (C) longitudinal section of a *Fusarium*-infected tomato stem.

The organisms that cause wilt diseases may have broad host ranges as with *Pseudomonas solanacearum* and *Verticillium albo-atrum*, or may exhibit host specificity as with *Erwinia tracheiphila, Fusarium oxysporum*, and *Ceratocystis ulmi*. Most of them penetrate their hosts through wounds. Some enter wounds in roots, as with *Fusarium oxysporum* and *Verticillium albo-atrum*, whereas others enter through insect wounds, as with *Erwinia tracheiphila* and *Ceratocystis ulmi*. Those which enter primarily through wounds in roots are generally successful soil inhabitants, whereas those which enter primarily through insect wounds are not. Those which gain entry through wounds in roots are generally controlled by resistance whereas those which enter insect wounds are kept in hand more often by the control of the insect vectors which make the wounds and carry the inoculum.

The role of metabolites produced by organisms in the production of wilt is difficult to assess. Some of the discoloration, perhaps even some of the wilting, may be attributable to a poisoning by these toxic materials. In any event, toxicity of metabolites, plugging by mycelial or bacterial growth, plugging by formation of tyloses by the plant host, and destruction of the vascular continuity collectively result in restricted flow of water. Hence, a deficit of water in the aerial portions which lose water by transpiration and a wilting results.

In FUSARIUM wilt, the causal organism is found typically in the xylem, as with most true vascular wilt diseases. Here it grows and the passage of water is restricted by the organism (Fig. 41-4), its products, residues, and host responses as described above. The fungus is known to produce pectolytic enzymes, fusaric acid, lycomarasmin, and ethylene. Presumably, the pectolytic enzymes cause the degradation of pectic materials which then collect in gummy masses. The fusaric acid and lycomarasmin are dumped into the xylem stream like so many pollutants in a river and may play some role in poisoning the tissues. Some suggest that the turgor pressure is altered by a change in permeability of the protoplastic membrane caused by the products of the pathogen and/or host responses.

Symptoms. The first symptoms are vein clearing and epinasty. **Vein clearing** is a loss of the green color of chlorophyll along the veins. **Epinasty** is a dropping of the petioles to a lower angle—not a wilting. Plants tend to become stunted and yellow and begin to wilt during the heat of the day, but recover during the cool of the evening. Eventually, the wilt becomes permanent (Fig. 41-3). Before the permanent wilt sets in, brown streaks can be seen in the vascular bundles when the stem is cut (Fig. 41-3C). On closer examination, hyphae can be seen in the xylem (Figs. 41-3B, 41-4).

Disease Cycle. The fungus is transported long distances with transplants as infections in them or contamination in clinging soil particles. This is probably insignificant now that soils are generally contaminated in most tomato-growing regions of the United States and the fungus survives indefinitely in most soils. Infected transplants serve, however, as sources of primary infection along with primary infection of the roots from overwintering mycelium in the soil. Seed transmission is rare in the FUSARIUM wilt of tomato but common with the similar wilt of pea.

Spread within a field is by rain and implements. Secondary infection usually does not occur until the plants die. The fungus is mostly in the vascular tissue and grows very little in the cortex until the plant dies. Thereafter, it sporulates on the dead tissues and produces the secondary inoculum. Most of the infections, for several

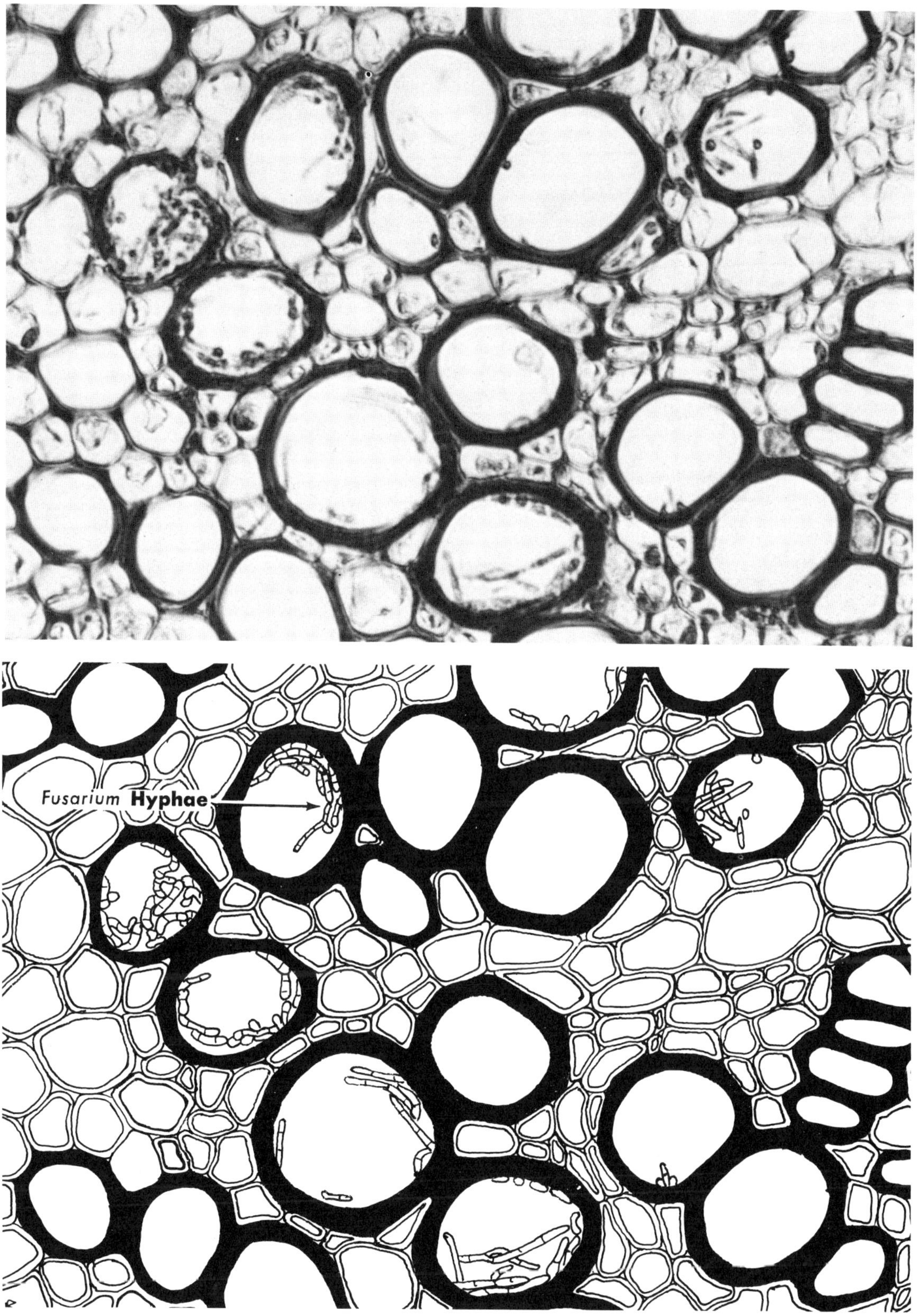

Fig. 41-4. Photomicrograph and representative drawing of hyphae of *Fusarium oxysporum* f. *lycopersici* in the xylem elements of an infected tomato plant (commercial preparation).

weeks at the start of the growing season, are a series of primary infections. Many of the first infections to appear are from infected transplants, while those showing symptoms later are predominantly the result of infection by inoculum in the soil.

The FUSARIUM wilts are generally favored by warm weather and the disease of tomato is less severe in Canada and the northern states. Cotton wilt is a serious disease and the Panama disease, or FUSARIUM wilt of banana, is crippling the industry and causing the abandonment of large acreages in Central and South America.

Control. With FUSARIUM wilts of tomato and cabbage control is limited mostly to the use of resistant varieties. Pea wilt is reduced with the use of clean seed. No satisfactory control is currently available for the disease of bananas.

Experiment XVI

Fusarium Wilt of Tomato
Isolation and Observation

Procedures for isolation of organisms from diseased tissues vary depending on the type of organism, type of tissue, and the condition of the tissue at the time of isolation. They are commonly refinements of three basic techniques:

(1) dilution of materials containing propagules
(2) tissue transplants from infected tissues containing propagules
(3) isolation of spores from diseased tissues.

Tissue transplants may be taken from the center of a lesion, or from the advancing margin where contaminants are less frequent and sometimes the pathogen more readily recovered. The success of the technique depends on the elimination of surface contaminants by sterilization, commonly with 10% Chlorox (active ingredient: sodium hypochlorite) for 10 to 20 minutes. The technique is especially useful for isolation of pathogens from fleshy tissues since a thorough surface sterilization for elimination of surface contaminants can be achieved without elimination of the pathogen. Whole sections of fleshy parts can be surface sterilized and then implanted in agar. In the case of apples, potatoes, melons, and large woody parts, the entire organ or a section of it can be surface sterilized, the outer parts removed with sterile instruments, and the inner, infected tissues implanted directly without additional treatment. Tissue transplants are also good for isolation of organisms causing systemic diseases.

EXPERIMENT

1. *Isolate* the causal organism of FUSARIUM wilt of tomato from infected plants (Fig. 41-3 A). Cut roots and stems into 1 inch segments and surface sterilize in enough 10% Clorox, with a couple drops of Tween 20, to completely cover the pieces. Tween 20 reduces surface tension and eliminates air bubbles around hairs and other small projections. The air bubbles prevent the sterilant, Chlorox, from reaching the surface, and contaminants survive. Pick up the sterilized pieces with forceps which have been sterilized in the Bunsen burner. Implant 2 pieces of root and 2 pieces of stem in the peripheral areas in each of two petri dishes of potato-dextrose agar by pushing the pieces firmly into the agar. If all are placed in proximity, contaminants surviving on one piece could overgrow the others. Implant pieces of healthy tomato plants (controls) in two other plates to demonstrate the absence of *Fusarium*.

Fig. 41-5A. A simple microtome.

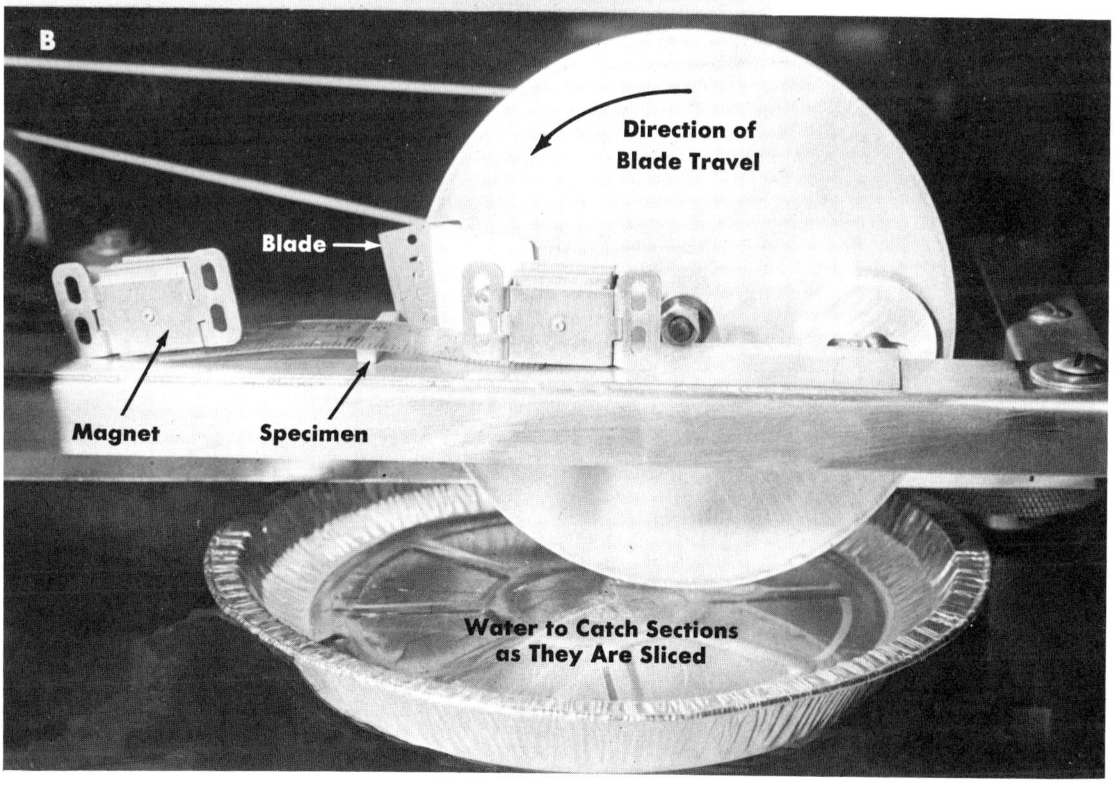

Fig. 41-5B. Side view of simple microtome setup.

2. *Observe* darkening of the vascular tissues by sectioning stems (Fig. 41-3 C). Cross sections, made by hand or with a microtome for fresh material (Fig. 41-5), allow for observation of the darkening and the presence of causal organism hyphae in the xylem elements (Figs. 41-3 B, 41-4).

Expendable Supplies × Total Number of Students or Groups

1 healthy tomato plant (Fig. 41-3 A)
1 *Fusarium*-infected tomato plant (Fig. 41-3 B)
4 petri dishes of potato-dextrose agar

General Supplies Available

Clorox
1 graduate cylinder, 100 ml.
a few 250-ml. beakers
Tween 20

DISCUSSION QUESTIONS

1. What are the advantages and limitations of transplant techniques for isolation of pathogens? For what kinds of diseases is this technique useful?
2. What is the value of implanting sterilized pieces of healthy tomatoes in the experiment?

Experiment XVII

Fusarium Wilt of Tomato: Effect on Transpiration

The systemic vascular diseases characteristically exhibit wilting caused by vascular dysfunction. The degree of dysfunction can be measured by the degree of decrease in transpiration.

EXPERIMENT

Measure the transpiration rate of a healthy and a *Fusarium*-infected tomato plant with a potometer (Fig. 41-6). Cut the stems below water and retain the cut ends under water for a few minutes. Insert the cuttings into a water-filled potometer. Be sure that the bottom of the cutting is below the surface of the water. Mark a line with a wax pencil near the end of the capillary tube and adjust the meniscus to this line by turning the clamp. Record the distance that the meniscus moves each 1 or 2 minutes and plot graphs on graph paper provided at the end of this experiment.

Expendable Supplies × Total Number of Students or Groups

1 healthy tomato plant
1 *Fusarium*-infected tomato plant

Nonexpendable Supplies × Total Number of Students or Groups which are Seated in a Laboratory Section

1 potometer

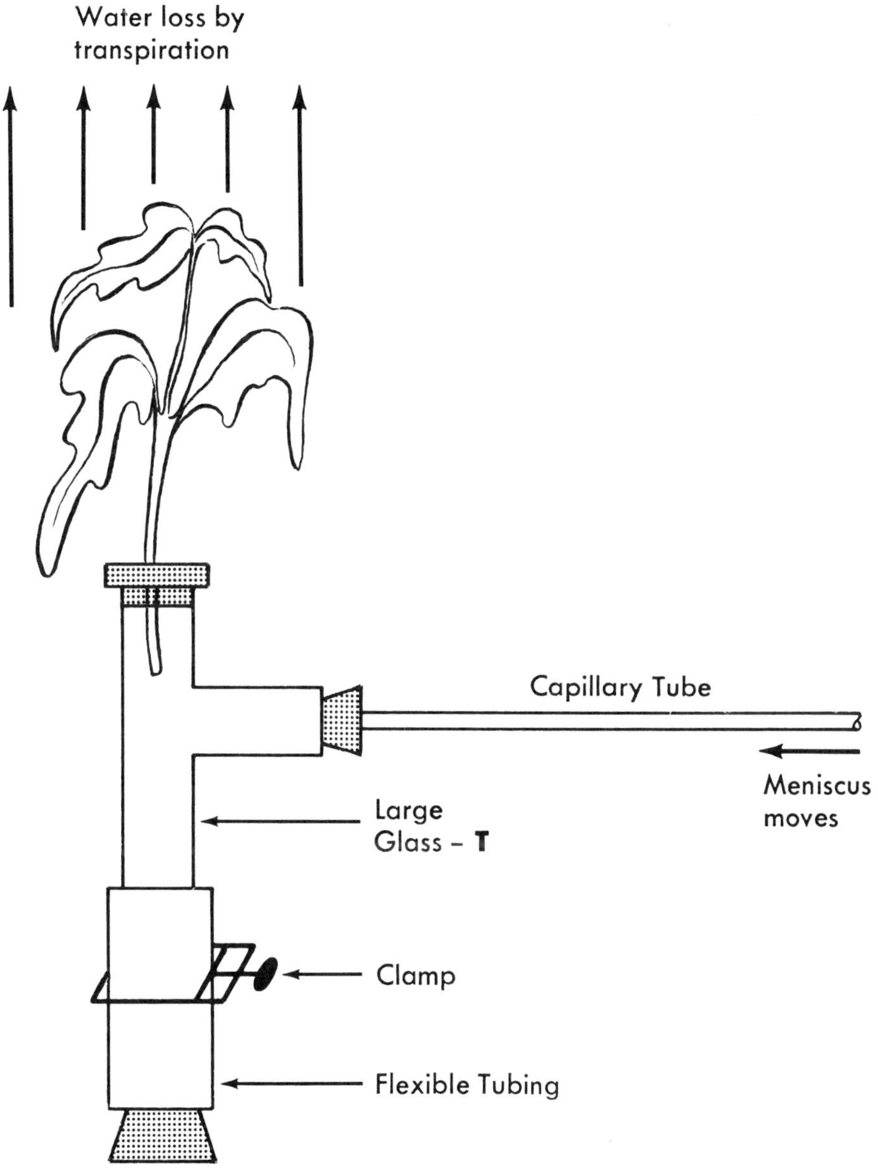

Fig. 41-6. Potometer for measurement of transpiration rate.

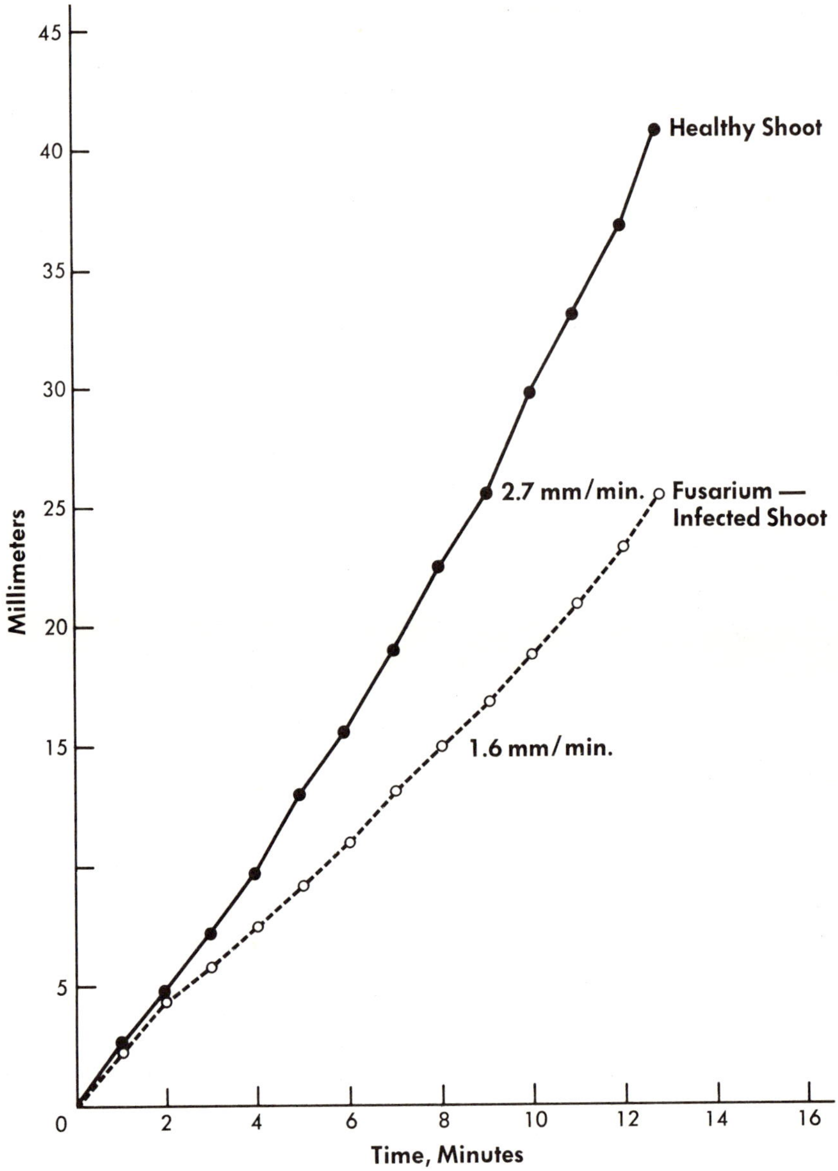

Fig. 41-7. Graph of transpiration rate of both a healthy and a *Fusarium*-infected shoot.

DISCUSSION QUESTIONS

1. Why should the plants be cut below the surface of water?
2. Why must the cut stem of the plant be below the surface of the water in the potometer?
3. What would the graphs look like if you raised the temperature? Why? What would they look like if you placed a fan nearby? Why?
4. What is the significance of a reduced rate of transpiration?
5. What happens to the meniscus if the lamp heats the water in the potometer?

42

Verticillium Wilt

VERTICILLIUM wilt is an example of a true vascular wilt disease (Fig. 42-1,-2). (See Chapter 41 for additional discussions on wilt diseases.)

Causal Organism. *Verticillium* has no known perfect stage. The life cycle is an endless repetition of an asexual reproductive cycle (Fig. 42-5). The genus, *Verticillium* is recognized by the hyaline, branched conidiophores which are verticillate, or whorled. To be whorled, some nodes must have three or more side branches. The conidia are small, hyaline, one-celled, and are borne successively on the tips of the conidophore branches. Balls of conidia occur in a sticky, water soluble matrix at the tips (Fig. 42-3). Some forms of *Verticillium* produce **microsclerotia** (Fig. 42-4) which are small (*micro*) compared to the large sclerotia of *Sclerotinia, Botrytis, Monilinia,* and *Claviceps.* Some authorities distinguish *albo-atrum* and *dahliae* as two distinct species, primarily on the basis of interpretation of sclerotia and daurmycelium (durable of perennial mycelium) in the original research papers and an application of these differences. The more liberal interpretation of lumping them into one species, *albo-atrum,* is accepted here with the understanding that some may or may not produce microsclerotia daurmycelium.

Disease Cycle. Penetration is usually through wounds in roots. Large numbers of parasitic nematodes, especially the dagger nematodes, increase the incidence of VERTICILLIUM wilt, presumably by providing increased numbers of penetration points. The fungus invades the xylem elements of the vascular system and progressively grows upward into aerial portions of the plant—even if its a tree. Rapid upward movement may be achieved in some plants by the passage of the small conidia up the xylem with the transpiration stream. They lodge farther up, produce small colonies which produce more conida which move quickly upward again. Secondary cycles are not dominant since the secondary inoculum is generally not released into the soil until the plant dies. Technically, infections higher up in the plant resulting from conidia produced and released farther down are secondary infections. Typically, the series of infections observed in a field during a season represent a series of primary infections from inoculum in the soil.

VERTICILLIUM wilt is not necessarily a disease of warm climates but symptoms are more pronounced with the onset of warm weather since this accentuates the wilting.

Verticillium albo-atrum is a good soil inhabitant and overwinters as dormant mycelium and microsclerotia. Microsclerotia are viable for many years in the soil. It has long been assumed that the source of inoculum first be a solanaceous crop such as tomato, potato, or eggplant. Indeed, these crops are very susceptible and tend to increase the inoculum level in the soil. However, it appears that *Verticillium* occurs in uncultivated areas and that the organism is native to our soils. We should view the population of *Verticillium* in the native, uncultivated areas as a great reservoir of variability and potential pathogenicity to crops.

Fig. 42-1. VERTICILLIUM wilt of strawberry showing response of a susceptible and resistant variety and the typical symptoms.

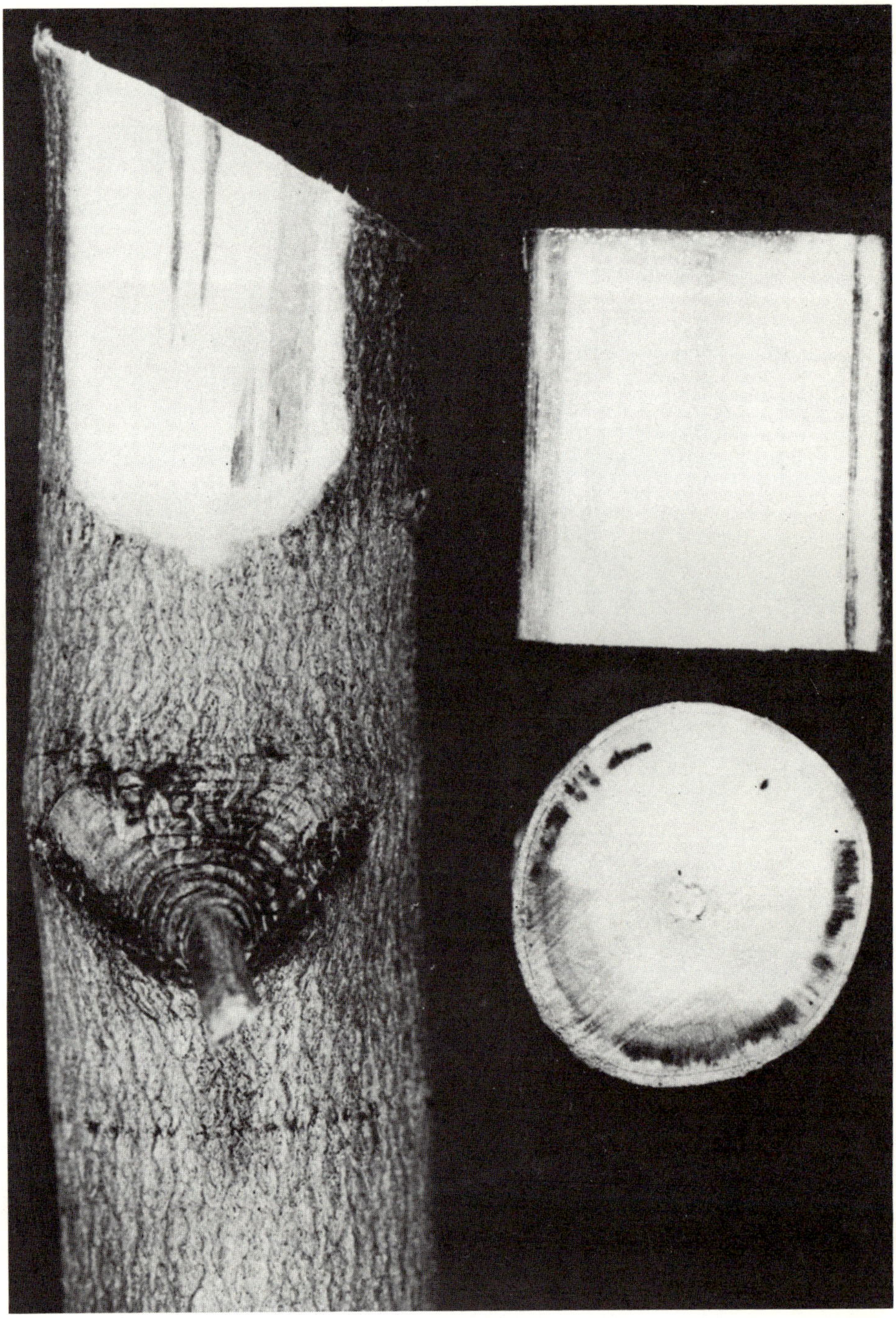

Fig. 42-2. VERTICILLIUM **wilt of maple showing the green and brown streaks in longitudinal and cross sections of the wood.**

Diagnosis. In none of the diseases caused by *Verticillium albo-atrum* can a positive diagnosis be given on symptoms alone. Isolation of the organism from diseased tissue is required in order to offer a positive diagnosis (Fig. 42-4 E). We should otherwise describe a diagnosis, unsupported by isolation attempts, as provisional.

Control: achieved with disease-free plants, resistance, and avoidance of planting susceptible crops where solanaceous crops have been repeatedly cultivated. For high value crops, chemical fumigation of the soil can be profitable.

Fig. 42-3. Conidia and conidiophores of *Verticillium albo-atrum*.

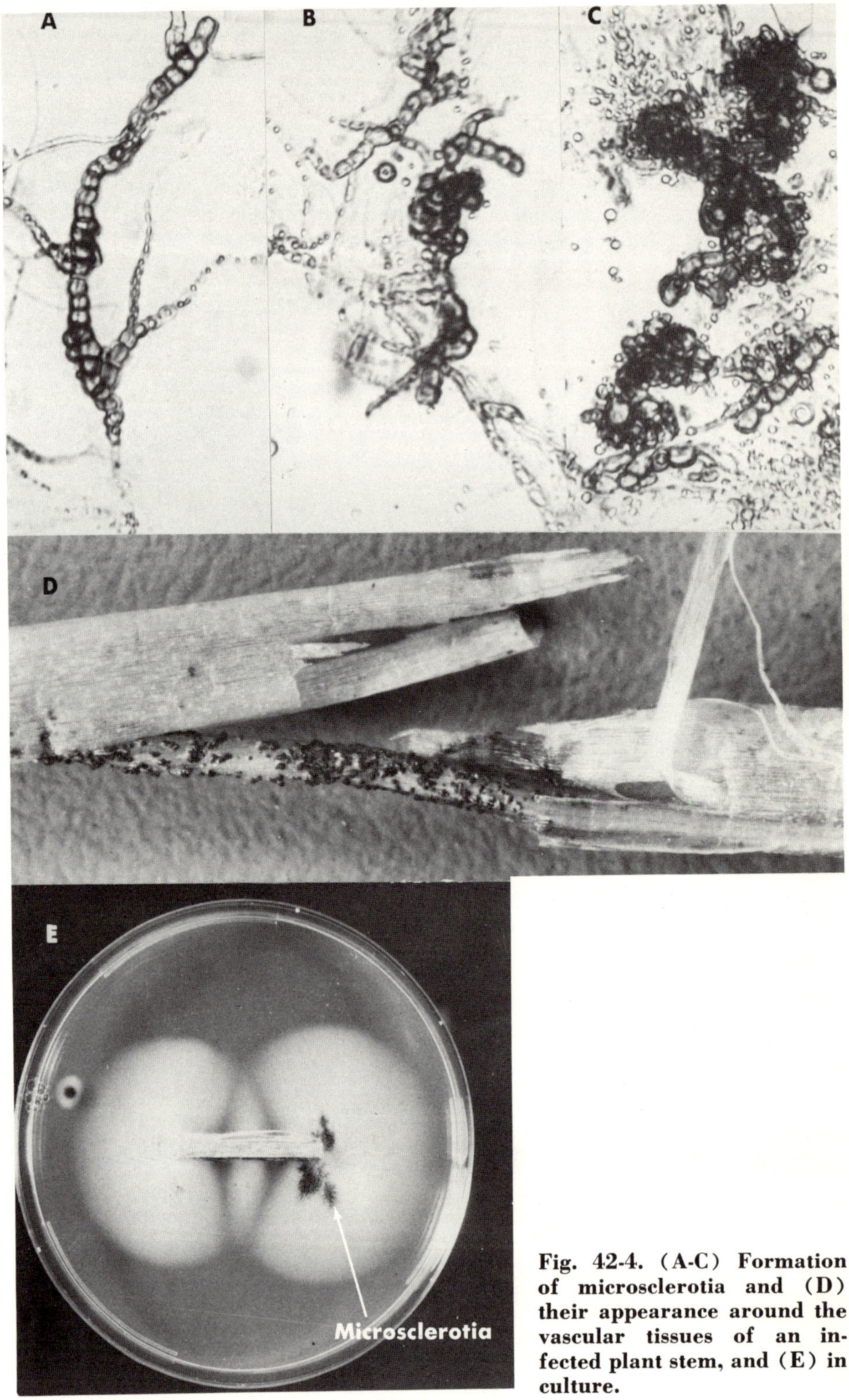

Fig. 42-4. (A-C) Formation of microsclerotia and (D) their appearance around the vascular tissues of an infected plant stem, and (E) in culture.

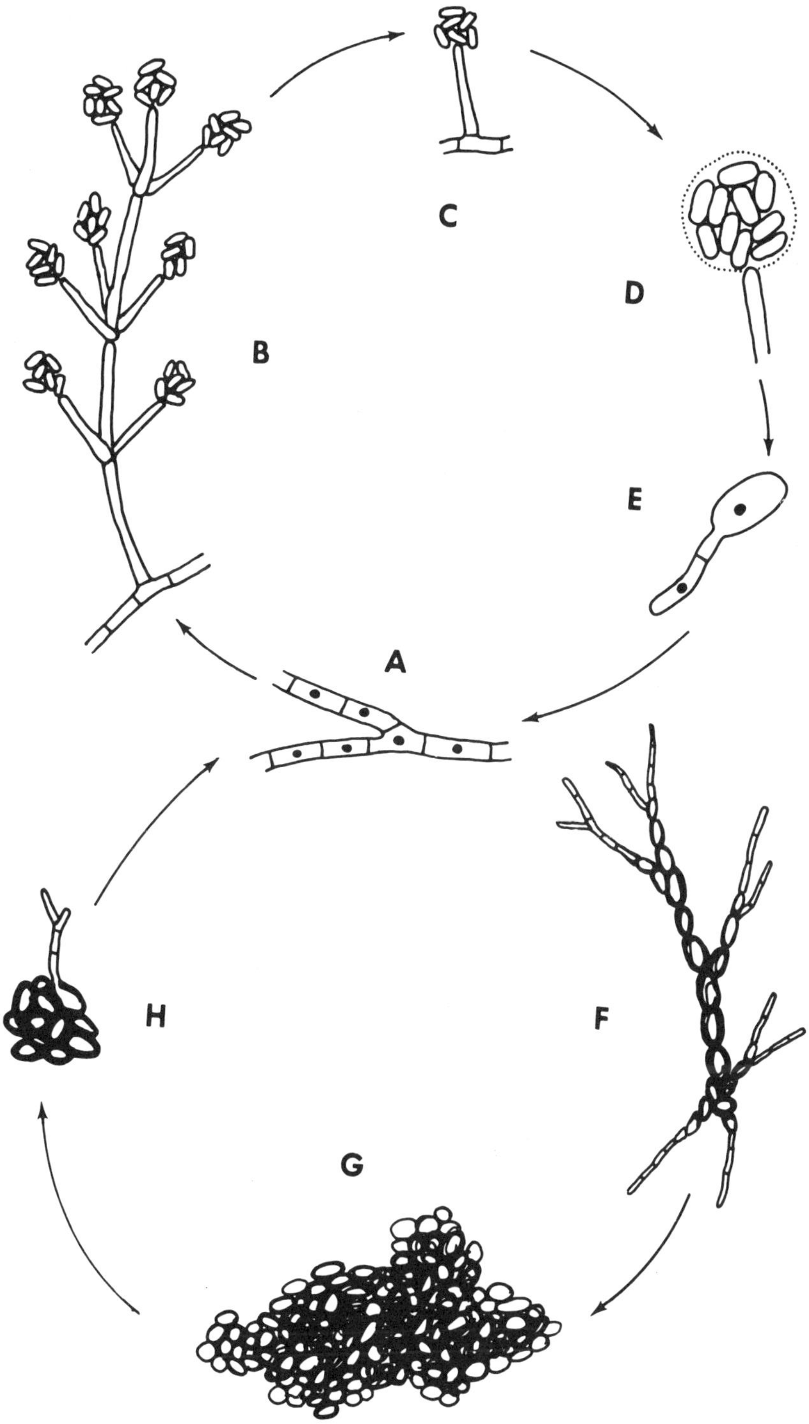

Fig. 42-5. Life cycle of *Verticillium albo-atrum*.

43

Anthracnose

The true anthracnose diseases are caused by species of *Colletotrichum* or *Gleosporium* (Fig. 43-1,-2,-3,-4). The only difference between these two genera is defined to be the occurrence of dark setae (hairs) in the acervuli of *Colletotrichum* but not *Gleosporium*. This is a variable characteristic and probably not a valid difference and therefore the two genera should be combined. Species of *Gnomonia, Glomerella,* and *Elsinoe,* which are Ascomycetes, typically produce asexual spores in the form-genus *Colletotrichum* (*Gleosporium*). Perfect stages have not been found for many species of *Colletotrichum* (*Gleosporium*). Some examples are:

Imperfect Stage	Perfect Stage	Host	Figure
Gnomonia veneta	*Gleosporium nervisequum*	sycamore	30-1
Gnomonia caryae	*Gleosporium*	hickory	30-3
Glomerella lindemuthiana	*Colletotrichum lindemuthianum*	bean	43-1
Glomerella cingulata	....	horsechestnut	
....	*Gleosporium musarum*	banana	
Glomerella glycines	*Colletotrichum glycines*	soybean	
Glomerella gossypii	*Colletotrichum gossypii*	cotton	
Glomerella lagenaria	*Colletotrichum lagenarium*	gourd	43-3
Gnomonia quercina	*Gleosporium quercinum*	oak	
Physalospora tucumanensis	*Collectotrichum falcatum*	sugarcane	
....	*Colletotrichum phomoides*	tomato	43-4

Bean Anthracnose. *Colletotrichum lindemuthianum* exhibits physiological specialization: three forms, *alpha, beta,* and *gamma,* are recognized, and a fourth, *delta,* has been described. The disease is worldwide and of most significance in humid, temperate zones. It is absent, or nearly so, from most surface irrigated areas of the western United States. The use of disease-free seed and resistant varieties has greatly reduced the importance of this disease. The bacterial and viral diseases now appear to be more significant in the production of beans.

Causal Organism. Hyphae are branched and septate. Conidia are hyaline, oblong, and are borne on short, erect conidiophores in acervuli (Figure 43-2). The conidia occur in pink masses in the acervuli.

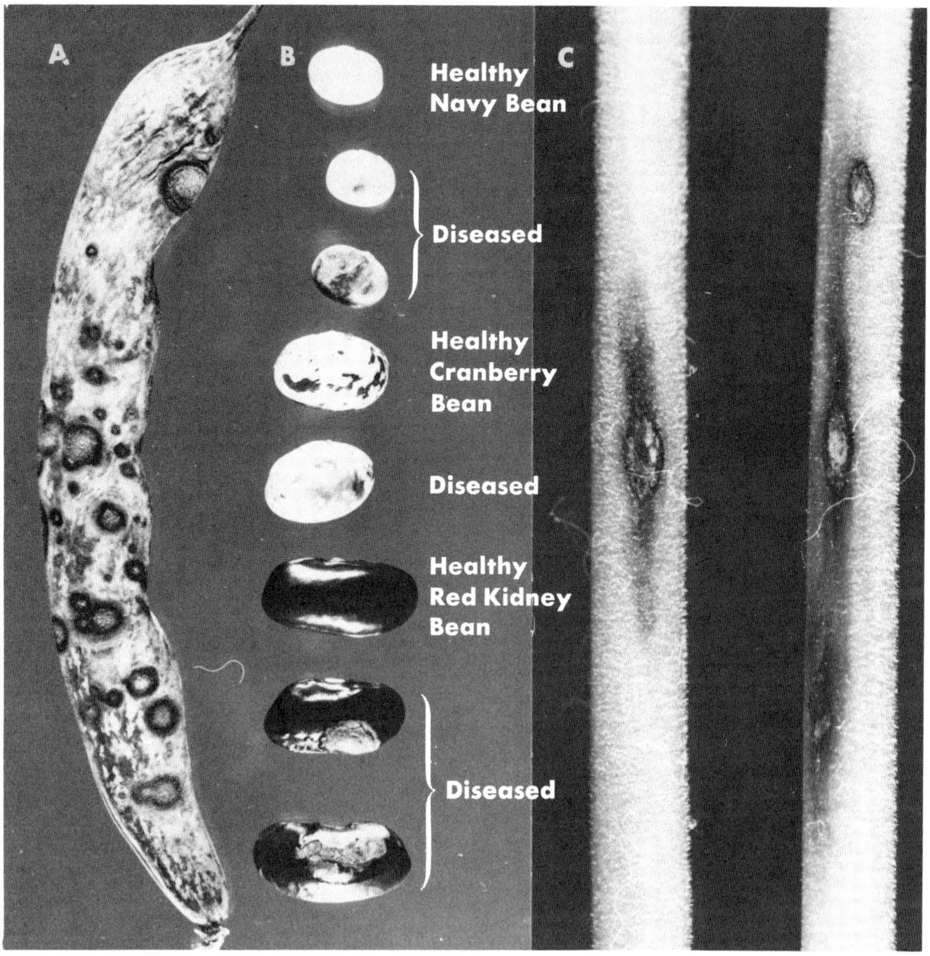

Fig. 43-1. Anthracnose of bean on (A) pods, (B) seeds, and (C) stems.

Disease Cycle. The fungus overwinters in infected plant debris and seed, and cannot survive in soil in the absence of bean refuse. Conidia are produced in great abundance in acervuli on overwintered refuse and lesions on cotyledons from infected seed. These conidia constitute the primary inoculum. Water is essential for release of the conidia from the matrix which holds them together. Conidia are spread by splashing and blowing rain and produce germ tubes which penetrate the host tissues directly.

Control: usually sufficient with disease-free seed and the use of resistant varieties. Three-year rotations and plowing under refuse can help.

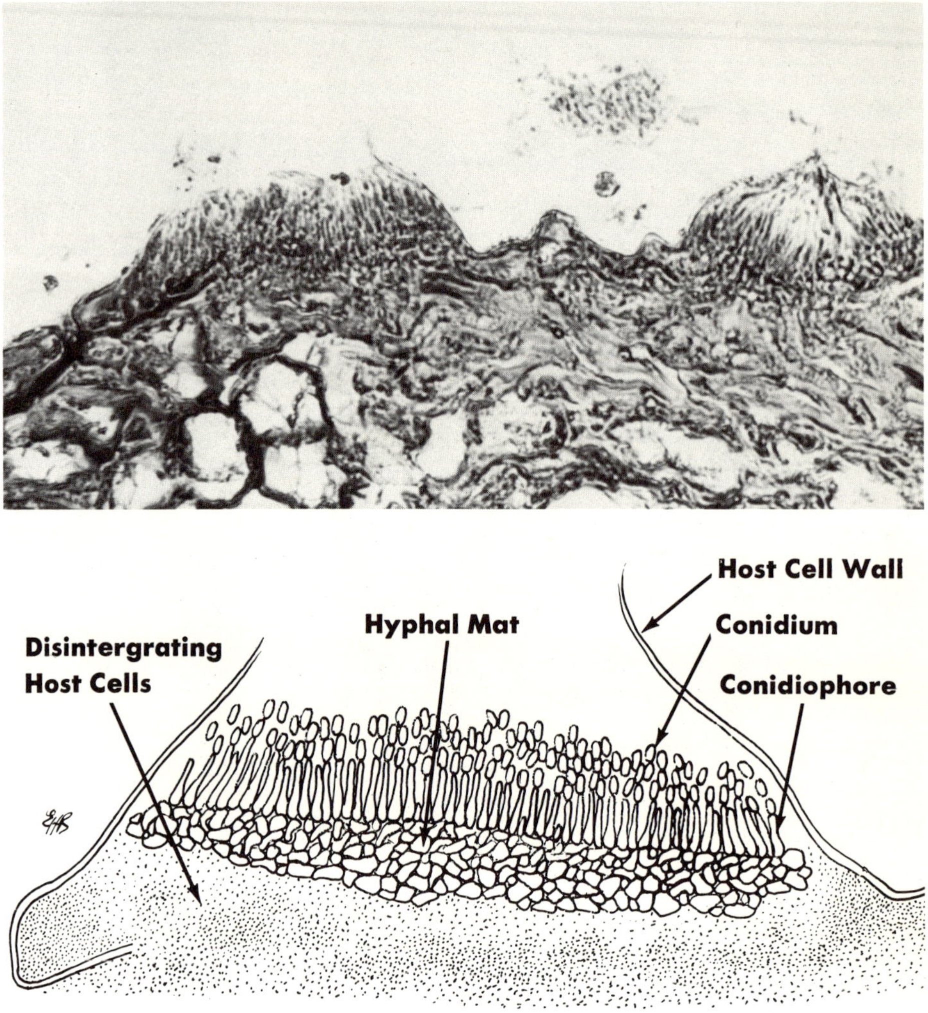

Fig. 43-2. Photomicrograph and reconstructed drawing of an acervulus of *Colletotrichum lindemuthianum*.

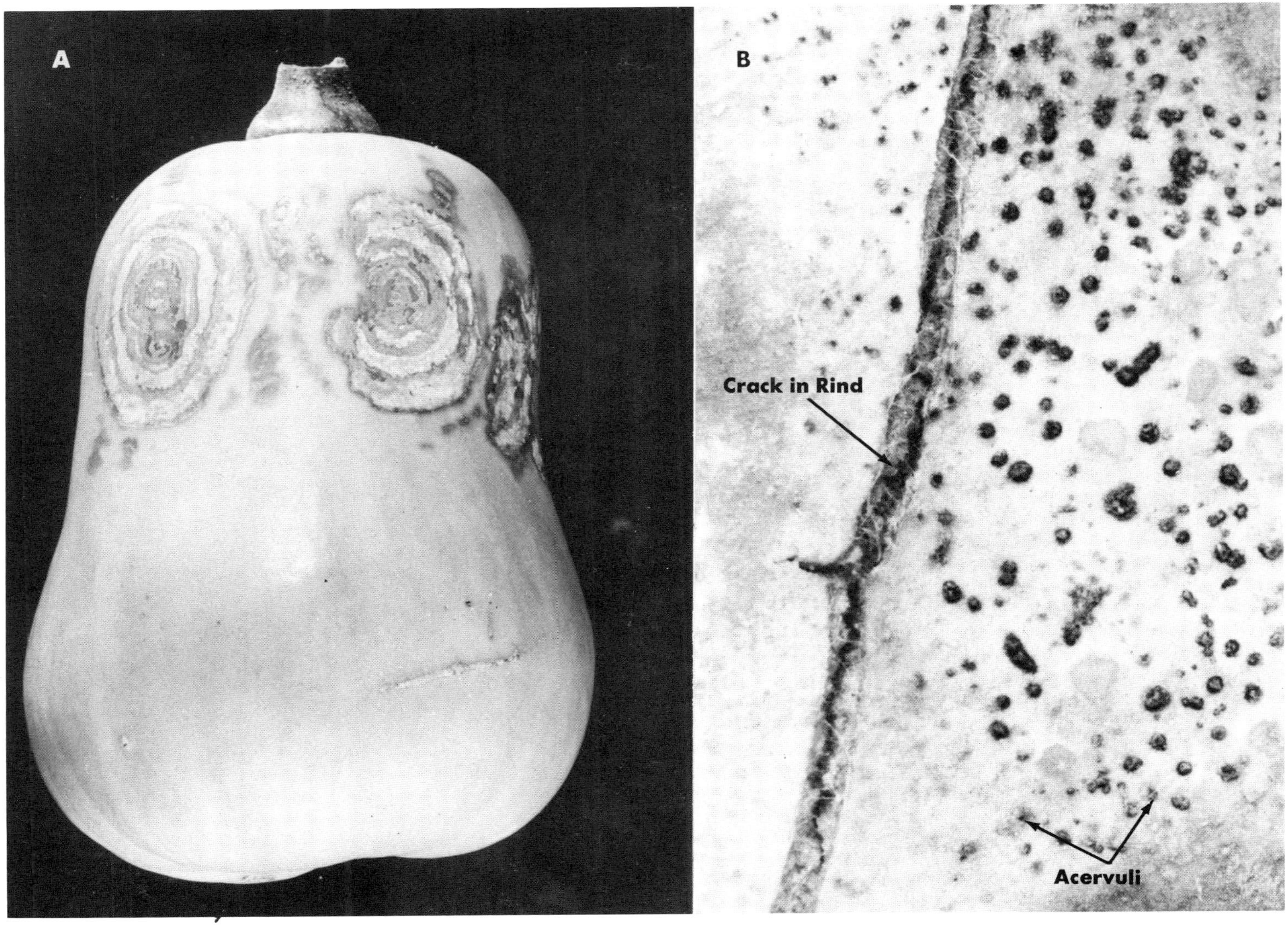

Fig. 43-3. (A) Anthracnose of gourd. (B) Closeup showing acervuli.

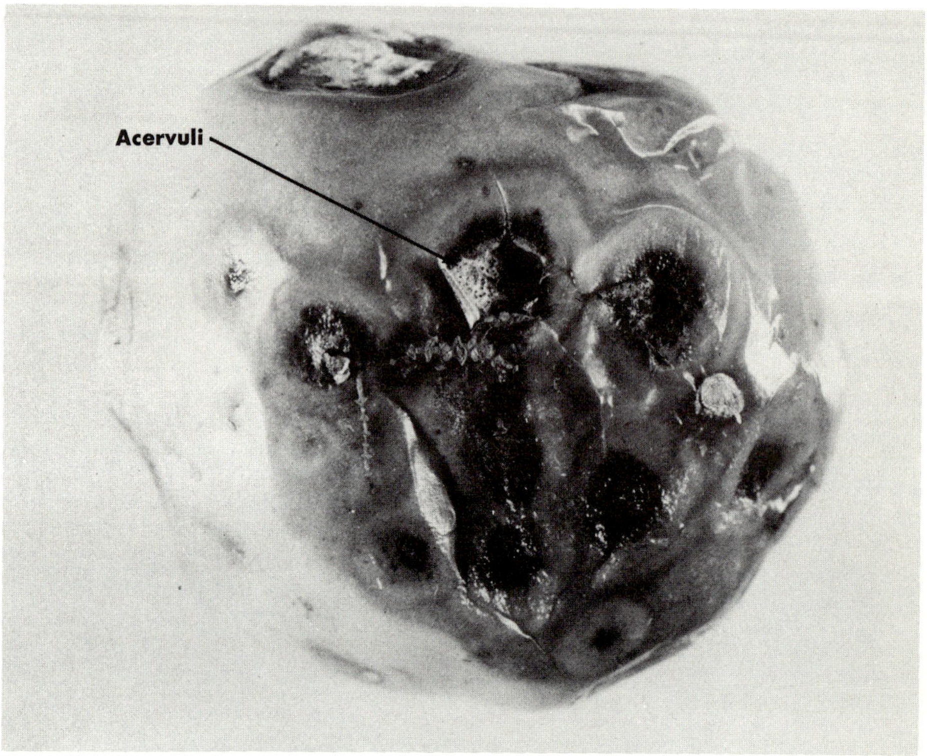

Fig. 43-4. Anthracnose of tomato.

44

The Basidiomycetes

The common mushrooms of the field, the golf course, and the steak platter are Basidiomycetes. So too are the puffballs, shelf fungi, rusts, and smuts. These fungi produce basidia and basidiospores. Characteristic of most species are clamp connections, secondary and tertiary mycelium, and the dolipore septum. The life cycles are generally dominated by the dikaryotic phase.

The sub-class Heterobasidiomycetidae includes the rust fungi (Uredinales) and the smut fungi (Ustilaginales). These exhibit few clamp connections and no tertiary mycelium. The Homobasidiomycetidae includes the pore fungi (Polyporales) and the gill fungi (Agaricales).

Outline of Genera Treated

- Heterobasidiomycetidae (*Sub-Class*)
  - Uredinales (*Order*)
    - Pucciniaceae (*Family*)
      - *Puccinia, Gymnosporangium*
    - Melampsoraceae (*Family*)
      - *Cronartium*
  - Ustilaginales (*Order*)
    - Ustilaginaceae (*Family*)
      - *Ustilago*
    - Tilletiaceae (*Family*)
      - *Tilletia, Urocystis*
- Homobasidiomycetidae (*Sub-Class*)
  - Polyporales (*Order*)
  - Agaricales (*Order*)

The Rusts

The rust fungi are obligate parasites expressing both simple and complex life cycles. Haustoria are frequent, but clamp connections are uncommon. These fungi typically produce five distinct reproductive stages with five different spore forms, some of which are parasitic on one host while the others are parasitic on an alternate host. This alternation of hosts can be obligate for completion of the life cycle. All produce teliospores and basidiospores. These five stages, often referred to by roman numeral, and their nuclear condition are:

Stage	Structure	Spore	Nuclear Condition
0	Spermagonia bearing	spermatia and receptive hyphae	haploid
I	Aecia bearing	aeciospores	dikaryotic
II	Uredia bearing	uredospores	dikaryotic
III	Telia bearing	teliospores	dikaryotic, later diploid
IV	Promycelia bearing	basidiospores	haploid

There are two general life cycle patterns: (1) long-cycled or **macrocyclic** species and (2) short-cycled or **microcyclic** species. Long-cycled rusts produce at least one binucleate spore in addition to the teliospore, whereas the teliospore is the only binucleate spore produced by the short-cycled rusts. Those which are long-cycled but lack the uredial stage are called **demicyclic.** They can be outlined as follows (stage IV can be written as superscript to stage III since basidiospores are produced in the rust fungi only by germination of the teliospore):

Stages	Type of Cycle
0, I, II, III^{IV}	macrocyclic
0, I, III^{IV}	macrocyclic (demicyclic)
0, III^{IV}	microcyclic
III^{IV}	microcyclic

Spermagonia were once called pycnia and the spermatia, pycnospores. Now that the function of the spermatia in the fertilization process is known, spermatia is the proper term. Two general types of spermagonia occur. Four distinct kinds of aecia occur and are used as form-genera in an artificial classification system for the rust fungi when the sexual stage, the telia, are not known or found. The system is parallel to that used for the Fungi Imperfecti. They are ***Aecidium*** as in *Puccinia* (Fig. 46-2,-3), ***Roestelia*** as in *Gymnosporangium* (Fig. 48-2,-3), ***Peridermium*** as in *Cronartium* (Fig. 49-1,-2) and ***Caeoma.***

The rust fungi which complete their life cycles on a single host are **autoecious** (same household) and those which require two different, or alternate hosts for different spore forms are **heteroecious.** The plant on which the spermagonia and aecia develop is the **aecial host** and the plant on which the uredia and telia develop is the **telial host.** All microcyclic rust fungi are autoecious since the basidiospores are the only infectious spores produced during the life cycle. A relatively obscure precept, called **Tranzschel's Law** after its originator, states that a short-cycled species occurs on the aecial host of the long-cycled parent. This recognizes that the most reduced rust fungi retain teliospores and basidiospores and that the basidiospores are the only infectious spores of the two kinds. Thus, the only host that can be infected must be the host on which they are produced. In the long-cycled rusts, these infections give rise to aecia. In the short-cycled rusts, no aecia are produced, only telia. In some of these short-cycled rusts, the telia look like aecia and the teliospores like aeciospores. Only by germination to determine if basidiospores are formed can one recognize that they are, in reality, teliospores and not aeciospores. Tranzschel's Law embodies the concept that the short-cycled species mimic their long-cycled parent by occuring on the aecial host, sometimes looking like aecia, and are systemic if the parents were.

Control: provided with host resistance (stem rust of wheat), chemical sprays (cedar-apple rust), removal of alternate hosts (white pine blister rust), and destruction of overwintering inoculum (hollyhock rust).

46

Stem Rust of Wheat

An interesting question to pose is: What is the most important plant disease? Your answer may depend on whether you live in Kansas City or Hong Kong, eat rice or wheat, or make money selling cotton. Of greatest significance are the diseases which cause the greatest losses of our most important food crops such as rice and wheat. History records this with early investigations on diseases of rice in Japan, cabbage in Russia, potatoes in Ireland, and wheat and barley in Europe. Important plant diseases were noted and respected for their significance in even earlier times. For example, pagan celebrations were held in honor of the God Rubigo during the days of the Caesars in Rome, for this god brought about the rust of wheat.

Causal organism. Stem rust diseases of wheat and other grass species are caused by *Puccinia graminis.* Although morphologically uniform, the fungi which cause these diseases are restricted in pathogenicity to a single species of host or closely related hosts. These pathogenically distinct groups are designated by trinomials such as *Puccinia graminis* f. *tritici* on wheat and *Puccinia graminis* f. *hordei* on barley. Further specialization on pathogenicity is exhibited by each form. These further restrictions in host range are designated as races. There are hundreds of different races of *Puccinia graminis* f. *tritici.* These races and the resistance to them in the host are determined by numerous genes in both host and parasite. It is believed that most all highly specialized host-parasite combinations are governed by a gene-for-gene relationship as elucidated with the flax rust.

Disease Cycle. In northern latitudes, the fungus overwinters as teliospores. They germinate in the spring to produce a hypha which is not unlike a germ tube of limited length and called the promycelium. Cross walls are laid down which divide the promycelium into four cells, each of which produces a sterigma on which a single hyaline, one-celled, uninucleate basidiospore is formed (Fig. 46-6 A, E). The basidiospores infect the barberry plants by direct penetration and can infect only the barberry. They constitute primary inoculum for the barberry and for the barberry only. There is no secondary inoculum for this host. Infection of the barberry produces spermagonia (Fig. 46-1) on the upper surface and aecial initials (beginnings) on the lower surface (F). The fungus is heterothallic so that spermatia can fertilize only receptive hyphae of the opposite and compatible strain (G). The nucleus of the spermatium enters the receptive hypha (H) and multiplies and migrates downward into the aecial initials (I). Some of these cells are dikaryotized, or made $(n + n)$ (J). The aecium develops into a cup-shaped structure (K) with chains of aeciospores (Fig. 46-2, 46-3). The aeciospores and most of the cells of the aecium are dikaryotic but some of the basal cells of the aecium remain uninucleate.

Aeciospores (L) can infect only the wheat host and constitute primary inoculum for the wheat. Penetration is not direct as with the basidiospores and the barberry,

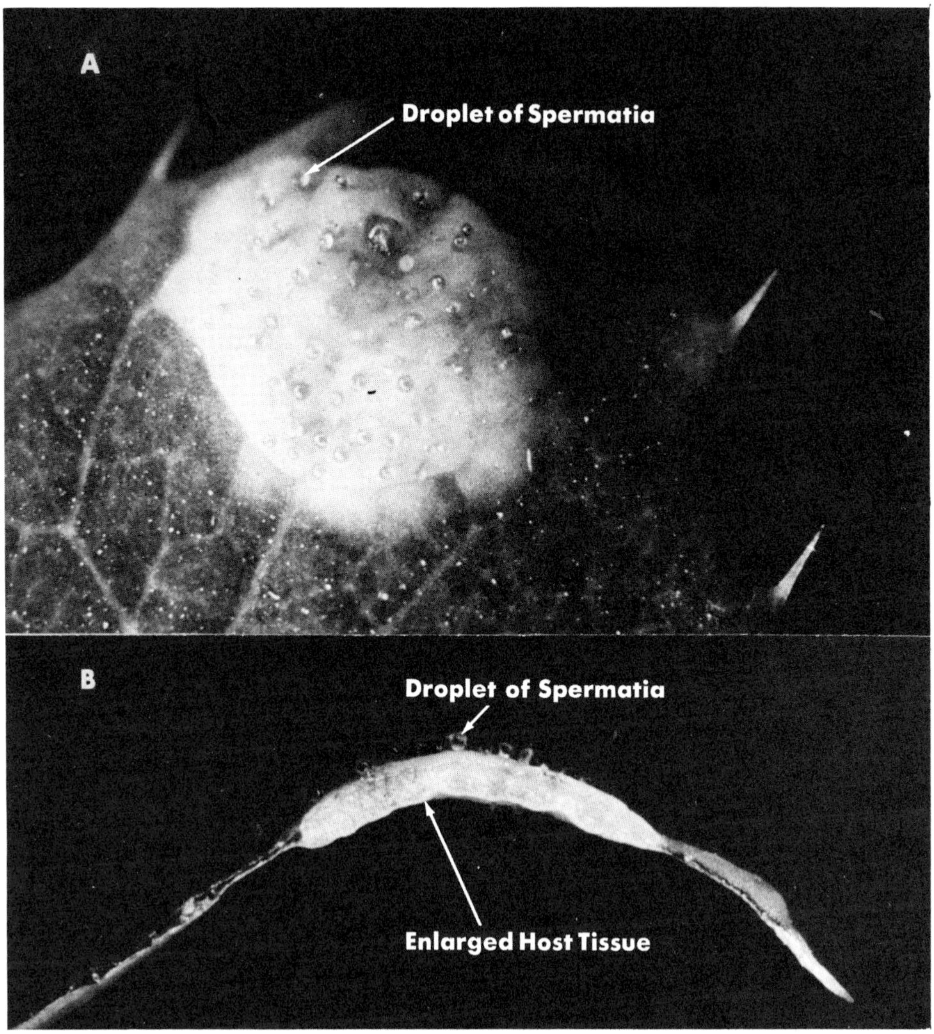

Fig. 46-1. (A) Top and (B) side views of spermagonia of *Puccinia graminis* on barberry.

but through the stomata. Soon after infection, uredia (M) are produced (Fig. 46-4, B, C). Uredia produce uredospores (N, O) which can infect only the wheat host. They reinfect the wheat and constitute secondary inoculum. Several generations of uredospores can be produced in a single growing season creating a rapid repeating secondary cycle for multiplication of the disease.

In southern climates of the United States and Mexico, uredospores survive the warm dry summers and summerover. They constitute overwintering inoculum and primary inoculum for the wheat. Both aeciospores and uredospores, therefore, may constitute primary inoculum depending on the climate of the area and if the uredospores can overwinter. Later in the season, teliospores are produced among uredospores in some uredia and in telia which contain only teliospores (Fig. 46-4 A, C, 46-5). They have a required dormancy period and are unable to germinate after formation until a cold period has passed. Teliospores do not infect any host. They

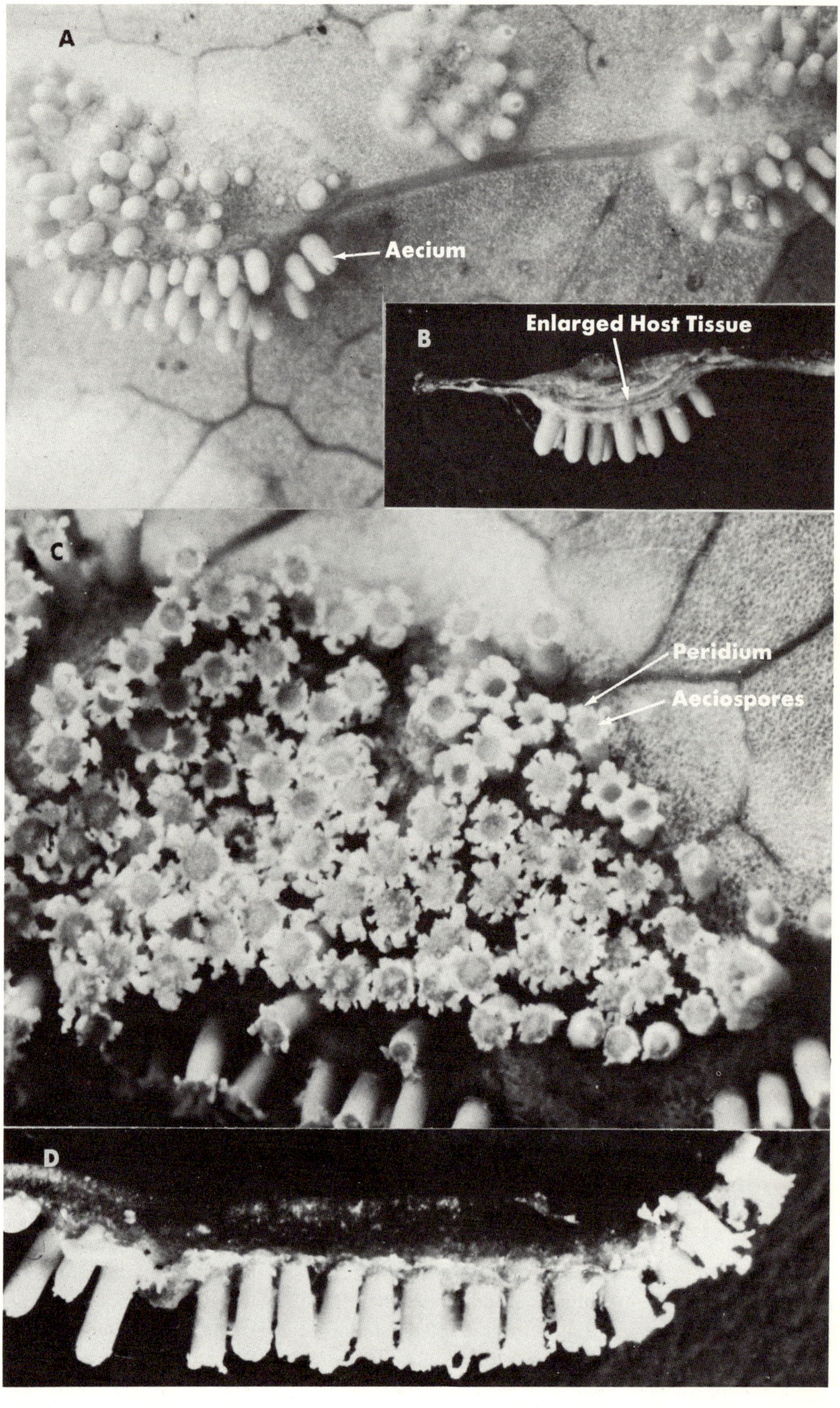

Fig. 46-2. Top (A) and side (B) view of immature aecia and top (C) and side (D) view of mature aecia of *Puccinia graminis*. See Fig. 46-3 for photomicrograph of a cross section through a specimen such as Fig. 46-2 B.

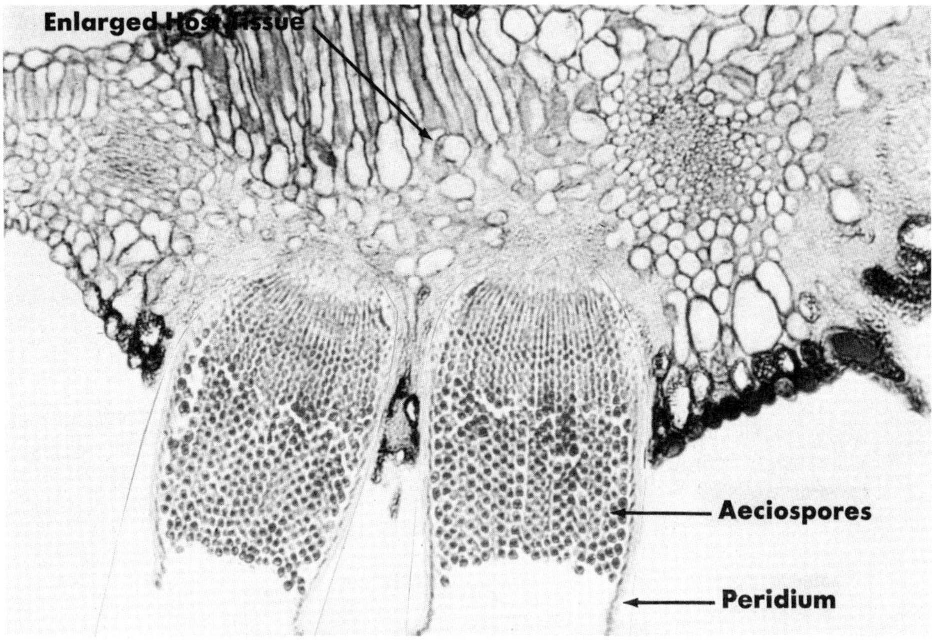

Fig. 46-3. Photomicrograph of a cross section of two aecia of *Puccinia graminis* on barberry. This is a cross section through a specimen such as exhibited in Fig. 46-2 B. Note the greatly enlarged host tissues, especially the upper palisade cells.

germinate after the dormant period by issuing a promycelium which gives rise to basidiospores which infect the barberry host.

Control. The disease is controlled by the use of resistant varieties. Because uredospores produced in southern areas are blown northward and produce new infections which produce more uredospores which are blown northward, there is a leap-frogging effect from Mexico, through the great plains of the United States, into Canada. When a race of the fungus occurs which can attack the previously resistant varieties, epidemics occur. These so-called new races are brought about by genetic recombinations and mutations. Once they occur, they tend to increase rapidly since their host, the wheat, is resistant to other infections of the same organism. These resistant varieties, in other words, select the new races which can attack them. This has been called man-guided evolution. The suggested reading is along this line.

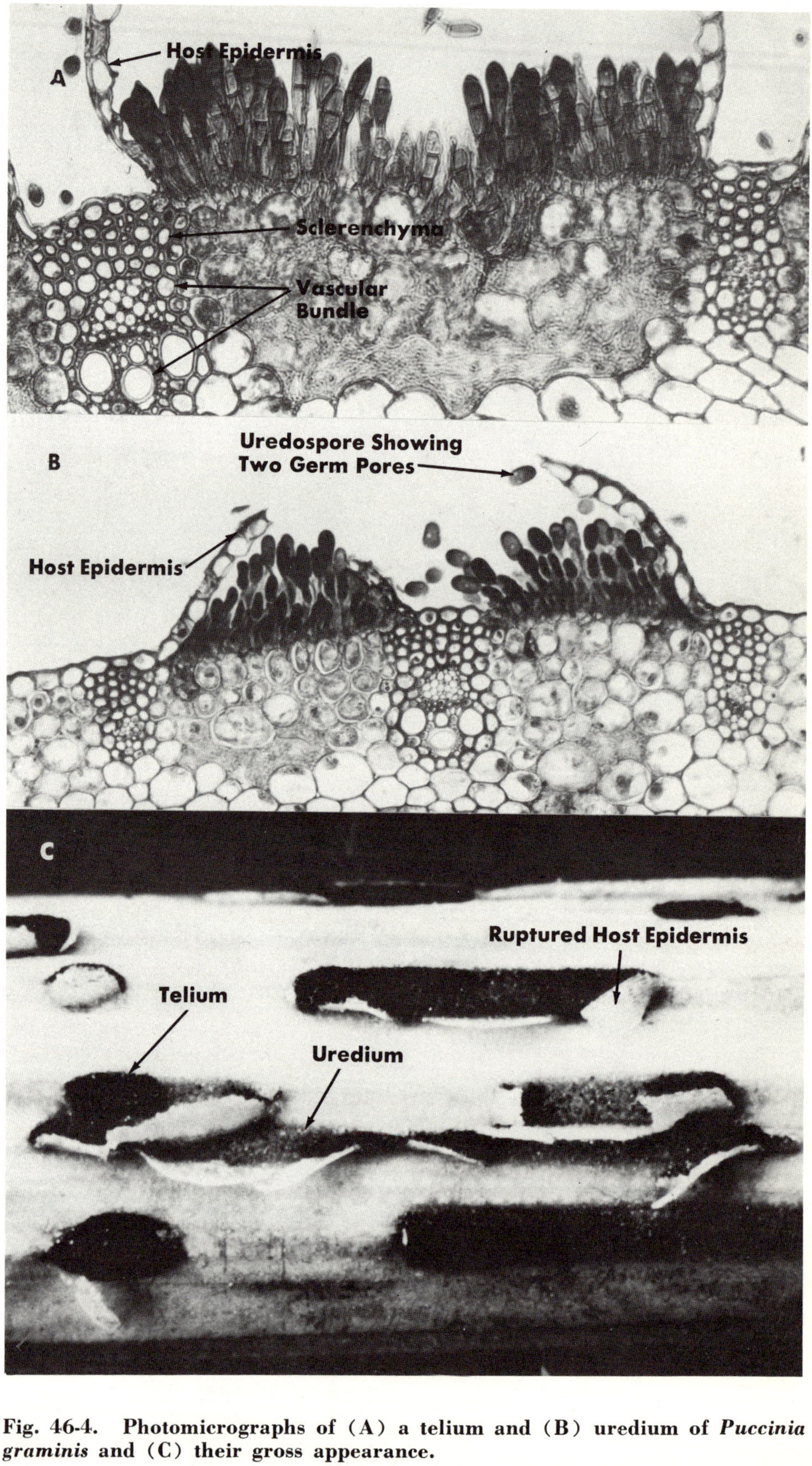

Fig. 46-4. Photomicrographs of (A) a telium and (B) uredium of *Puccinia graminis* and (C) their gross appearance.

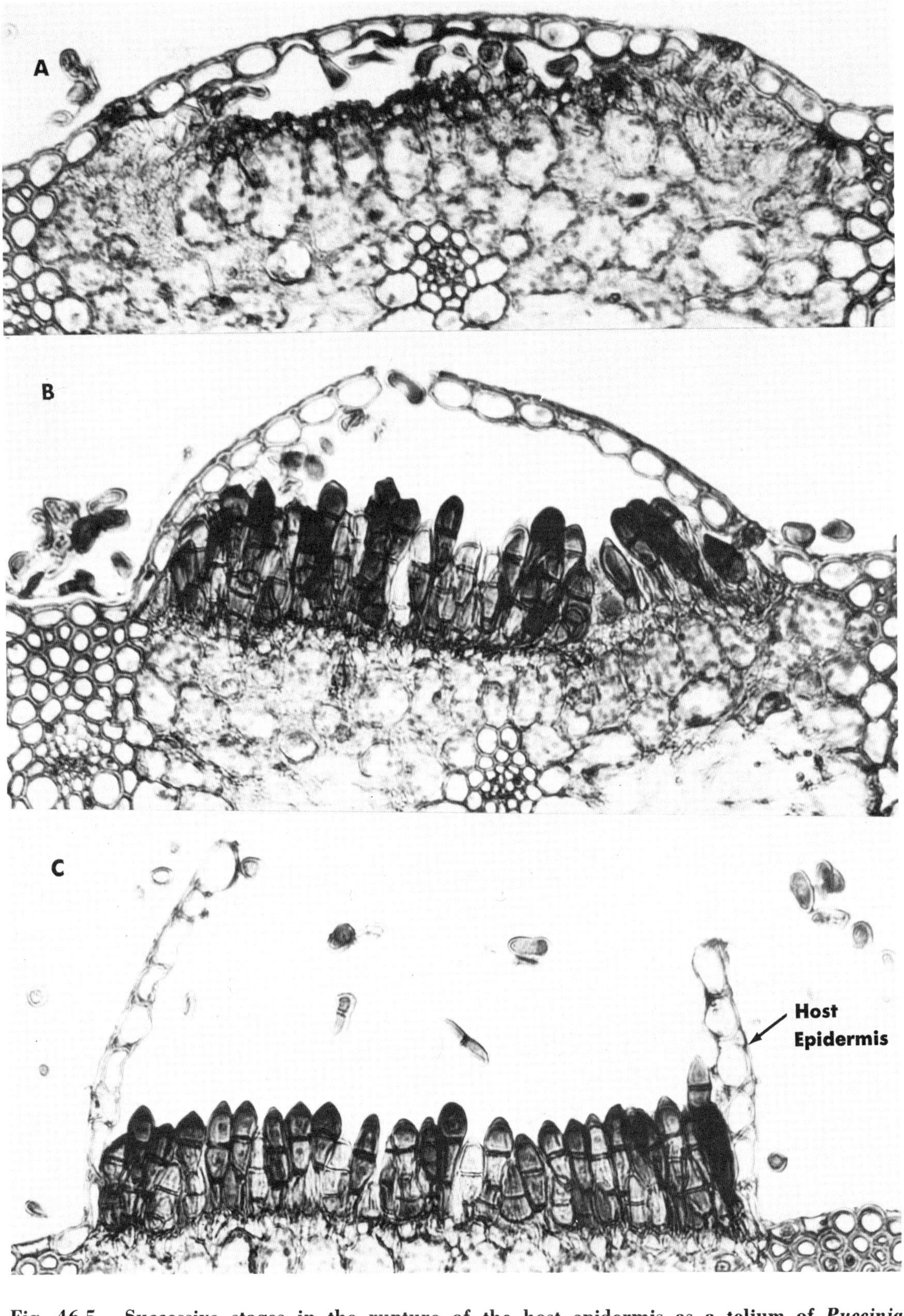

Fig. 46-5. Successive stages in the rupture of the host epidermis as a telium of *Puccinia graminis* matures.

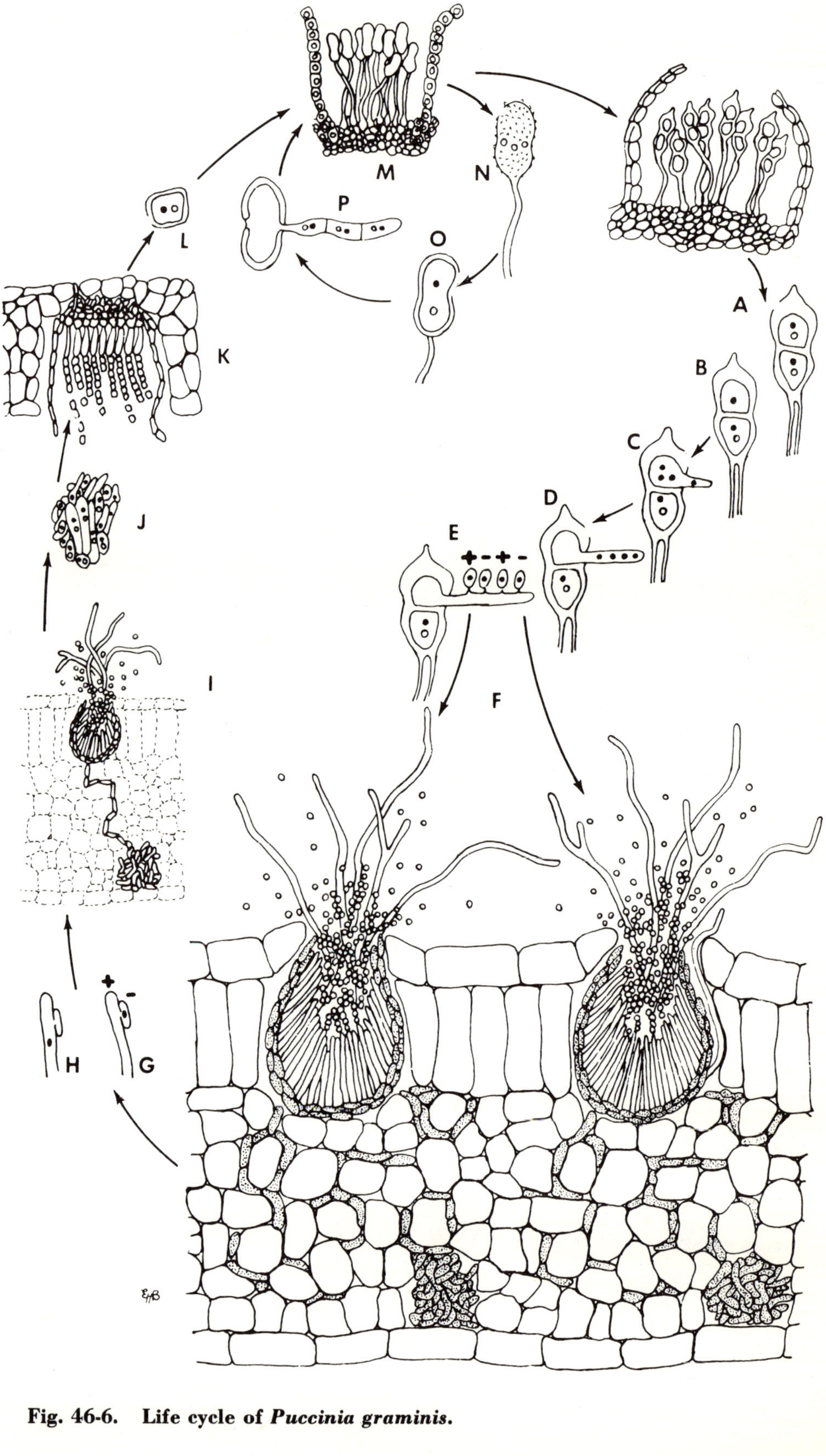

Fig. 46-6. Life cycle of *Puccinia graminis*.

SUGGESTED READING

Flor, H. H. "Host-Parasite Interaction in Flax Rust"—Its Genetics and Other Implications." *Phytopathology,* Vol. 45 (December, 1955), pp. 680–685.

Johnson, T., "Man-Guided Evolution in Plant Rusts." *Science,* Vol. 133 (February, 1961), pp. 357–362.

47

Hollyhock Rust

Hollyhock rust (Fig. 47-1, 47-2) offers a number of contrasts to stem rust of wheat. Whereas *Puccinia graminis* is heterothallic, heteroecious, and macrocyclic, *Puccinia malvacearum,* (the cause of hollyhock rust) is homothallic, autoecious, and microcyclic.

Microcyclic rust fungi, in addition to mycelium, consist of teliospores and basidiospores. Spermagonia may or may not be present. No aecia or uredia occur in microcyclic species. Sometimes the telia and teliospores present a morphological similarity to aecia and aeciospores of macrocyclic species. On germination, however, they produce basidiospores. The teliospores are the only binucleate spores produced. Since the basidiospores are the only infectious spores, there obviously can be no alternation of hosts and all microcyclic species are autoecious.

Puccinia malvacearum produces only teliospores and basidiospores. It overwinters as teliospores and dormant mycelium in overwintering, perennial plant parts. Teliospores are not infectious. They germinate and produce the basidiospores which infect the hollyhock and its relatives. The infections in overwintering plants do not represent primary infections since they are continuations of last year's infections. Primary infections must be new infections of the current growing season, resulting

Fig. 47-1. Hollyhock rust on lower (A) and upper (B) leaf surface.

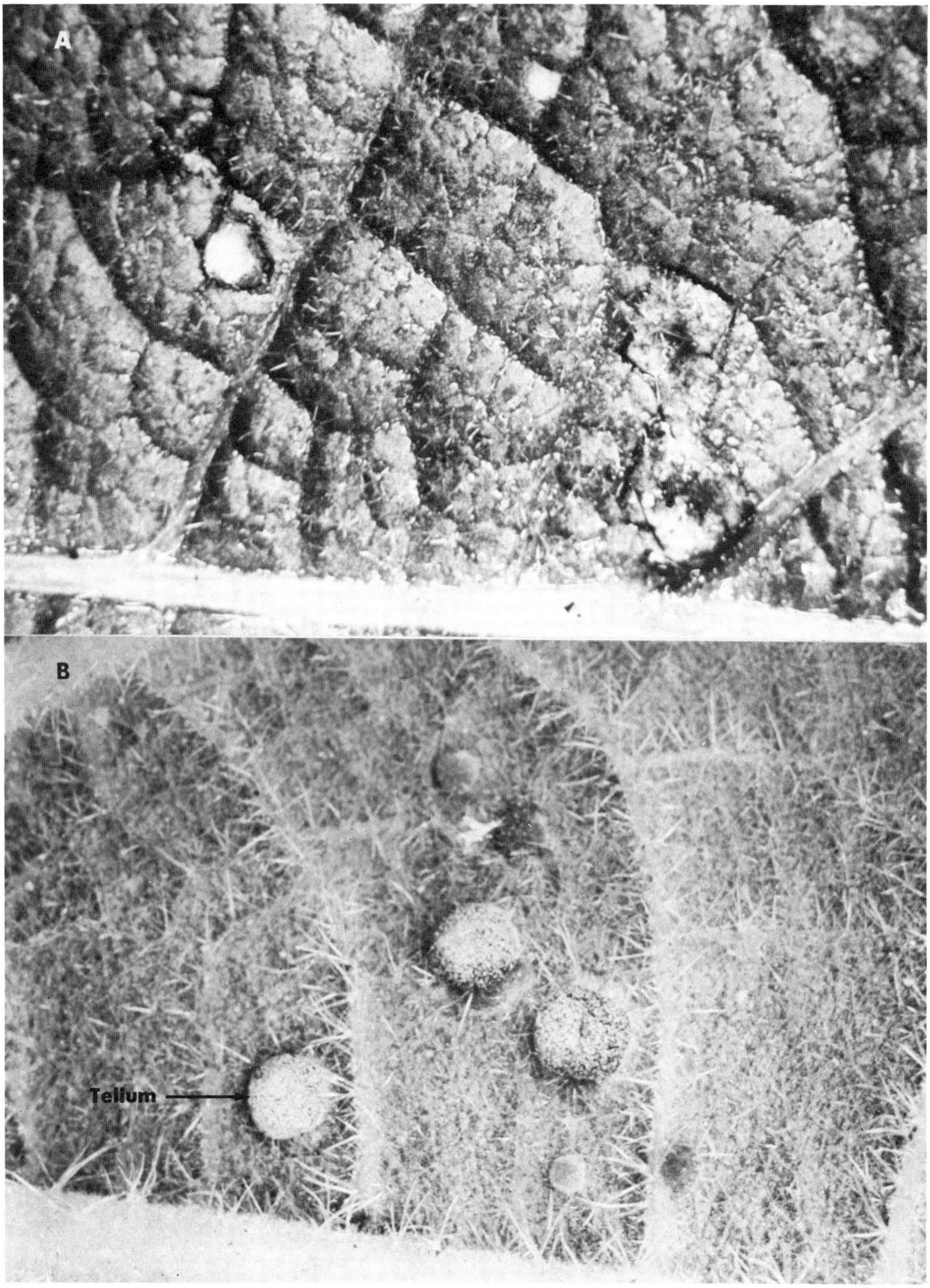

Fig. 47-2. (A) Close-up of the lesion of hollyhock rust on the upper surface of the leaf and (B) a close-up of telia on the lower surface.

from inoculum derived from overwintered structures. The basidiospores which are produced by the overwintered teliospores which develop on the infected plants which overwintered constitute the primary inoculum. The second and successive crops of basidiospores represent repeating secondary inoculum. Structurally, then, the primary and secondary inocula are identical and the difference is solely their source. There is no dormant period for these teliospores and they may germinate promptly.

Typically, the disease is severe in the spring and again in the fall. The intervening period of relatively light infections is due to unfavorably low moisture and high temperature for the production of inoculum and infection.

The nuclear behavior is interesting. The nuclei in the developing basidiospores (Fig. 47-3 H) divide so that the basidiospore becomes binucleate when released (I-J). The developing mycelium, however, reverts to a uninucleate(n) condition. The hyphal tips that give rise to the telia become binucleate again (B). The two sister nuclei in each cell of the two-celled teliospore fuse (F) and undergo meiosis on germination (G).

Control. The disease is controlled by the removal and destruction of overwintering debris.

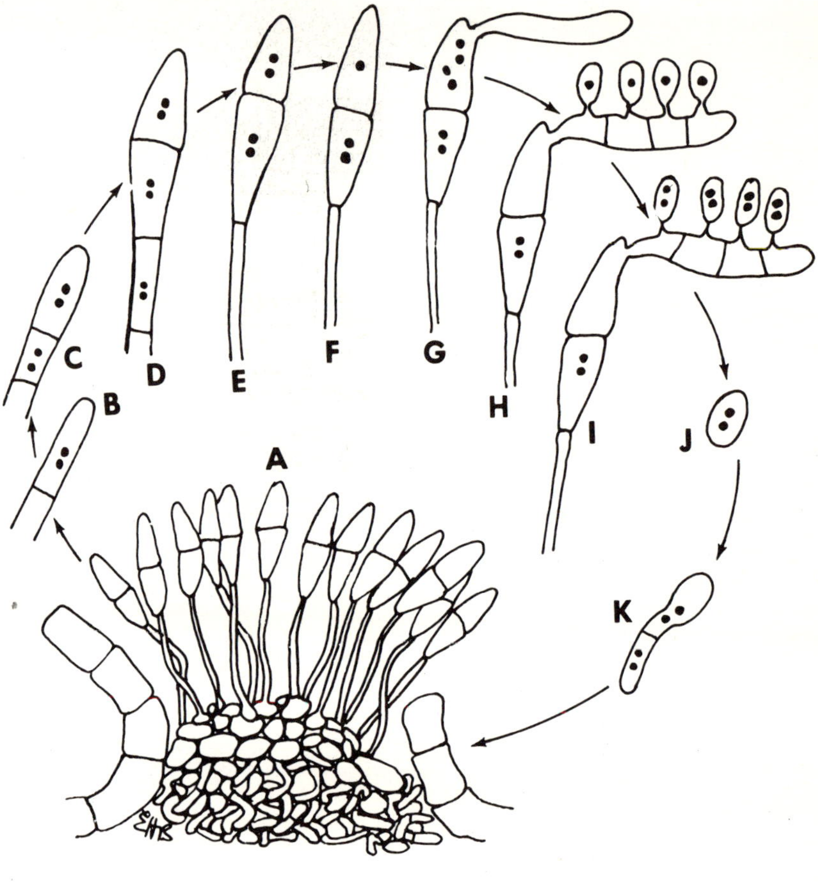

Fig. 47-3. Life cycle of *Puccinia malvacearum*.

48

Cedar Apple Rust

Cedar apple rust is an appropriate name for a disease caused by a fungus that attacks both the cedar and the apple. It is also fitting since the reddish galls produced on the cedar are called cedar apples! This disease is caused by *Gymnosporangium juniperi-virginianae.* Two similar diseases, Hawthorn rust and quince rust, are caused by closely related fungi, *Gymnosporangium globosum* and *Gymnosporangium clavipes,* respectively. These three fungi and related species are heteroecious and **demicyclic,** a form of macrocyclic species in which no uredial stages occur. No uredial stages occur in most species of *Gymnosporangium.*

Causal Organism. The fungus overwinters as binucleate mycelium in the galls on the cedar (Fig. 48-5A, 48-81). They are composed of host tissues and fungus. In the spring, the telia absorb water and expand into long, gelatinous horns (Figs. 48-4, 48-5B, 48-8A) in which the telia are embedded in great numbers. Many of the commercial preparations of microscope slides are cross sections of the telia when they first begin to swell and they appear as dome-shaped pustules (Fig. 48-6). These expand to become the long, slender, gelatinous horns.

The teliospores germinate by the production of a promycelium which forms four sterigmata and four basidiospores (Fig. 48-7). When the basidiospores form, the yellow protoplasm which flowed from the teliospore into the promycelium moves into the basidiospores. Thus, the basidiospores are yellow, whereas they are hyaline in the other rust fungi we are studying. The basidiospores infect only the apple and crab-apple by direct penetration and constitute the single source of primary inoculum for the apple. This is the only inoculum for the apple and therefore there are no secondary infections of the apple host. The basidiospores are wind-blown for a distance of two to three miles. Those which might be carried farther apparently become dehydrated and nonviable.

Spermagonia and aecia are produced on the apple host (Figs. 48-1,-2,-3 and 48-8F). Aeciospores constitute the primary inoculum for the cedar. This fungus is forced to alternate between apple and cedar since only two infectious spore forms are produced and each spore is able to infect only one host and not the other, and the organism is an obligate parasite.

Aeciospores (Fig. 48-8G) infect the cedar in late summer to early fall and small galls are visible the following summer (H). Telial horns are not produced until the following spring. Thus, if basidiospores infect an apple in 1967, telial horns are not produced until the spring of 1969—a two year cycle.

Control. easily effected by removal of apples from cedar orchards and cedars from apple orchards! The carbamate fungicide, Ferbam, is very effective in chemical control of infections of the causal organism on both hosts.

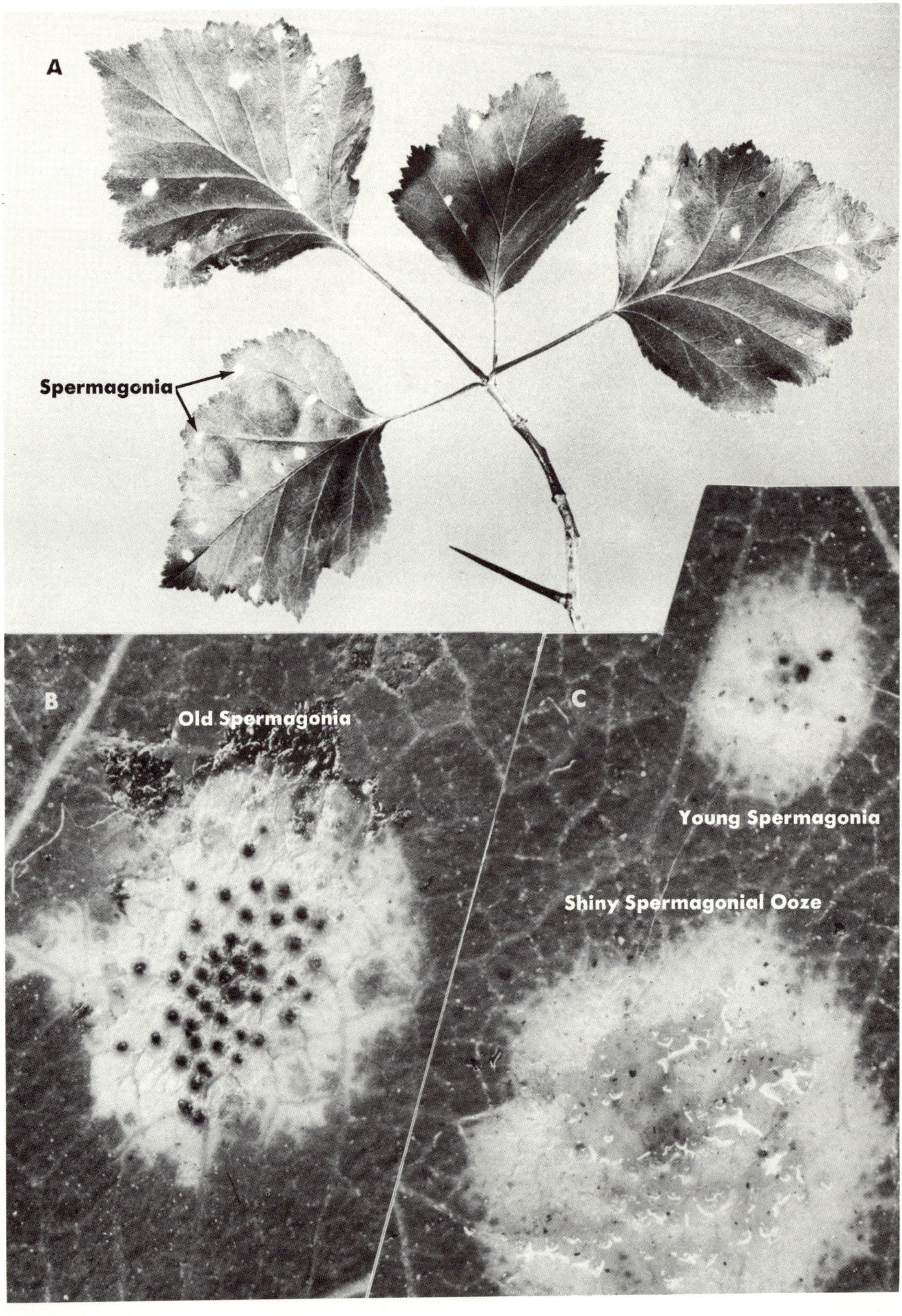

Fig. 48-1. Spermagonia of *Gymnosporangium globosum* on Hawthorne.

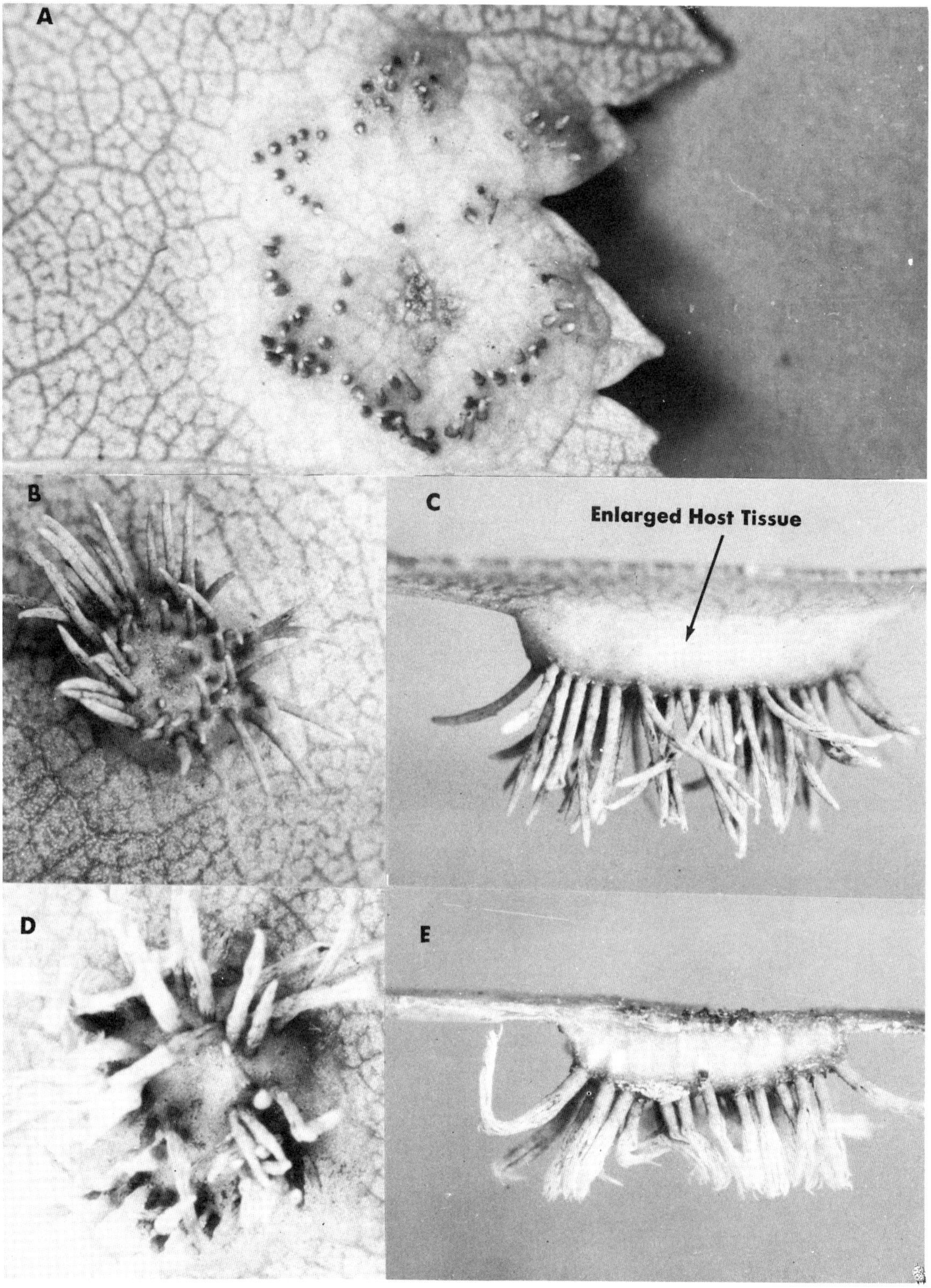

Fig. 48-2. Stages in the maturation of aecia of *Gymnosporangium globosum.* (A) Top view of young aecia; (B) top view of mature aecia; (C) side view of mature aecia; (D) top view dehiscent aecia; (E) side view of dehiscent aecia.

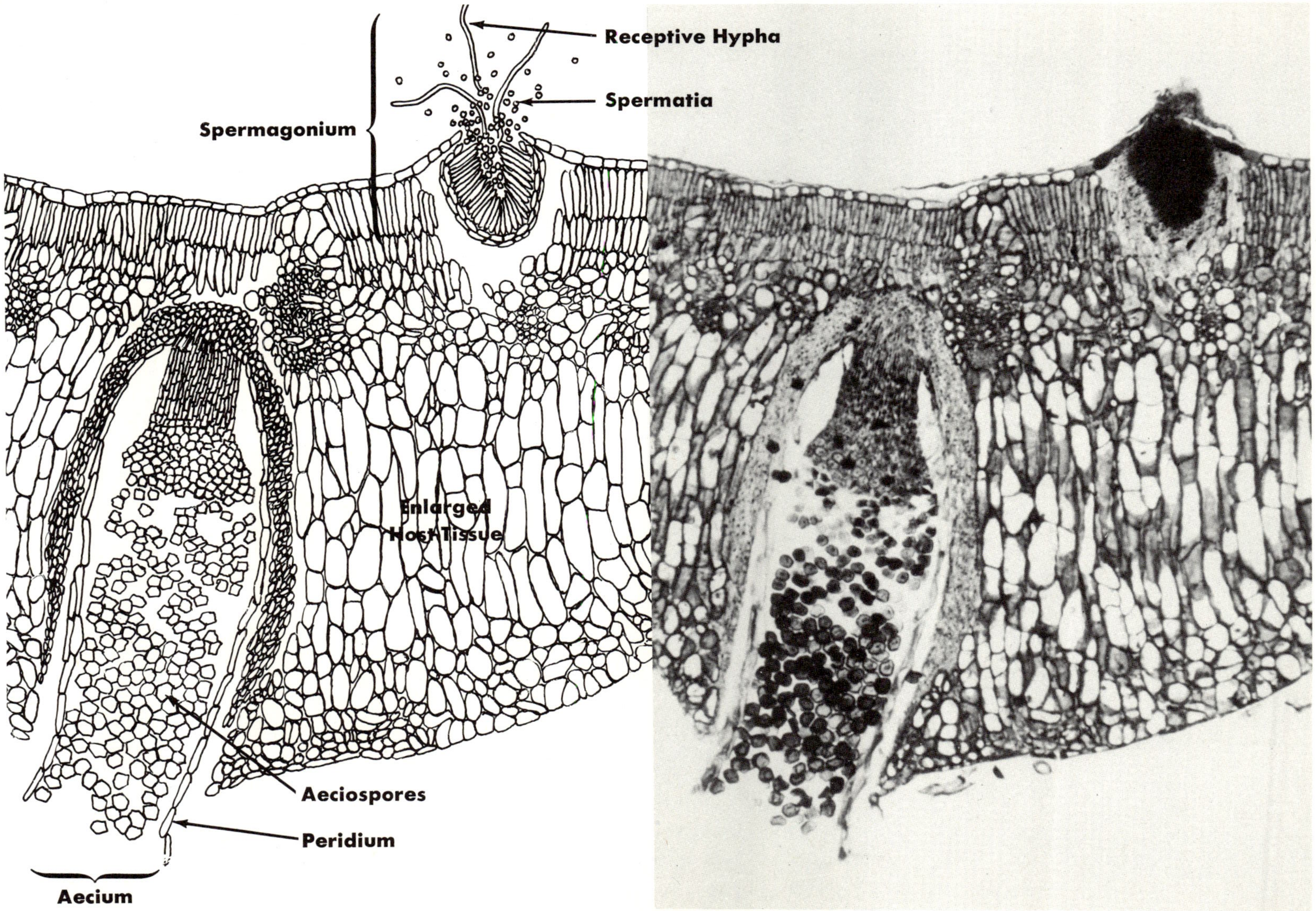

Fig. 48-3. Photomicrograph and reconstructed drawing of the spermagonium and aecium of *Gymnosporangium juniperi-virginianae*, and hypertrophied host tissues.

Fig. 48-4. Expanded telial horns on galls of *Gymnosporangium juniperi-virginianae.*

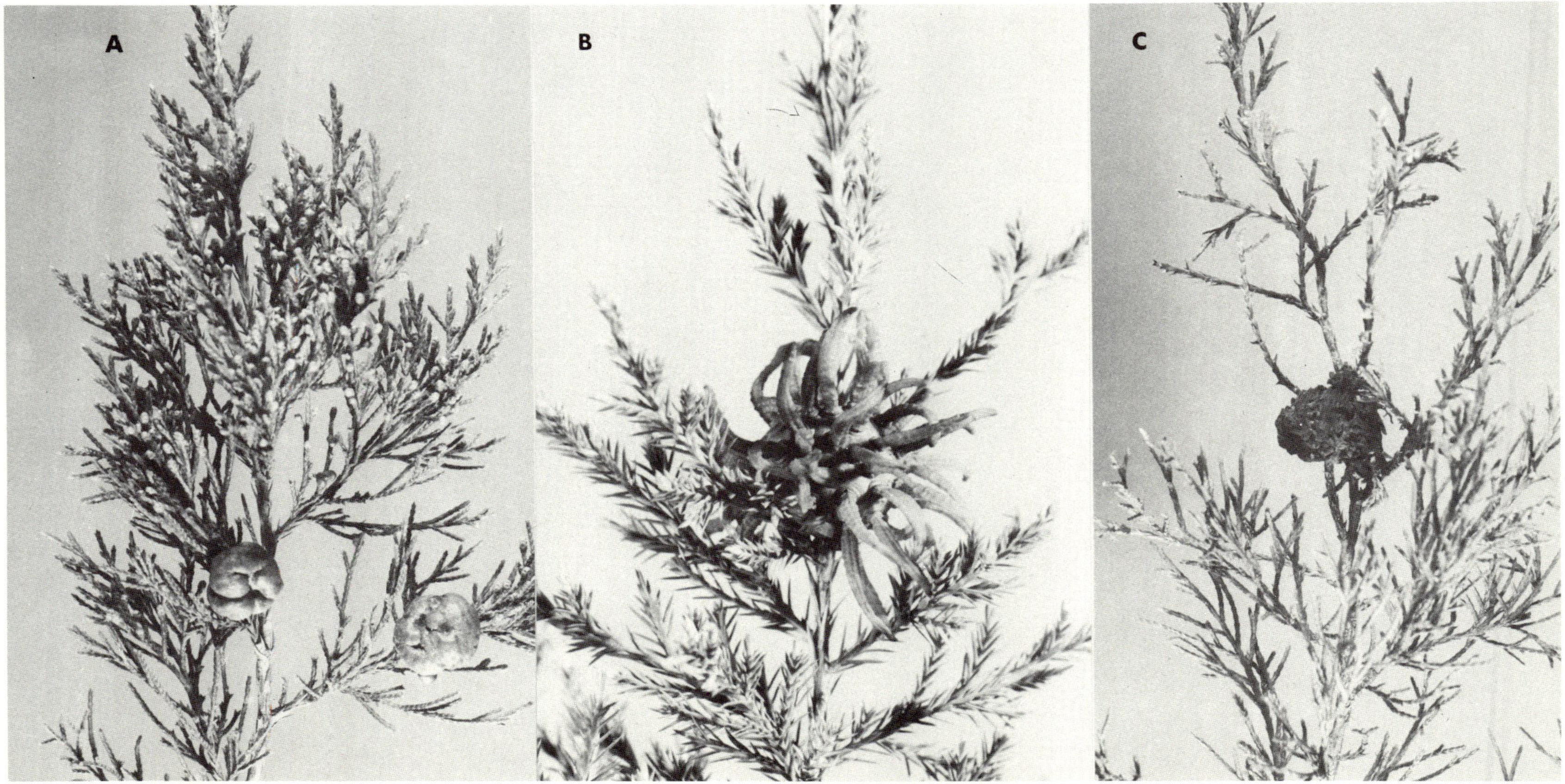

Fig. 48-5. **Telial galls of *Gymnosporangium juniperi-virginianae* showing (A) immature galls, (B) mature galls, and (C) old inactive gall.**

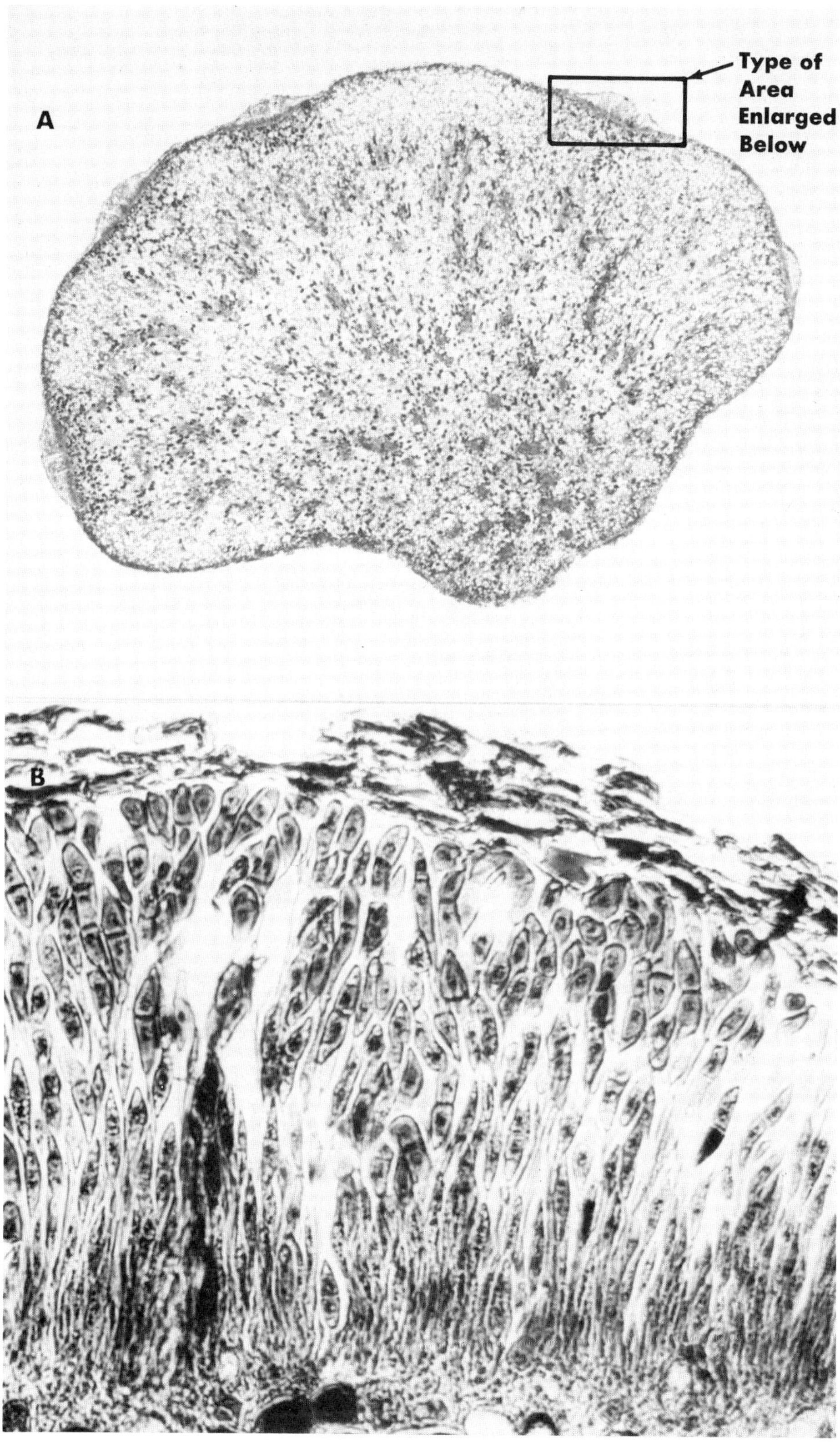

Fig. 48-6. (A) Photomicrograph of a cross section of a gall of *Gymnosporangium juniperi-virginianae* and (B) a close-up of the teliospores.

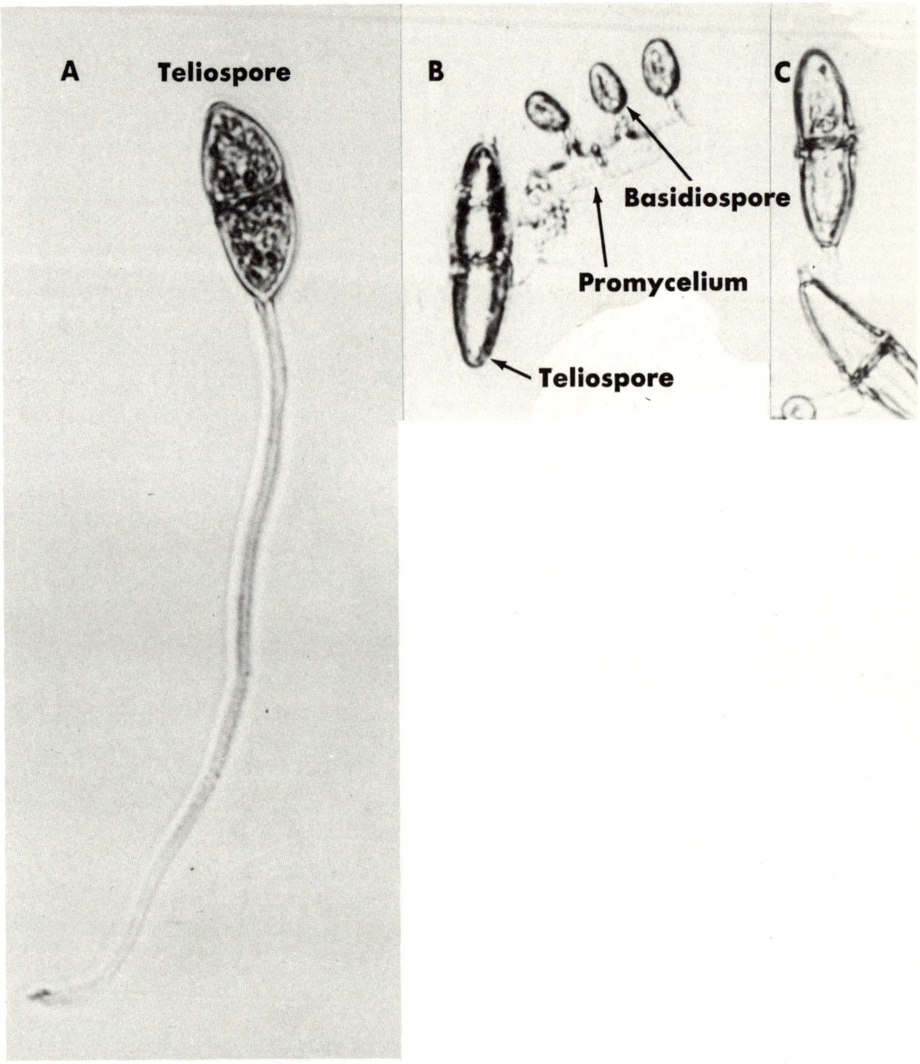

Fig. 48-7. Germinated (A) and ungerminated (B, C) teliospores and basidiospores of *Gymnosporangium juniperi-virginianae.*

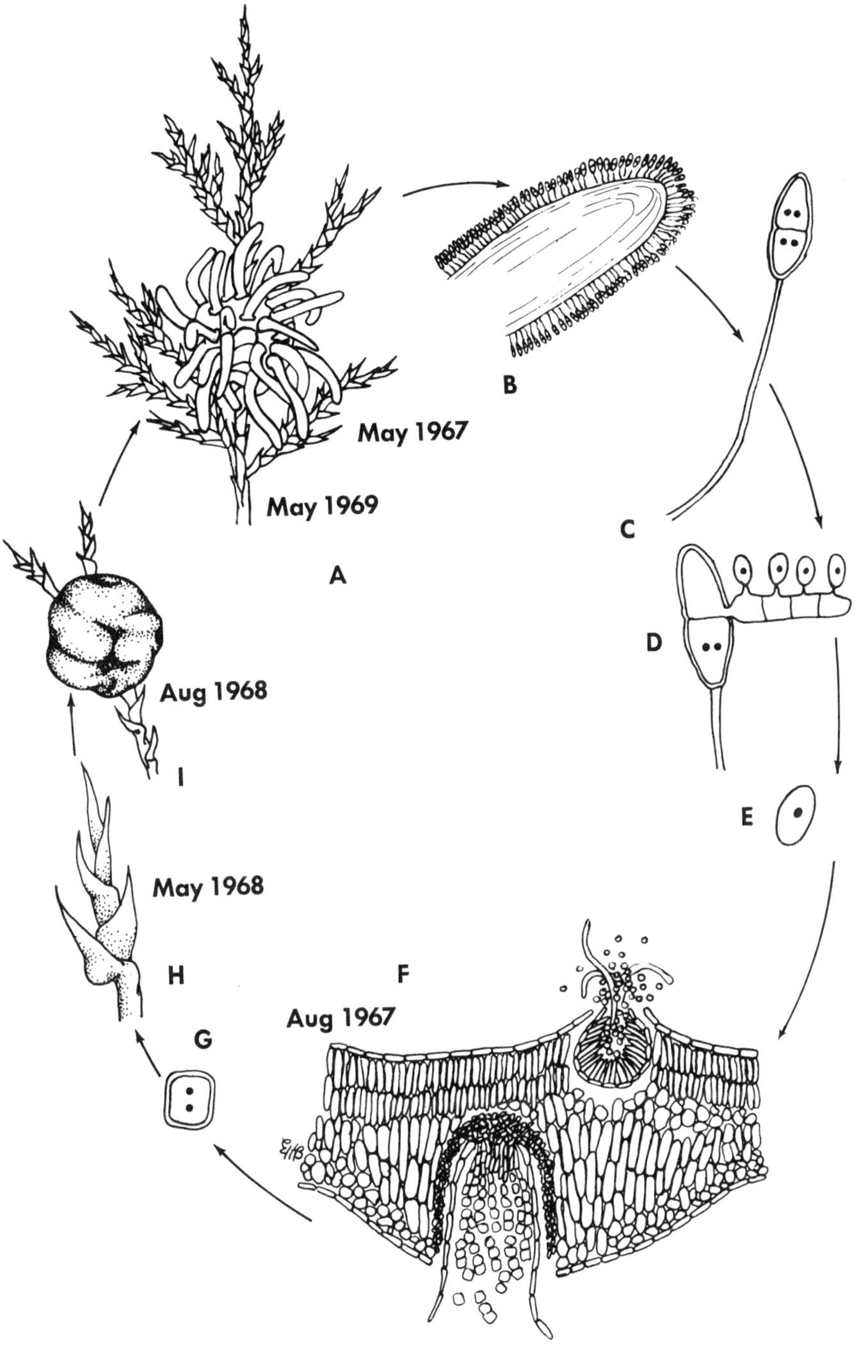

Fig. 48-8. Life cycle of *Gymnosporangium juniperi-virginianae.*

49

White Pine Blister Rust

The forests of New England were cleared for the plow to grow food and were stripped for logs to build cabins. And, as America grew, so did the need for lumber. The forests of Michigan filled this need with lumber to build new cities and to rebuild Chicago after the Great Fire. It made men rich for it was worth two billion dollars—or twice the estimated value of the gold resulting from the California gold rush of 1849. At the turn of the century, America sought to heal much of the scars from this unrestricted logging. Seed of native American white pine was shipped to Germany where seedlings were raised. On the return voyage, some of these seedlings carried infections of *Cronartium ribicola,* the fungus that causes white pine blister rust (Fig. 49-1). Presumably the fungus originated in Asia and spread to the European continent.

Disease Cycle. *Cronartium ribicola* is a long-cycled and heteroecious rust fungus and occurs on the majority of white, or five-needle, pine species and not on pines of fewer or greater numbers of needles per bundle. Currants and gooseberries, species of *Ribes,* are the telial hosts. Cultivated red currants are relatively resistant but not the European black currant. This species is prohibited from culture by law in some states. Native species of *Ribes* are abundant throughout most of the range of the five-needle pines.

Infections of the pine occur by stomatal penetration of the needles by germ tubes of the basidiospores. The developing mycelium enters the vascular system and grows rapidly through the twig. During the second or third year following infection of the needle, discoloration and spindle-shaped swellings occur (Fig. 49-1 A). Squirrels are fond of these tasty swellings before development of the fruiting structures occurs. Spermagonia appear in small blister-like areas which rupture, ooze droplets, and dry. The aecia usually do not appear until the following season (Fig. 49-1). The aecial blisters rupture and expose masses of yellow-orange aeciospores (Fig. 49-2). This infection of pine is systemic, and the production of aeciospores continues, thereafter, yearly. The infected areas progressively enlarge until the stem is girdled by the lesion. When stems are girdled and killed, the foliage turns brown and dies. This branch with dead leaves is called a flag.

Aeciospores are wind-blown and infect species of *Ribes* through the stomates of the leaves during moist periods. Uredia may appear as soon as seven days. Uredospores are wind blown, re-infect *Ribes,* and create a repeating cycle for the *Ribes.* Telia are produced in the fall in columns (Fig. 49-3). Teliospores are not readily detached from the column and the basidiospores are produced in place. They germinate promptly and do not overwinter. Overwintering is chiefly by means of perennial mycelium in systemic infections of pines and mycelium in *Ribes.*

Control: based on the fact that infection can occur only if basidiospores are available from telia produced on *Ribes* plants. Thus, the most important means of control

Fig. 49-1. Aecial blisters of white pine blister rust.

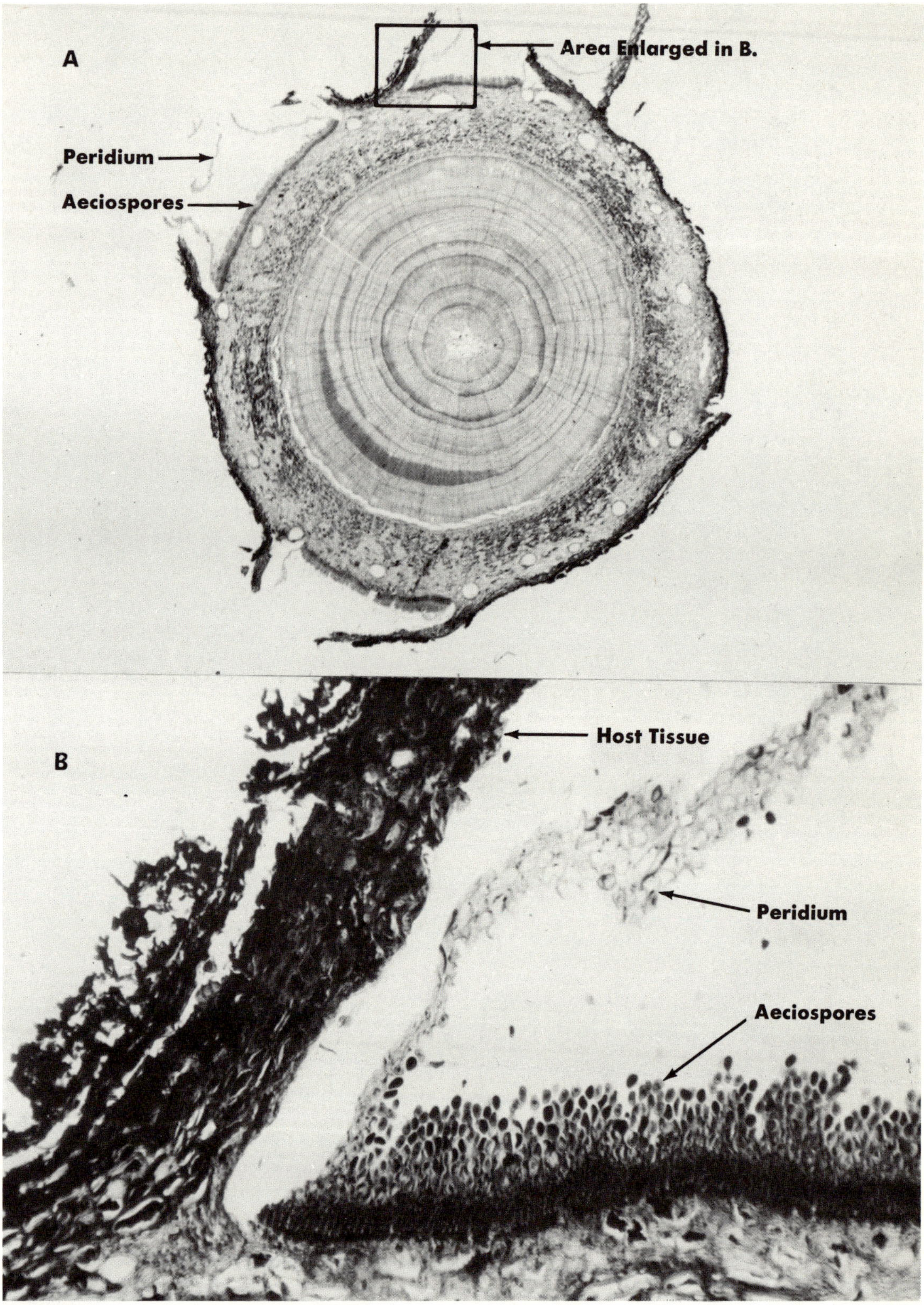

Fig. 49-2. Photomicrograph (B) (× *125*) and close-up (A) of a cross section of a pine branch with aecia of *Cronartium ribicola*. Make a comparison of the ruptured bark and peridium with Fig. 49-1C.

is eradication of the *Ribes*. This works well since no repeating cycles occur on the pine but only on the *Ribes*. Thousands of acres of pine forests have been scouted for *Ribes* and the plants have been mechanically and more recently chemically destroyed.

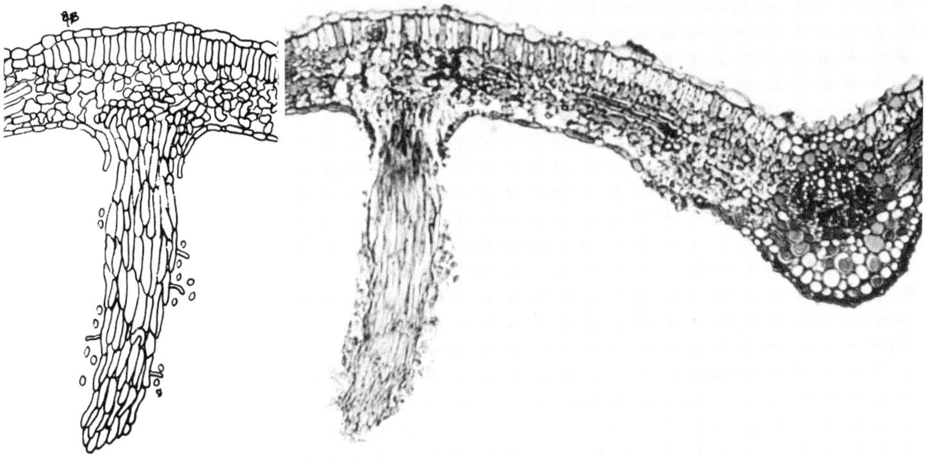

Fig. 49-3. Photomicrograph and representative diagram of a telium of *Cronartium ribicola* (× 125).

SUGGESTED READING

Offord, H. R., Quick, C. R., and Moss, V. D., "Blister Rust Control Aided by the Use of Chemicals for Killing *Ribes*." *Journal of Forestry*, Vol. 56 (January, 1958), pp. 12–18.

50

Needle Rusts

The needle rusts are diseases of the leaves of coniferous trees (Fig. 50-1). The most common ones are caused by long-cycled species of *Coleosporium* in the family Pucciniaceae. The spermagonia and aecia occur on needles of pines and the uredia and telia on ferns, flowering plants, and shrubs. A few short-cycled species occur with spermagonia and telia, or telia only, on the pines. (Note that this follows Tranzschel's Law.) Needle rusts of other conifers such as fir, Douglas fir, hemlock, spruce, and larch are caused, primarily, by species in the family Melampsoraceae.

None of the needle rusts are usually very significant economically except on young trees in nurseries. Control can be secured by destruction of alternate hosts for 1,000 feet. It is more practical to establish nurseries in areas where the alternate hosts are not prevalent.

The close similarity of the gross symptoms of needle rust and scale insects presents a good example of the difficulty of diagnosis for untrained personnel (Fig. 50-1 A, C). Close examination reveals the nature of the fungus-caused malady (Figs. 50-1 B, D and 50-2).

Fig. 50-1. (A, B) aecial blisters of needle rust on a 5-needle pine and (C, D) scale insects on a 2-needle pine.

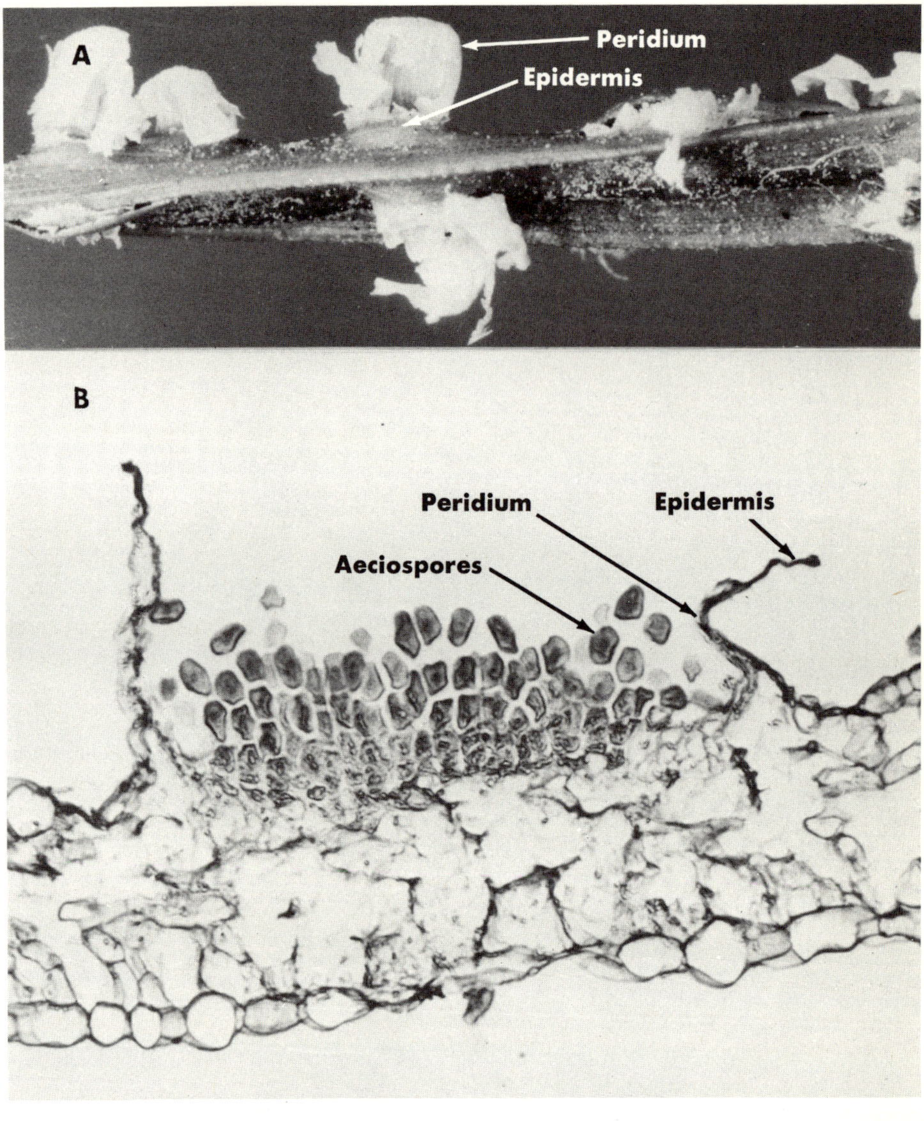

Fig. 50-2. (A) Close-up and (B) photomicrograph ($\times$ *250*) of a cross section of aecial blisters of needle rust.

51

The Smut Diseases

The smut diseases are generally characterized by black, dusty masses of spores. These spores are, in reality, teliospores but are frequently and incorrectly called chlamydospores. The causal fungi are not obligate parasites; even so, they are not usually found as saprobes. The corn smut fungus, *Ustilago maydis*, however, is uniquely adapted and persists as a saprobe in the soil and on refuse. The fungi which cause the smut diseases are very specific in their selection of hosts and host organs. Some smut fungi may attack only the stems, flowers, anthers, or ovules of their hosts and no other part and no other host. Some diseases are localized and some are systemic. Many smuts are either seedling or ovary infectors.

Wheat bunt is of special historical interest and significance. In the late 1700's, Tillet demonstrated the contagious nature of black smut, and Prevost, in the early 1800's proved the pathogenic nature of the fungus. This was the first demonstration of the pathogenic nature of any microorganism, but with the popularity of the Theory of Spontaneous Generation this was ignored for the next forty years. In their turn, the Tulasne brothers, Charles and Louis (1840), Anton deBary (1853), and Kühn (1860) experimented with wheat bunt and confirmed the earlier works and extended the horizons of knowledge on the nature of the smut diseases. Translations of these works appear in the Phytopathological Classics.

Classification. The smut fungi comprise the Order Ustilaginales. The Ustilaginaceae are characterized by the production of basidiospores in lateral positions on a septate promycelium. Species in the Tilletiaceae produce terminal basidiospores on a non-septate promycelium (Fig. 51-4). A third family, the Graphiolaceae, is obscure and not treated here. Terminology is adopted after Fisher and Holton cited at the end of the chapter.

Control: usually by seed treatment. This may be a hot water dip or chemical dusting or dip. Soil treatments and the use of resistant varieties are useful.

Life Cycles: in general consist of germination of the teliospore (Fig. 51-4 A) by the production of a promycelium (D,O). Meiosis (C,L) occurs during germination. The basidiospores are uninucleate and haploid (F,R). Plasmogamy occurs between two compatible haploid bodies. This may be between two germinating basidiospores while on the promycelium (Q) or after release (G). It may be between haploid hyphae, as in *Ustilago maydis*.

OAT SMUTS. Two very similar smut diseases (Fig. 51-1) occur world-wide, side-by-side on oats in the same fields. Similar diseases caused by similiar smut fungi occur on other small grains (Fig. 51-2). These diseases on the same crop are so similar in symptoms, that positive diagnosis can be made only by examination of the teliospores to determine if their cell walls are smooth or spiny (Fig. 51-3). Smooth spores are characteristic of *Ustilago hordei,* which causes covered smut of oats and barley, whereas finely-spined spores are characteristic of *Ustilago avenae,* which causes loose smut of oats and semi-loose smut of barley. Loose and covered refer to the rupture or persistance of the delicate membranes which cover the sporulating sori.

Teliospores do not overwinter in the soil or in crop refuse but are carried with the seed. They germinate in the spring at the same time as the seeds of the host. Basidiospores are produced which germinate, conjugate, and form dikaryons. Only the dikaryons are infectious. Systemic invasion follows with growth of the fungus along with the growing point. It enters the flower primordia and produces masses of the binucelate dikaryotic hyphae. The cells round-up, form cell walls, and become teliospores (Fig. 51-3). The spores are wind-blown to other flowers and contaminate the adhering bracts of the developing kernels. There are no secondary cycles.

LOOSE SMUT OF BARLEY AND WHEAT is caused by *Ustilago nuda* (Fig. 51-2) which is distinct from all other smut fungi in that it is the only one which is known

Fig. 51-1. Comparison of a loose smut diseased floret and panicle of oats with a healthy floret and panicle.

to infect the host via the embyro. Spores from smutted spikes are wind-blown to healthy flowers where germination, conjugation, and infection of the ovary occur. The dikaryotic hyphae is the primary inoculum. The ovary becomes resistant about a week after pollination and timing, and therefore, is important. The hyphae remain dormant in the scutellum, pericarp, and the remains of the integuments and constitute the overwintering stage. No secondary cycles occur.

CORN SMUT is caused by *Ustilago maydis* which is uniquely adapted as a saprobe. Spores persist in soil and refuse. Seed-borne spores are uncommon. Primary inoculum consists mostly of air-borne basidiospores produced saprophytically on corn refuse. All young meristematic tissues are susceptible. Systemic invasion of seedlings occurs only occasionally. Galls occur on tassels, stems, and ears (Figs. 51-5, 51-6). The basidiospores germinate and penetrate the host directly. A limited amount of haploid hyphae develops after infection. After conjugation, however, the dikaryotic secondary hyphae continue the infection. Only the dikaryotic hyphae are capable of producing infections which terminate in the production of teliospores. Single basidiospores have been found which can give rise to complete infections with teliospores. These basidiospores were found to be diploid.

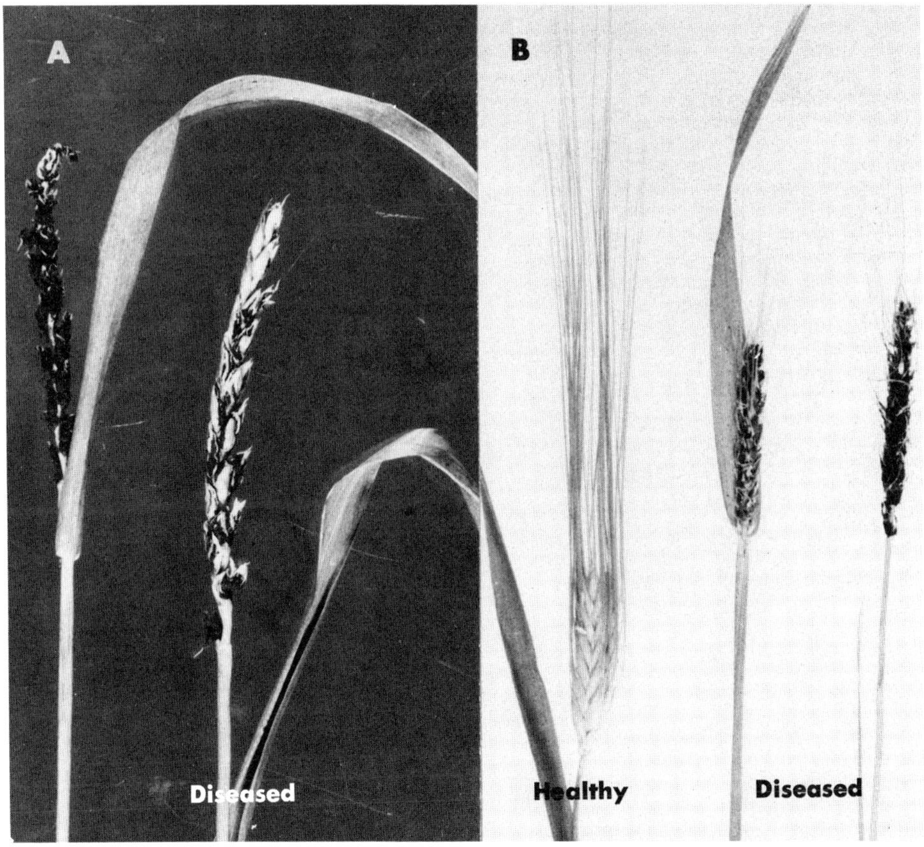

Fig. 51-2. Loose smut of wheat (A) and barley (B).

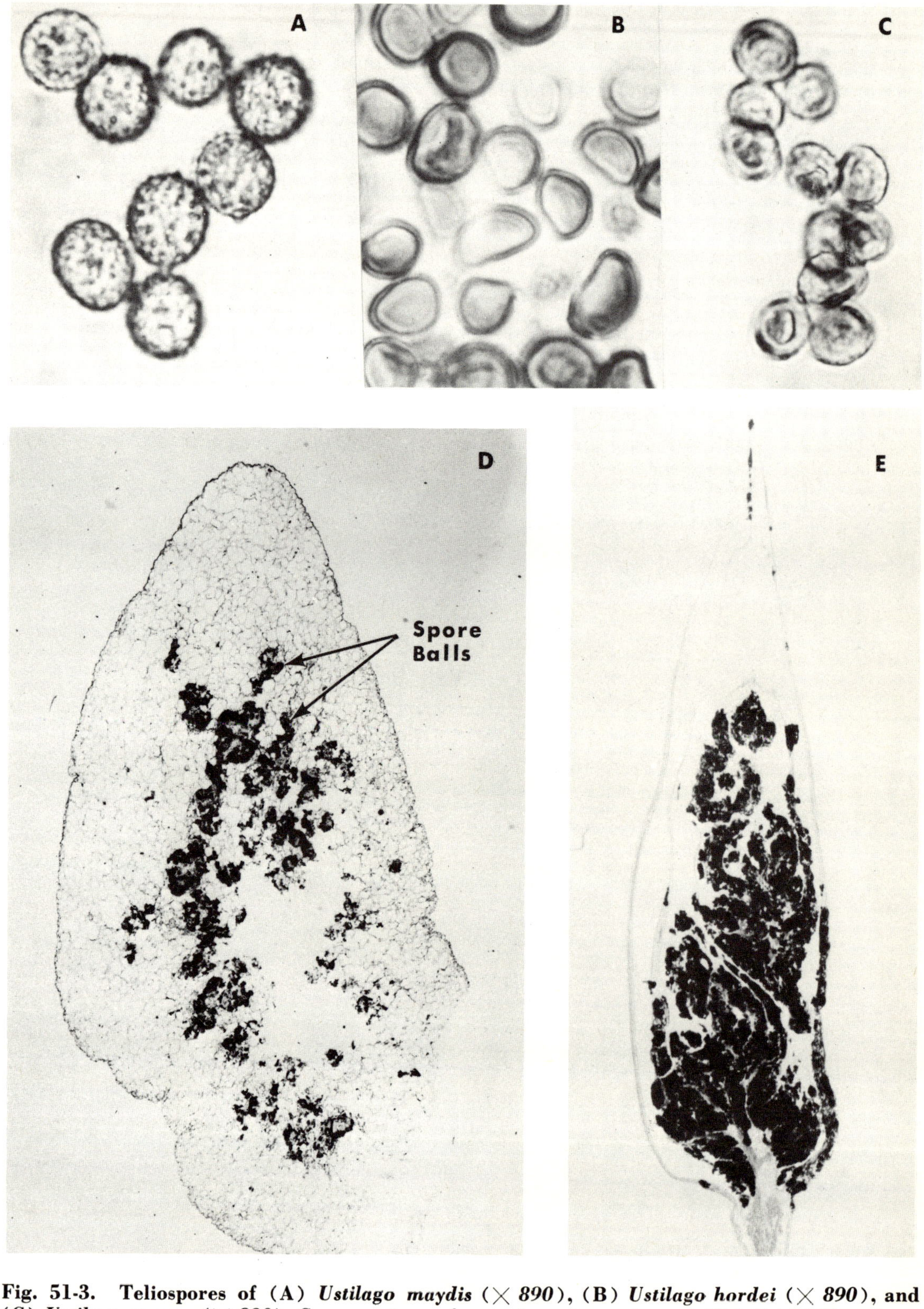

Fig. 51-3. Teliospores of (A) *Ustilago maydis* (× *890*), (B) *Ustilago hordei* (× *890*), and (C) *Ustilago avenae* (× *890*). Cross sections of an enlarged corn kernel (D) (× *10.7*) and oat floret (E) (× *10.7* to show the placement of spore balls. Almost the entire tissues of the oat floret have been replaced by fungus structures.

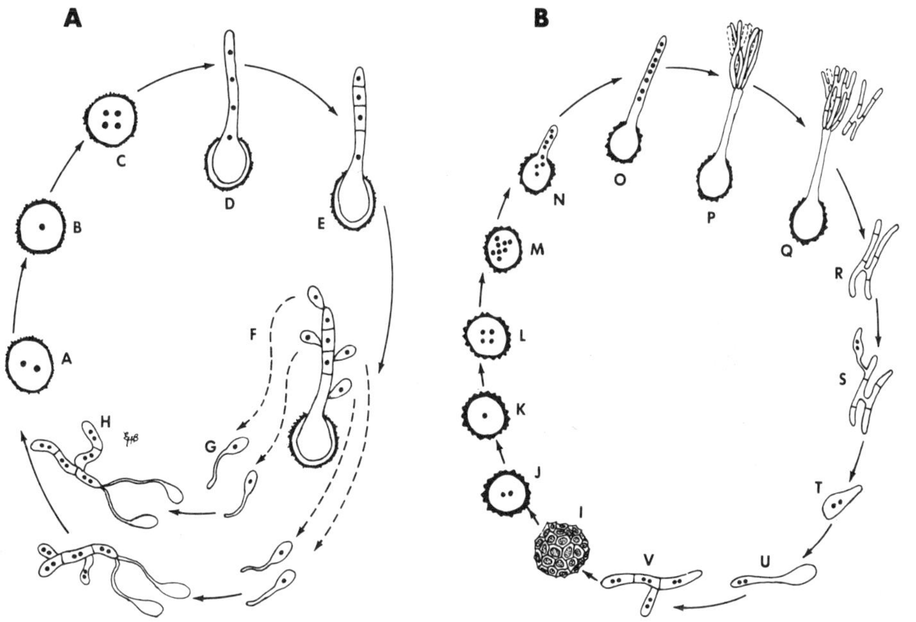

Fig. 51-4. Life cycles of (A) Ustilago and (B) Tilletia.

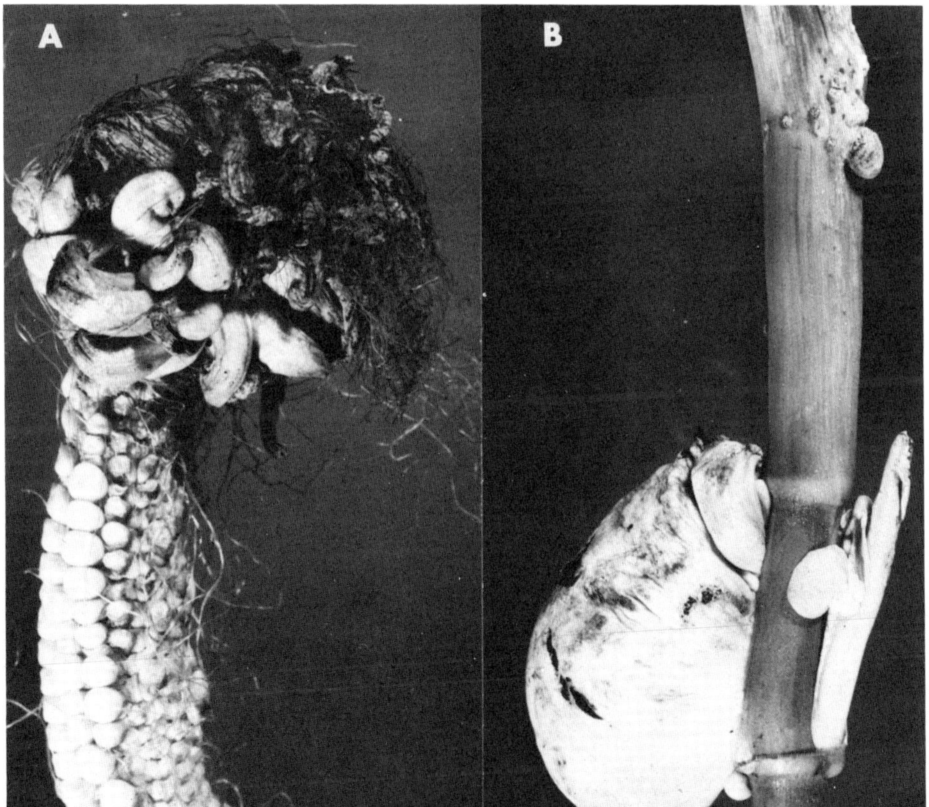

Fig. 51-5. Galls of corn smut on the (A) ear and (B) stem.

Fig. 51-6. Galls of corn smut on (A) the tassel and (B) a comparison of the outside and (C) inside of a smutted staminate floret with (D) a healthy floret.

SUGGESTED READING

Fisher, G. W., and Holton, C. S., *Biology and Control of the Smut Fungi.* New York, Ronald, 1957.

Experiment XVIII

Oat Smut: Germination

Teliospores germinate to give rise to promycelia and basidiospores. Observe germination of teliospores by dusting a small pinch on the surface of water agar in petri dishes or on glass microscope slides. If on the slides, place them on moist filter paper in petri dishes. (If lysis occurs incorporate 1% dextrose.)

Supplies × Total Number of Students or Groups of Students

1 petri dish of water agar or 1 glass slide coated with a thin layer of water agar
1 petri dish and filter paper

General Supplies Available

Teliospores of loose smut of oats, barley, and wheat

SUGGESTED READING

Popp, W., "A Comparative Study of Spore Germination of *Ustilago tritici* and *U. nuda.*" *Phytopathology*, Vol. 45 (November, 1955), pp. 585–590.

52

Wood Rots

Wood rots may be described as root, butt (Fig. 52-1 A), stump, trunk, or top rots to indicate that portion of the tree which is attacked. The diseases of these areas are not exclusive. The fungi which cause them attack most any part of the tree. Rots are also described as sap rots or heart rots (Fig. 52-3) to indicate the tissues attacked within the tree. Some organisms which cause a rot of the dead heartwood cannot attack the living sapwood. Such organisms are pathogenic but are they also parasitic? Some of the fungi which cause wood rots, such as *Polyporus versicolor* (Fig. 52-1 B), are called **slash fungi** and occur primarily on discarded timber, diseased wood, trash from cutting operations, and fallen trees. They are significant in the destruction and reduction of such slash and are, therefore, beneficial.

Wood rots are also described as white rots and brown rots. The **white rots** (Figs. 52-2, 52-3 C) are fairly firm, tend to form small pockets (52-2 C), and usually appear white but may be yellow to tan. All components of the wood, including lignin, are decomposed. Cellulose, which is white, is exposed by the rot. Paper is a pressed form of cellulose. The **brown rots** (Fig. 52-3) are dry, powdery, brown in color, and tend to form crumbly, cubical structures. Cellulose is decomposed by these fungi but the dark lignin remains more or less unaffected and gives the characteristic brown color to the rot. Brown rots sometimes follow white rots in the same tree, but white rots seldom follow brown rots.

The chemical reactions of decay are primarily reactions of oxidation and hydrolysis. The chemical make-up of wood is not fully understood and thus the chemistry of its degradation is also incompletely understood.

Zone Lines. An interesting and incompletely understood feature of wood rot is the appearance of zone lines, or decay lines (Fig. 52-4). These are dark brown to black lines which run through the wood with no relationship to the grain of the wood. They are most common in white rots and the decay of hardwoods, and rarely occur in the brown rots. They are formed most commonly by *Fomes pini, Fomes ignarius,* and *Xylaria polymorpha,* an Ascomycete.

Position of the zone line is marked by an aggregation of hyaline hyphae. These swell, fill the lumen of the cells, and become dark brown in color. These enlarged cells are not continuous but occur in clumps with interconnecting strands of hyphae. Some pigment escapes, stains the cell walls, and fills the pits between the ray cells. The zone lines may be the width of a single cellular element of the xylem tissue.

A controversy continues as to the cause of formation and the functions, if any, of the zone lines. One convincing story is that the zone lines represent a cross-section of the walls of a sclerotium-like body which is immersed within the host tissues. Where these lines come to the surface of the substratum, they become contiguous with a black structure with a white central core composed of hyphae surrounded by a ring of swollen cells. It seems that this sclerotium-like body protects the fungus body within from desiccation.

On microscopic examination, hyphae can be seen ramifying through the tissues

Fig. 52-1. (A) Butt rot of beech caused by *Fomes applanatus* and (B) the rainbow conk, *Polyporus versicolor*.

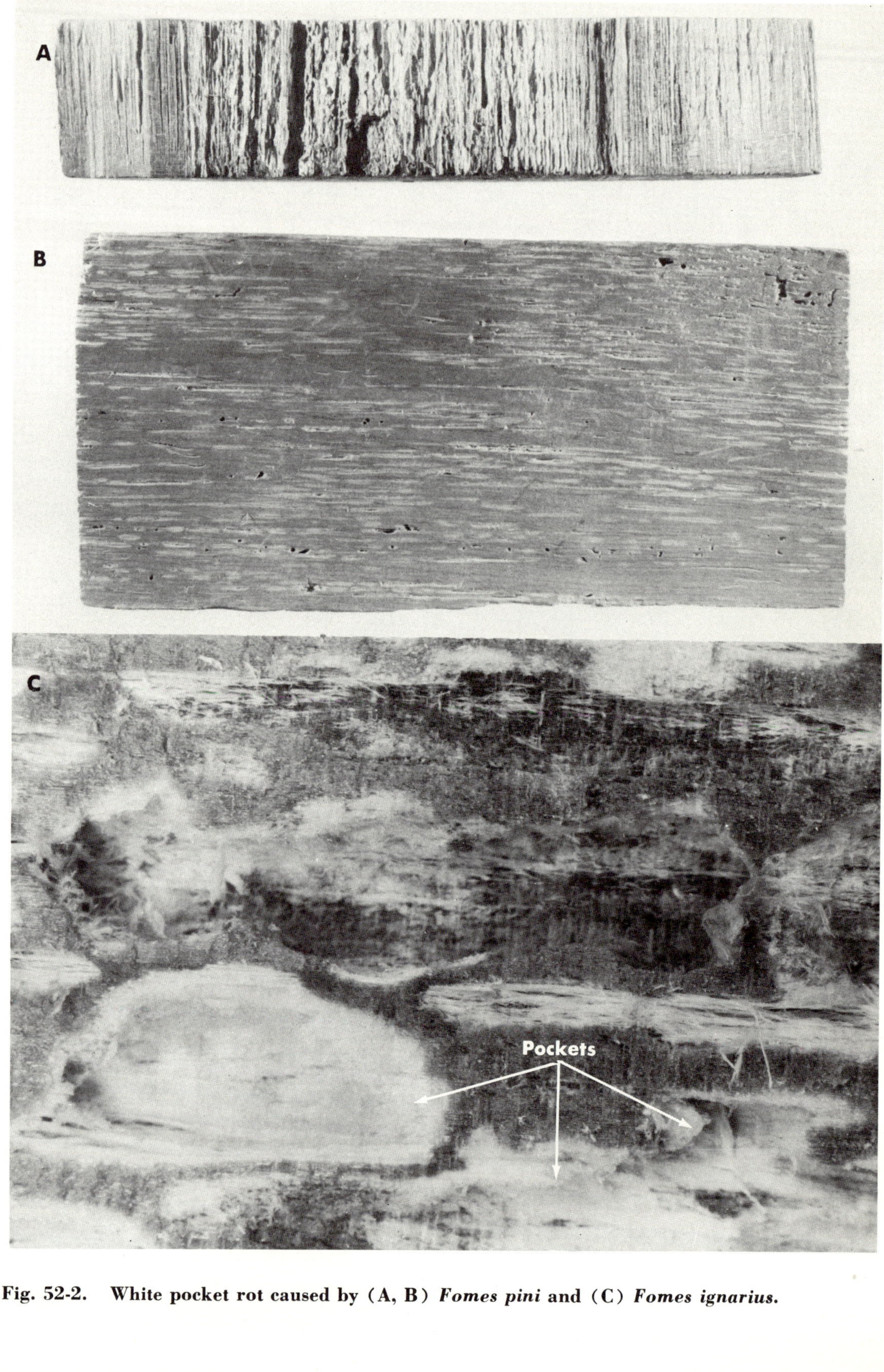

Fig. 52-2. White pocket rot caused by (A, B) *Fomes pini* and (C) *Fomes ignarius.*

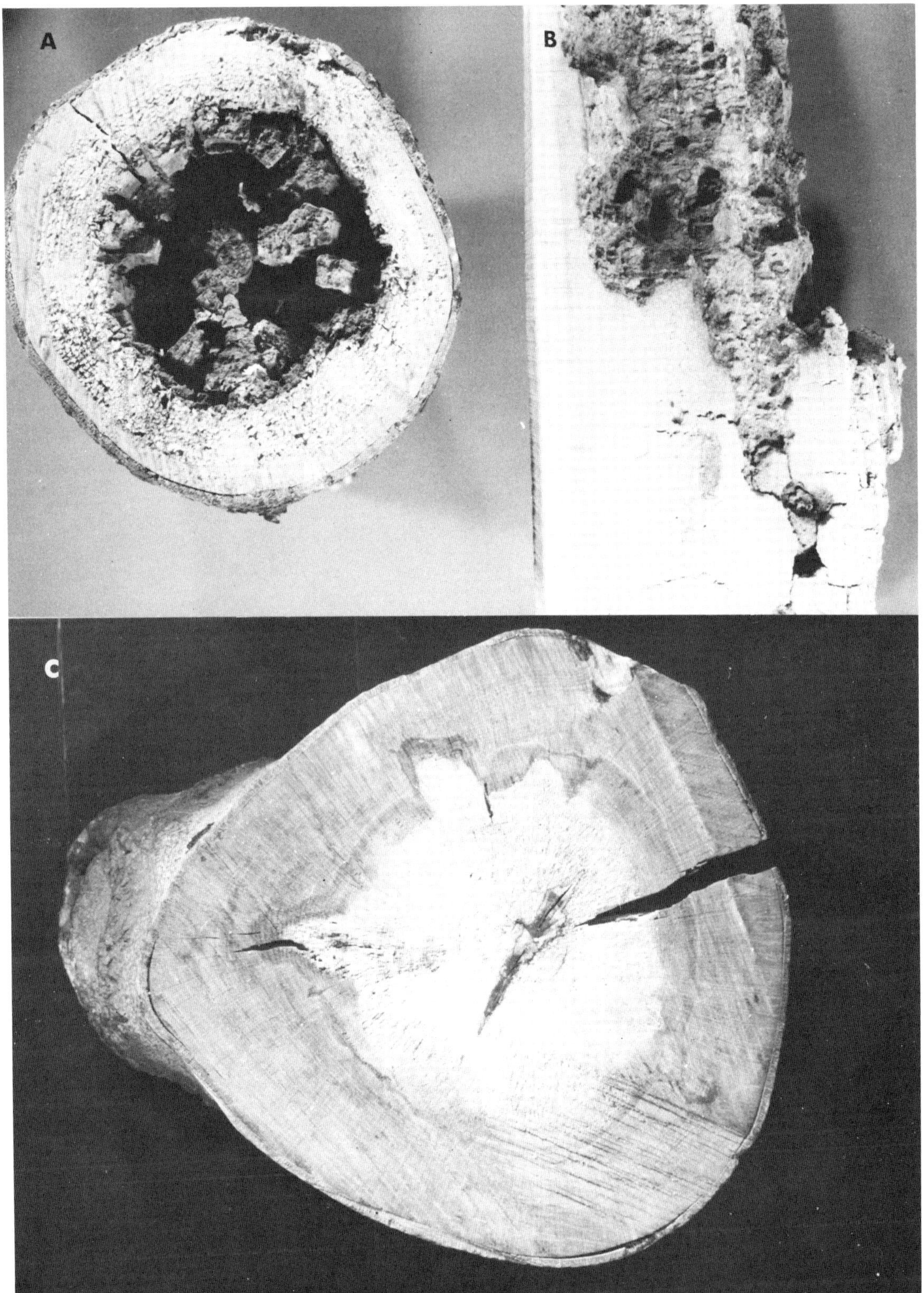

Fig. 52-3. Red brown rot of balsam fir caused by (A, B) *Polyporus schweinitzii* and (C) white rot of beech caused by *Fomes applanatus*.

Fig. 52-4. Zone lines caused by (A) *Fomes ignarius* and (B) *Xylaria polymorpha*. (C) Photomicrograph of zone lines (× *10.4*).

(Fig. 52-5). In older decay areas, the hyphae may have disintegrated, but **bore holes** (where the hyphae penetrated through cell walls) and the accumulation of decomposition products indicate their former presence.

Characteristics. Wood rot is recognized by the appearance of fruiting bodies, commonly called **conks** by foresters and **sporophores** by mycologists (Figs. 52-1 B, 52-13 A). These are sure signs of decay. Punky knots, swollen knots, unhealed branch stubs, and hollowed-out basal or trunk areas with punk (Fig. 52-1 A) are also common indications of decay.

Causal Organism. Wood rot is brought about by a large number of fungi, which, with only a few exceptions, belong to four families of the Basidiomycetes. We consider *Armillaria* of the Agaricaceae, and *Fomes* and *Polyporus* of the Polyporales.

The agarics are the **gill fungi.** They have umbrella-like sporophores with a stalk (stipe) and a cap (pileus). Vertical plates (gills) are usually arranged in radial symmetry about the central core on the underside of the cap. These gills are lined on both sides with basidia and basidiospores.

The polypores describe themselves as bodies of many pores (Fig. 52-13). The undersurface of the sporophore is covered with minute pores. These pores are the terminal, open ends of vertically aligned tubes whose inner surfaces are lined with basidia. The sporophore, or **conk,** is annual in *Polyporus* and perennial in *Fomes.* The perennial conks add a layer of tissue with verticle tubes and pores each year and some may be 50 to 75 years old.

Some of the important pathogens of hardwoods include *Polyporus squamosus, Polyporus sulphureus, Fomes ignarius,* and *Fomes applanatus.* Others which are more significant on conifers are *Polyporus schweinitzii* and *Fomes pini.* Life cycles of these higher basidiomycetes are similar to the one diagrammed for *Armillaria mellea* (Fig. 52-12).

Disease Cycles. Wounds, dead branches, and branch stubs are the most frequent points of entrance for germinating basidiospores. Of these, wounds caused by fire and stumps from cutting and thinning operations are the most important. In time, fruiting bodies appear with an attachment to a woody food base. They occur on stumps, at the base of trees, in rodent holes, and on dead roots—usually where high humidity prevails. The basidiospores constitute primary inoculum. Overwintering is accomplished

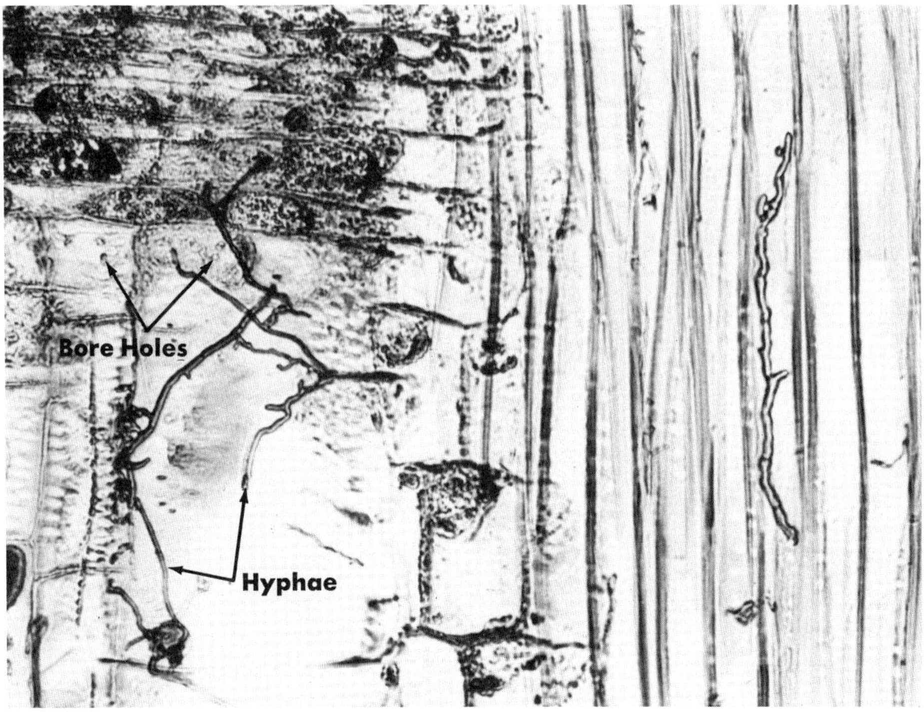

Fig. 52-5. Photomicrograph of wood tissue rotted by fungi (× *208*).

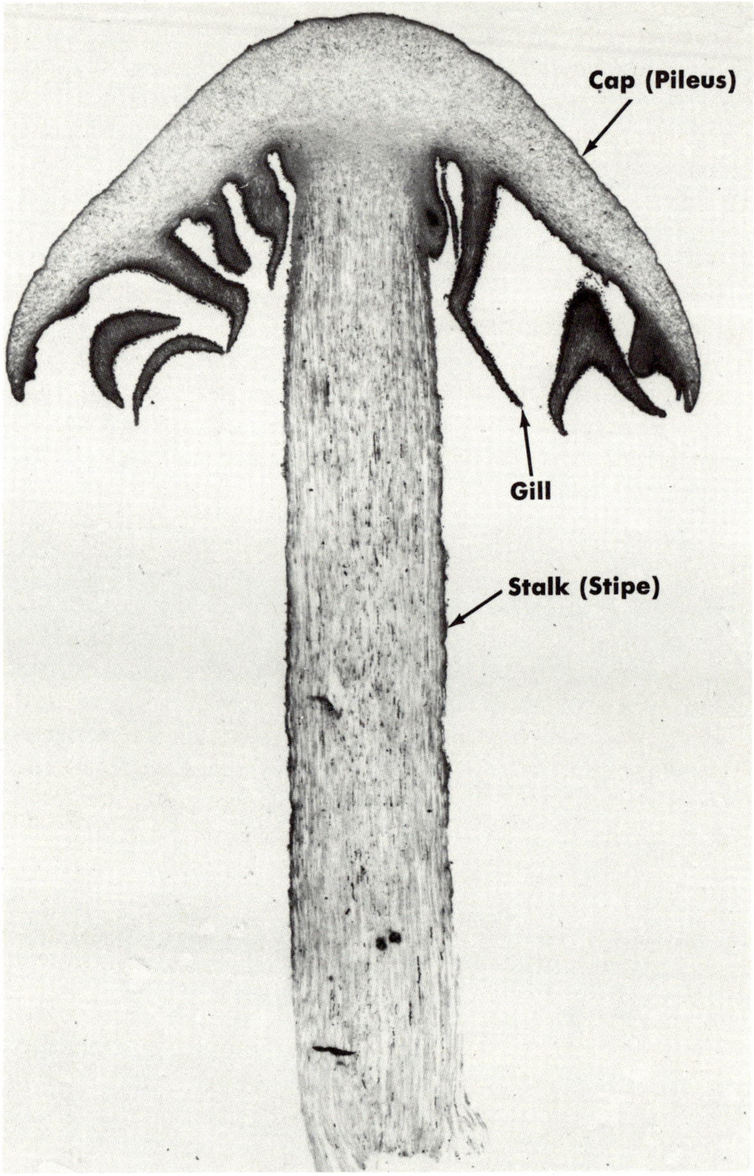

Fig. 52-6. Macrophotograph of a longitudinal section through an agaricus type of sporophore (× *8*).

by dormant mycelium in the rotting wood, dormant rhizomorphs, and perennial sporophores.

Control. Whereas cankers, blights, spots, and wilts are more important on ornamentals and shade trees in the urban communities, the wood rots are more important, economically, in timber production. The rate of decay in a timber stand is slow—about an increase of 0.3% to 0.6% per annum. In an individual tree, rot may progress one, or at the most, two feet per year.

Because the disease spreads and develops slowly, techniques of sanitation are useful in control. These diseases are relatively unimportant in a clear-cut area with new, even-aged stands. The most severe problems with wood rots arise in those stands which are mixed, uneven-aged, or thinned. Here the inoculum is more prevalent as a carry-over in older trees, and the trees in the understory bcome weaker and vulnerable to attack. If selective cutting is practiced, diseased trees should be removed and care should be exercised to avoid wounding the remainder. Another principle of control is to cut and harvest timber at that point in time when the ratio of sound wood to rotten is the greatest. This varies with the species and the site. For aspen, this is 45 years in Minnesota and 120 in Northern Ontario, and 150 for white fir and 250 for western hemlock in Oregon.

Armillaria mellea is called the shoe string fungus or honey mushroom. It is common, plentiful, and edible. It's found on shade, forest, orchard, and ornamental trees, whether coniferous or deciduous, and in temperate or tropical climates throughout the world. It occurs, for example, on apple, ash, pear, pine, spruce, elm, and beech.

Armillaria is most prevalent where trees are predisposed to infection by insects, lightning, frost, diseases, poor soils, and poor sites. In keeping with this it is not surprising that it is most common on trees outside of their natural range or on trees planted in plantations. Natural stands tend to be vigorous and to occur on preferred sites.

Fig. 52-7. Spore print made by placing a cap on a sheet of paper. The color of spores is useful in identifying the fungus.

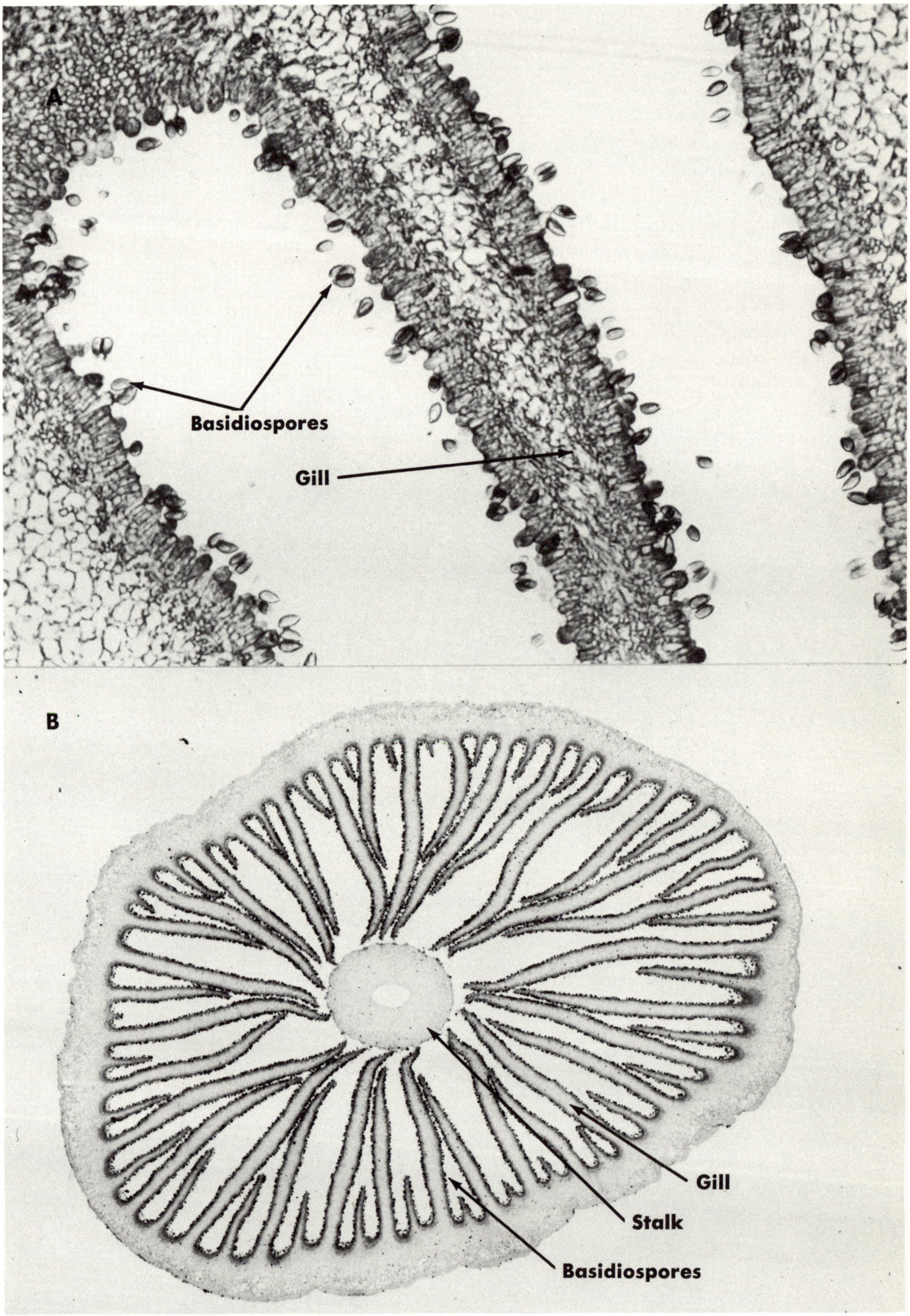

Fig. 52-8. Photomicrograph (× *250*) of cross sections of the gills (A) and cap (B) of an agaricus type of sporophore (× *25*).

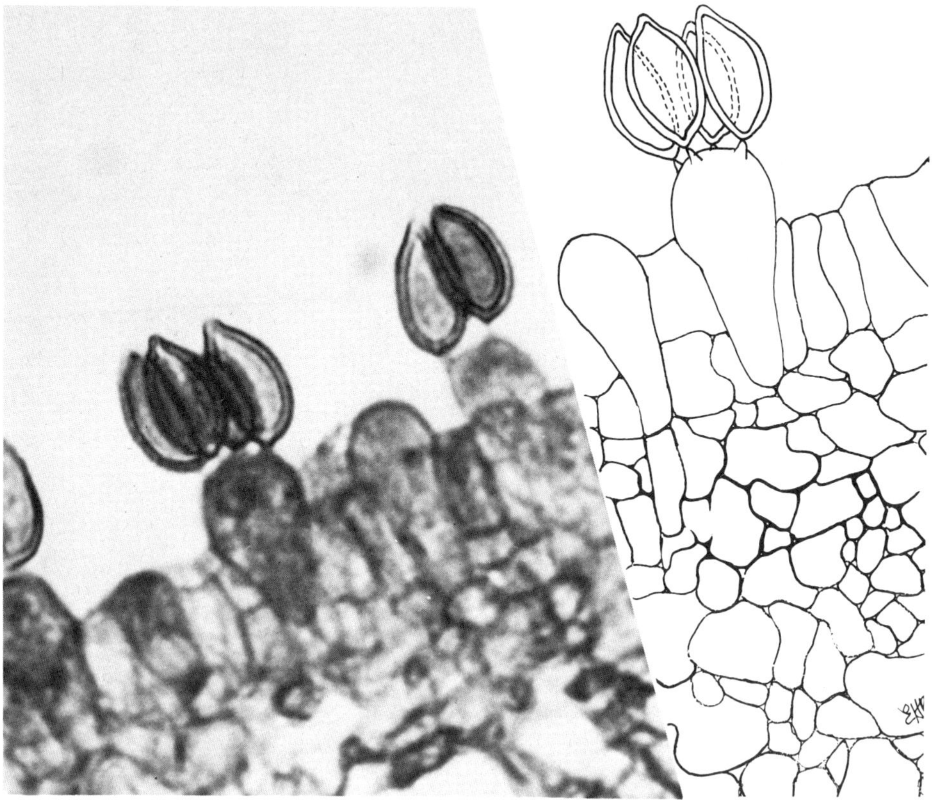

Fig. 52-9. Photomicrograph of basidia and basidiospores and representative drawing (× *1250*).

Basidiospores germinate and produce haploid, primary hyphae (Fig. 52-12 A,B). Two compatible haploid structures, either two germinating spores, two hyphae, or a germinating spore and a hypha, fuse to produce secondary, dikaryotic hyphae (C). This dikaryotic phase, the secondary mycelium, is the most significant portion of the life cycle in respect to proportion of time. This is the phase that rots the wood. The basidiospores are borne on basidia which are produced in and on fructifications called sporophores.

Most of the higher basidiomycetes exhibit heterothallism. Many are tetrapolar, or of four compatible types. These are determined by genes at two loci, A and B. Thus, compatible types are AB and ab, or Ab and aB. Many different alleles occur at both loci. Following dikaryotization, conjugate nuclear division and clamp formation (D-G) produces the secondary hyphae. The secondary hyphae is organized into the sporophore which is tertiary hyphae. In addition, the rhizomorphs (Fig. 52-11) are common under the bark of infected trees and about infected roots. They are strands of hyphae which are united and have a rind and a growing tip which reminds one of a root tip. Their shoe-string appearance gives the fungus its name.

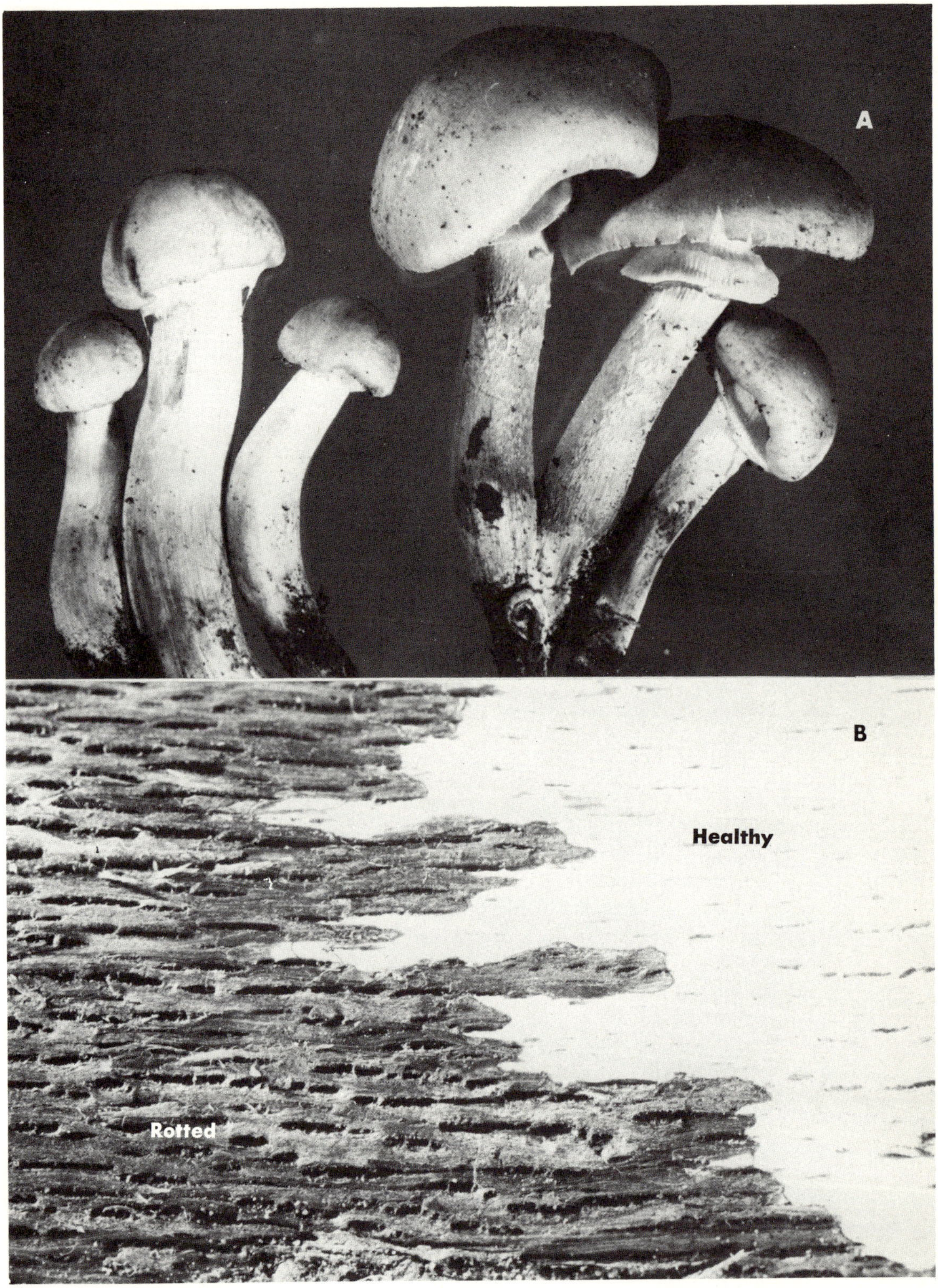

Fig. 52-10. (A) Sporophores of *Armillaria mellea* and (B) decay of oak caused by *Armillaria mellea.*

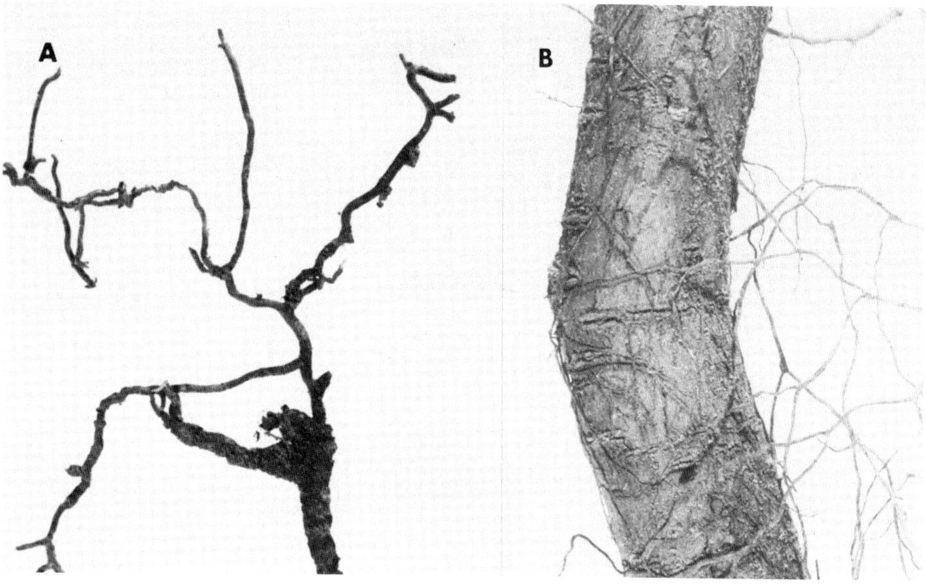

Fig. 52-11. Rhizomorphs of (A) *Armillaria mellea* alone and (B) about a main root of cherry.

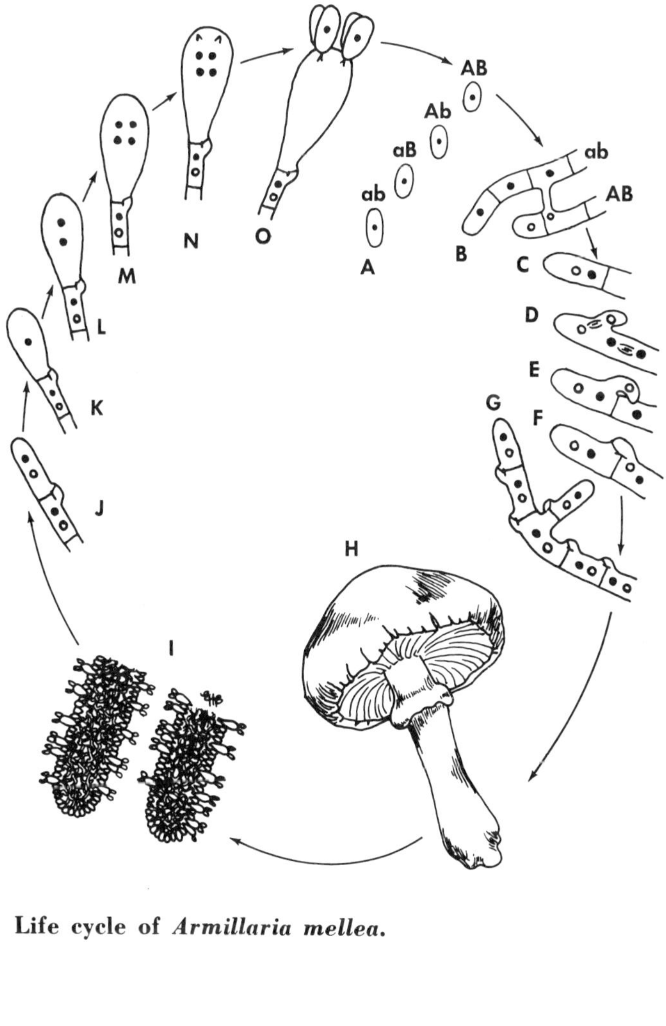

Fig. 52-12. Life cycle of *Armillaria mellea*.

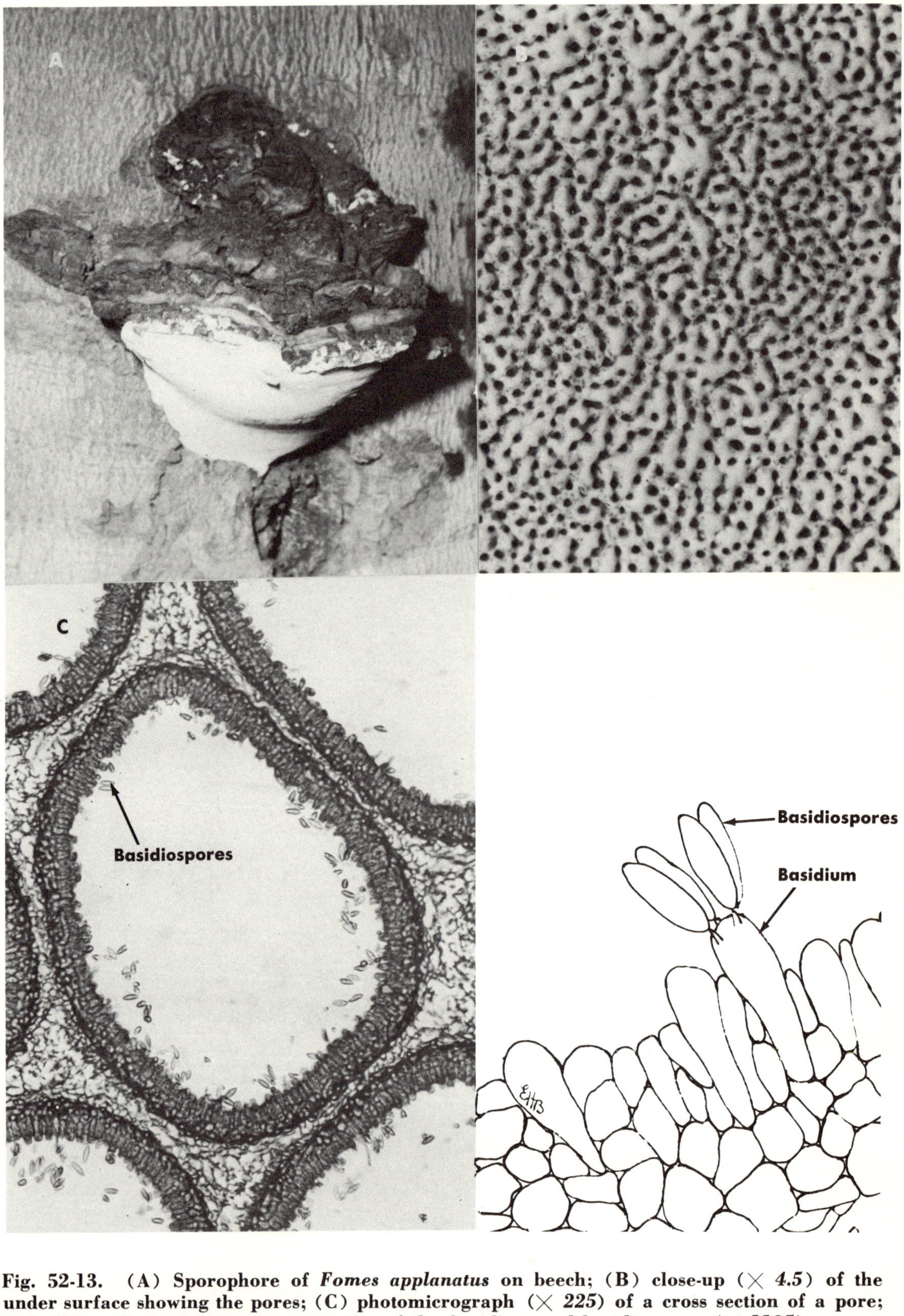

Fig. 52-13. (A) Sporophore of *Fomes applanatus* on beech; (B) close-up ($\times$ *4.5*) of the under surface showing the pores; (C) photomicrograph ($\times$ *225*) of a cross section of a pore; and, (D) a diagrammatic representative of the basidium and basidiospores ($\times$ *1125*).

53

Mistletoes

The mistletoes are seed plants. Their many genera constitute the family Loranthaceae. They are found primarily in the tropics throughout the world. Two genera occur in Europe and two other genera, *Phorodendron* and *Arceuthobium*, occur in North America. Species of *Phorodendron*, or tree thief, are the true mistletoes and are restricted to the Western Hemisphere. They occur primarily on hardwoods but also a few conifers. No serious diseases, from the standpoint of the forester, result except for the disease of the incense cedar. Ornamentals attacked by the true mistletoes are of concern to the homeowner and nurseryman. Our discussion centers on the other mistletoes, the dwarf mistletoes, caused by species of *Arceuthobium*. These are found exclusively on conifers and in the greatest variety in the American Northwest (Figs. 53-1, 53-2). That they are the most important problem in applied forest pathology in that region attests to their abundance and significance. Spruce and tamarack are attacked but not other eastern conifers by forms of *Arceuthobium* different from the more destructive western forms. The large stands of beautiful Ponderosa pine in the isolated Black Hills region are free of mistletoes. Continued care should be exercised to preclude their introduction.

Arceuthobium is unique in all the Angiosperms. The embryo is an undifferentiated mass of cells with an epidermal layer and a notch at the cotyledonary node. Seeds are disseminated by an explosive dehiscence for as far as ten yards. The seeds are mucilagenous and tend to stick to whatever obstruction they encounter. They germinate immediately or the following spring. A radicle-like tip grows until there is an obstruction in the bark and then produces an irregular mass of undifferentiated cells to form the foot, which inserts a haustorium into the host tree. By this time the radicle-like structure has deteriorated. A true plumule, embryonic shoot, and radicle never differentiate in the seedling.

The haustorium gives rise to the endophytic system—that portion "within the plant." The endophytic system is composed of two major types of strands, cortical strands and sinkers. Cortical strands are at first uniseriate, undergo divisions in longitudinal planes, and form tiers of four cells. Continued development into a cylinder of cells occurs with the inner ones differentiating into vascular elements. A continually functioning cambium and phloem appear to be absent.

Young sinkers do not have a tiered arrangement but grow centripetally and usually contiguous to one of the numerous phloem ray cells of the host until the cambium is reached. Cambial cells are not destroyed but usually initiate an aberrant ray cell around the young sinker. Scalariform vascular elements then differentiate in the central zone of the sinkers.

The endophytic system grows for two years before it produces branches and explains why mistletoe branches are not visible or observed on wood less than three years old. Plants are dioecious (two households) and, therefore, bear staminate (Fig. 53-2 A) and pistillate (Fig. 53-2 B) flowers on different plant bodies. Leaves are vestigial. Some species flower in the spring and early summer months whereas others, such as the group classified as *Arceuthobium campylopodium*, flower in the fall. *Arceuthobium* has a single sessile anther without a vascular bundle and is again unique in all the Angiosperms in having no locules in the anther but a continuous ring about a sterile column. The ovary has a single locule.

Control. Several features are important in controlling mistletoes: (1) they are confined to aerial parts, (2) they die with the death of their hosts, (3) seeds are not carried long distances. The greatest spread is from the overstory (highest tree crowns) to the understory (lower tree crowns). Uneven-aged stands, therefore, are more severely attacked. Clear cutting, to maintain even-aged stands, provides some measure of control. In light infections, the swellings which indicate infected areas can be pruned out along with a generous amount of adjacent tissues. Chemical control has not yet been devised but hopefully, a selective herbicide can be developed.

Four fungi occur as parasites on the dwarf mistletoes. One of them, *Colletotrichum gleosporioides,* frequently prevents the bearing of fruit. Perhaps such **hyperparasites**, or parasites of parasites, can be used to develop control measures.

Fig. 53-1. Dwarf mistletoe, *Arceuthobium campylopodium* f. *typicum*, on Ponderosa pine.

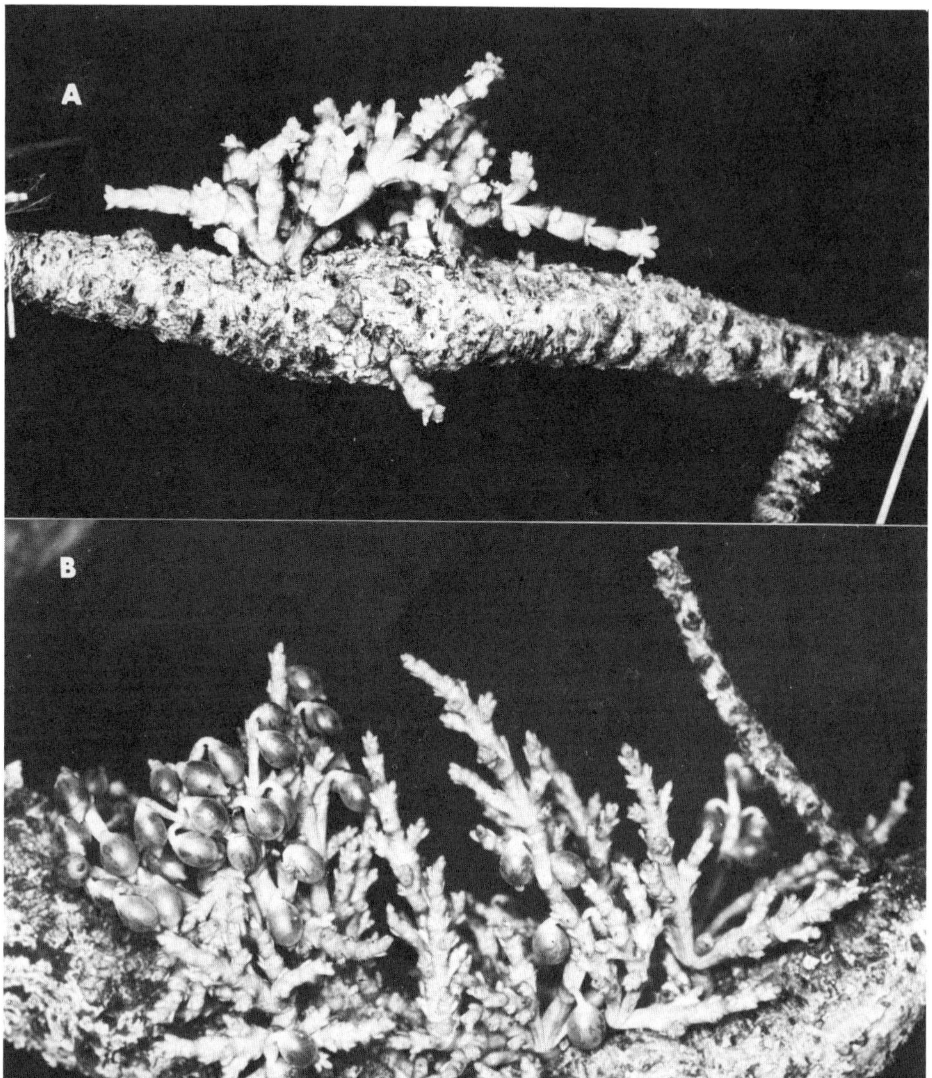

Fig. 53-2. **(A) Staminate and (B) pistillate plants of dwarf mistletoe, *Arceuthobium campylopodium* f. *typicum*.**

SUGGESTED READING

Gill, L. S. "Arceuthobium in the United States." Trans. Conn. Acad. Arts and Sci., Vol. 32 (July, 1935), pp. 111–245.

Kuijt, J., "Dwarf Mistletoes." *The Botanical Revue*, Vol. 21 (December, 1955), pp. 569–626.

54

Dodder (Cuscuta)

Higher plants cause injuries and diseases of other higher plants by shading, strangling, and parasitizing the others. Epiphytes, such as Spanish moss and other bromeliads, shade their living supports whereas grape vines and bittersweet strangle as well as shade theirs. Some, such as the mistletoes and dodders, are true parasites (Figs. 54-1, -2). The dodders are of special interest as vectors of viruses as well as obligate parasites and are useful in experimental transmission of viruses.

Seeds, which are viable in the soil for many years, germinate and produce small, incomplete seedlings. These die in a few weeks if contact is not made with suitable hosts. On contact, the seedling begins to twine and promptly issues **haustoria**, or suckers, which penetrate into the vascular tissues of the host (Figs. 54-2;3). Soon a vine-like growth develops (Fig. 54-1 B) which proliferates about its host (Fig. 54-1 A). Later, flowers are produced in abundance (Fig. 54-2 B).

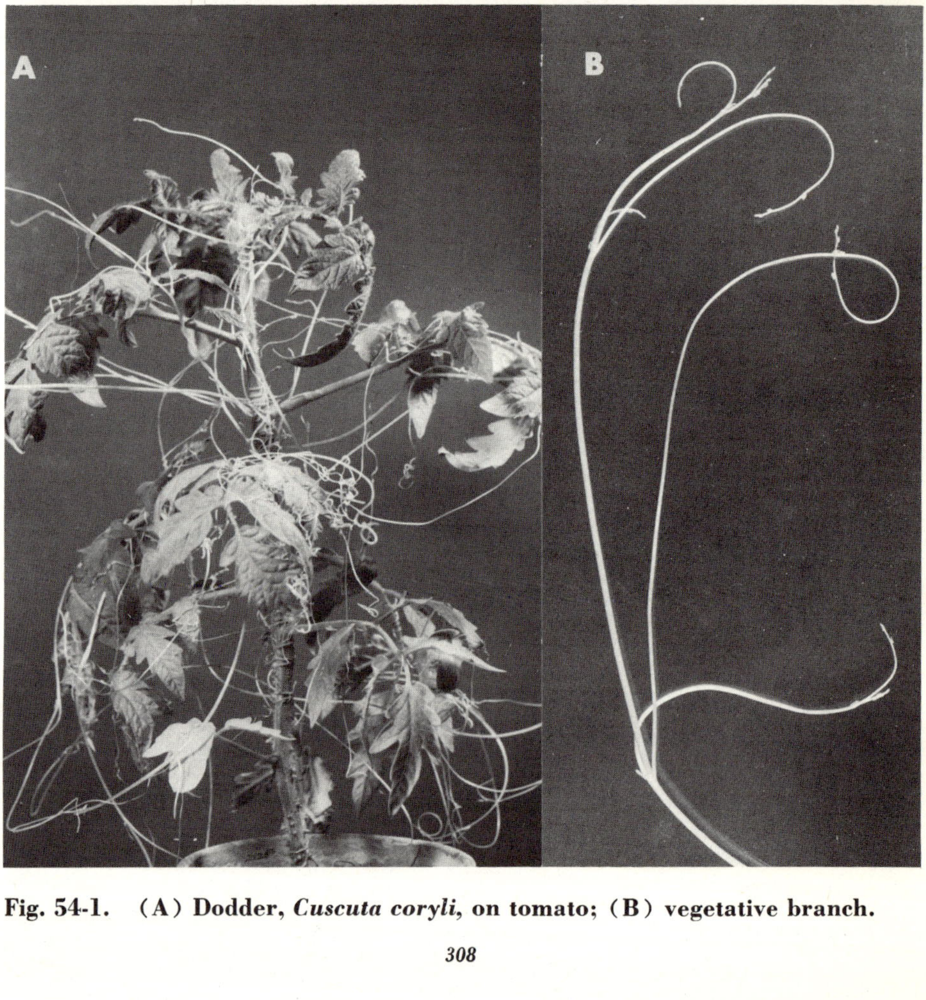

Fig. 54-1. (A) Dodder, *Cuscuta coryli*, on tomato; (B) vegetative branch.

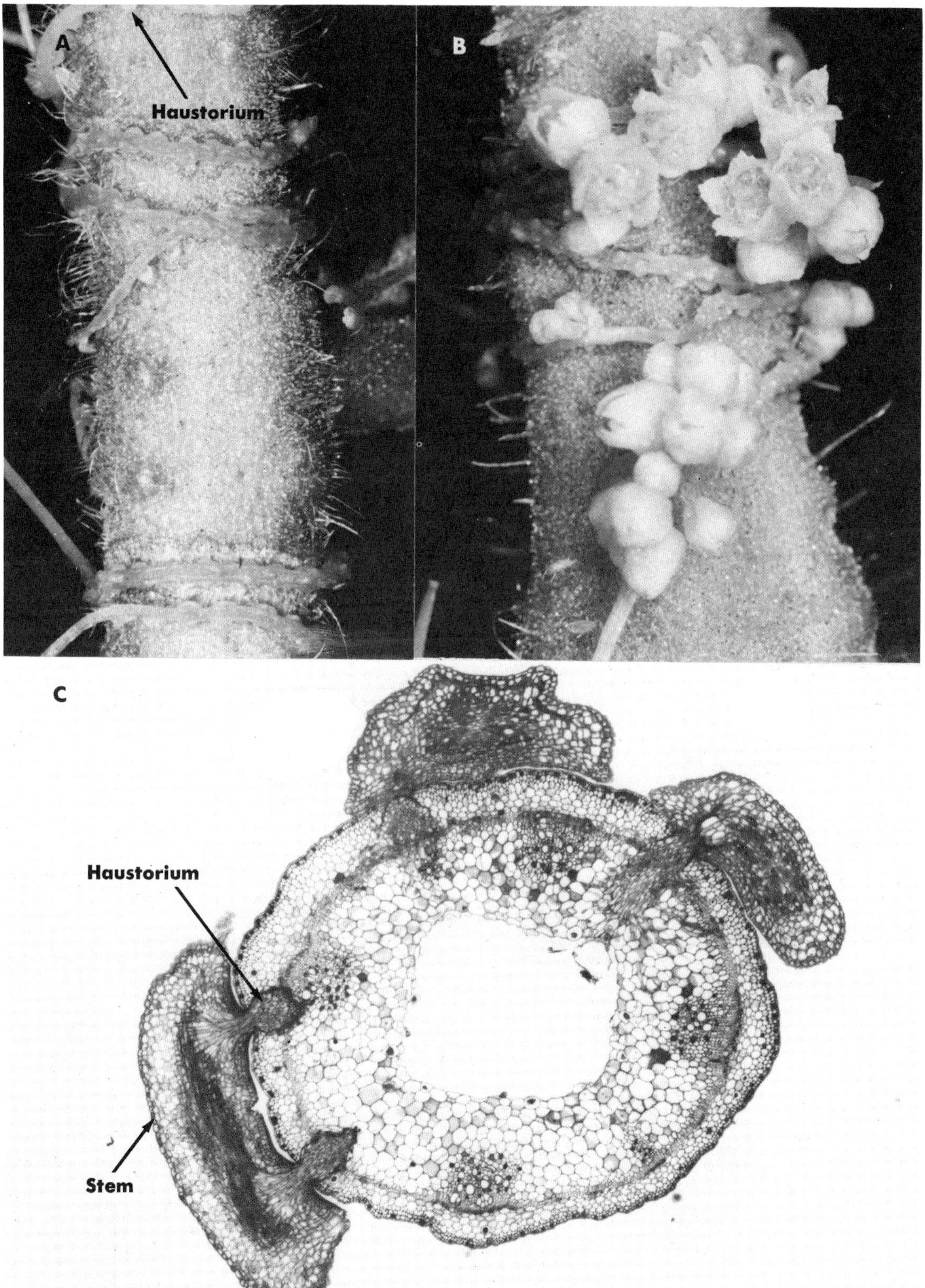

Fig. 54-2. (A) Dodder stem with haustoria; (B) flowers; and, (C) photomicrograph ($\times$ *19.2*) of a cross section of a cucumber stem parasitized by dodder showing cross sections of dodder stems and haustoria.

Control. On the whole crop seed must be free of dodder seed, but other practices are useful. Affected areas can be mowed or plowed to prevent maturation of the dodder and production of seed. Blow torches and burning straw can destroy many of the seeds on the ground.

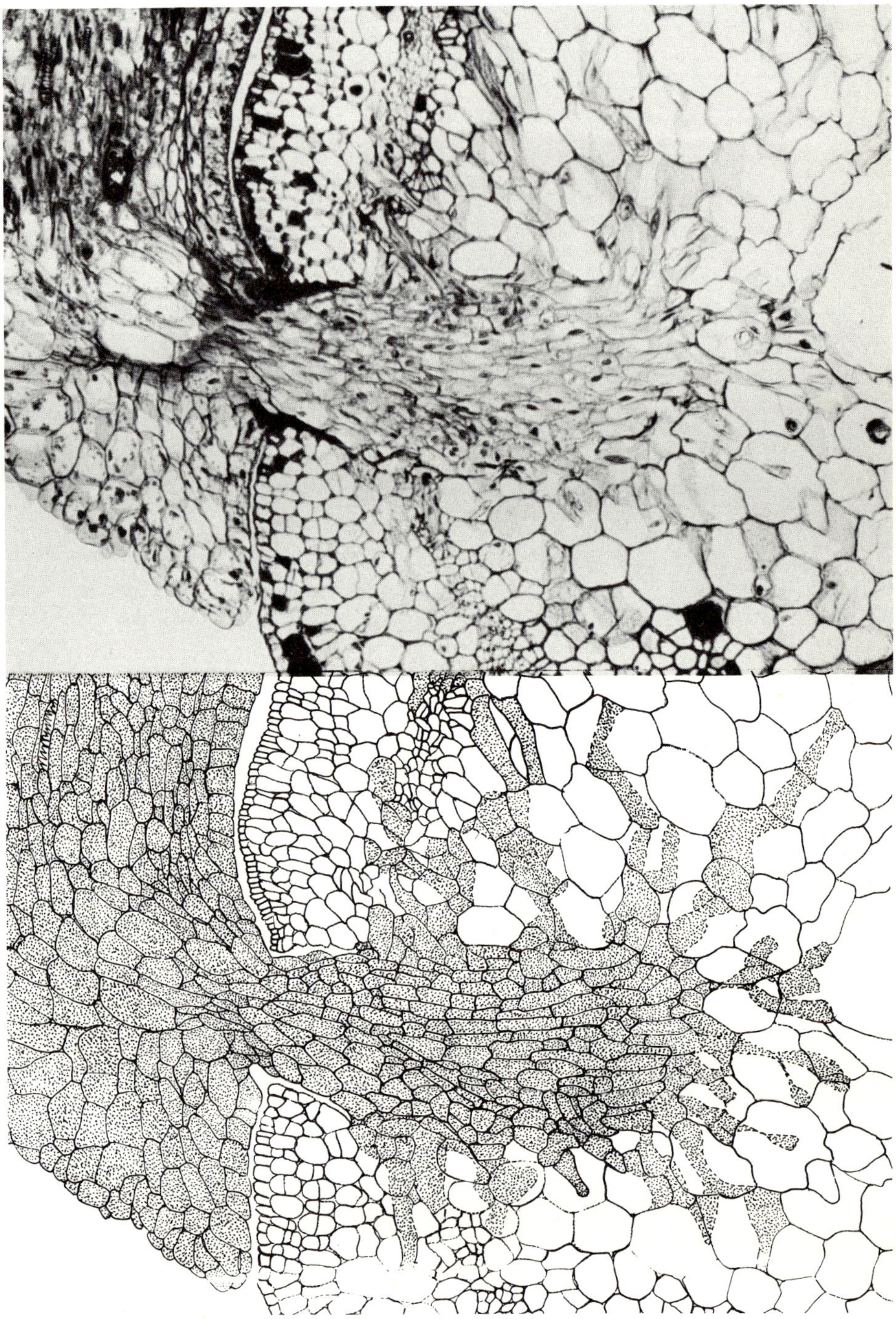

Fig. 54-3. Photomicrograph and representative drawing of a dodder stem and haustorium (stipuled) (× *100*).

55

The Nematodes: (Eelworms)

The eelworms, or nematodes, are common, widespread invertebrates which inhabit salt water, fresh water, soil, and refuse. Some are harmless whereas others parasitize plants and animals and cause serious diseases. They are not related to earthworms or flatworms, but are treated with the round worms.

Those which occur in the soil are 1/8 to 1/64 inch in length. They possess a stylet (spear) (Fig. 55-2 A) if phytophagus (plant-eating) or predacious on other eelworms and very small animals in the soil, but are spearless if saprobic. Secretions are injected through the stylet to paralyze prey, dissolve chitin, or make the contents of host cells more fluid.

Plant parasitic nematodes are generally obligate parasites and are mobile and ectoparasitic or sedentary and endoparasitic or ectoparasitic. The mobile and vagrant ectoparasites, such as *Xiphenema,* the dagger nematode (Fig. 55-3), are sometimes vectors of viruses, especially those that cause the ring spot diseases (See Chapter 11). The sedentary nematodes, such as *Meloidogyne,* the root-knot nematode (Fig. 55-5), are often associated with stimulated cell growth of the host plant (Fig. 55-4).

Life Cycles. The nematode life cycles are relatively simple—from egg, through four molts, to the adult form. Some, such as the root-knot nematode, undergo the first molt in the egg. All of the organs, except those of reproduction, are in an advanced stage at hatching time. Sexes differ by the possession of spicules (Fig. 55-2 B) and bursa—the copulatory organs of the male. In many species, the sexes are very similar. The greatest differences occur with the sedentary parasites such as the root-knot nematode in which the female becomes greatly enlarged and pear-shaped (Fig. 55-5).

Characteristics. Symptoms of diseases caused by nematodes may include dead buds, distorted and yellow stems and leaves, necrosis, leaf spots and lesions, root galls, excessive branching of the roots, and Y-shaped root tips (Fig. 55-4).

Outline of Some Important Plant Parasitic Genera

Enoplida (*Order*)

Dorylaimidae (*Family*)	*Longidorus*—needle nematode
	Xiphenema—dagger nematode
Diphtherophoridae (*Family*)	*Trichodorus*—stubby root nematode

Tylenchida (*Order*)

Tylenchidac (*Family*)	*Ditylenchus*—stem and bulb nematode
	Pratylenchus—lesion nematode
Heteroderidae (*Family*)	*Heterodera*—golden nematode
	Meloidogyne—root-knot nematode
Aphelenchidae (*Family*)	*Aphelenchoides*—bud and leaf nematode

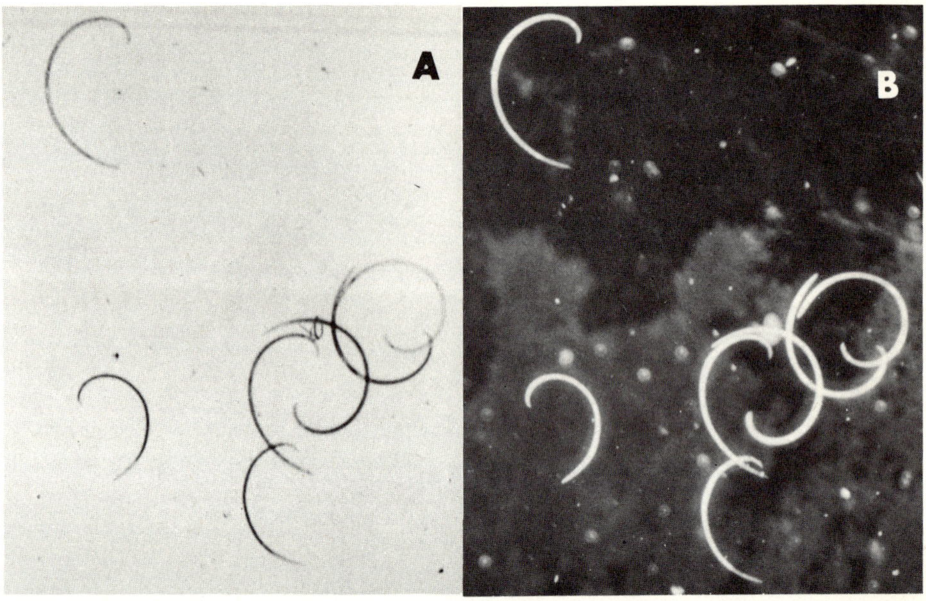

Fig. 55-1. Appearance of nematodes in a dissecting microscope with illumination from (A) below and (B) above (× *50*).

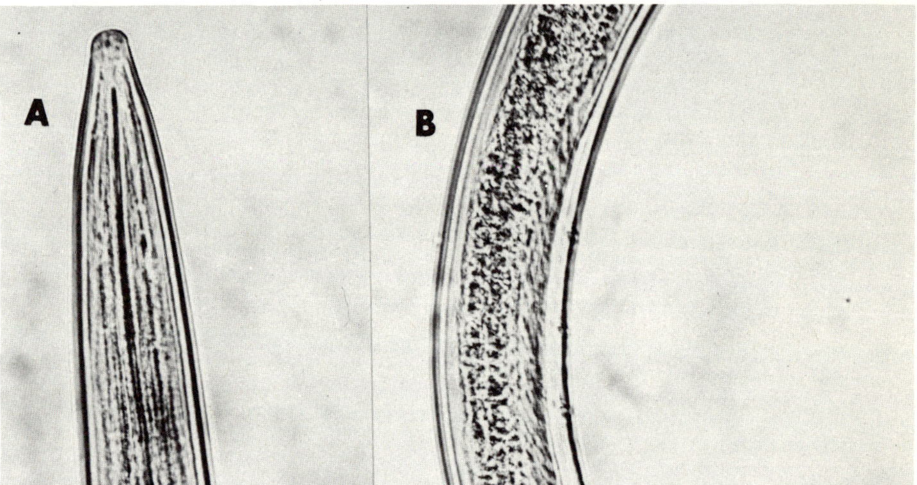

Fig. 55-2. (A) Head with spear and (B) tail with spicules.

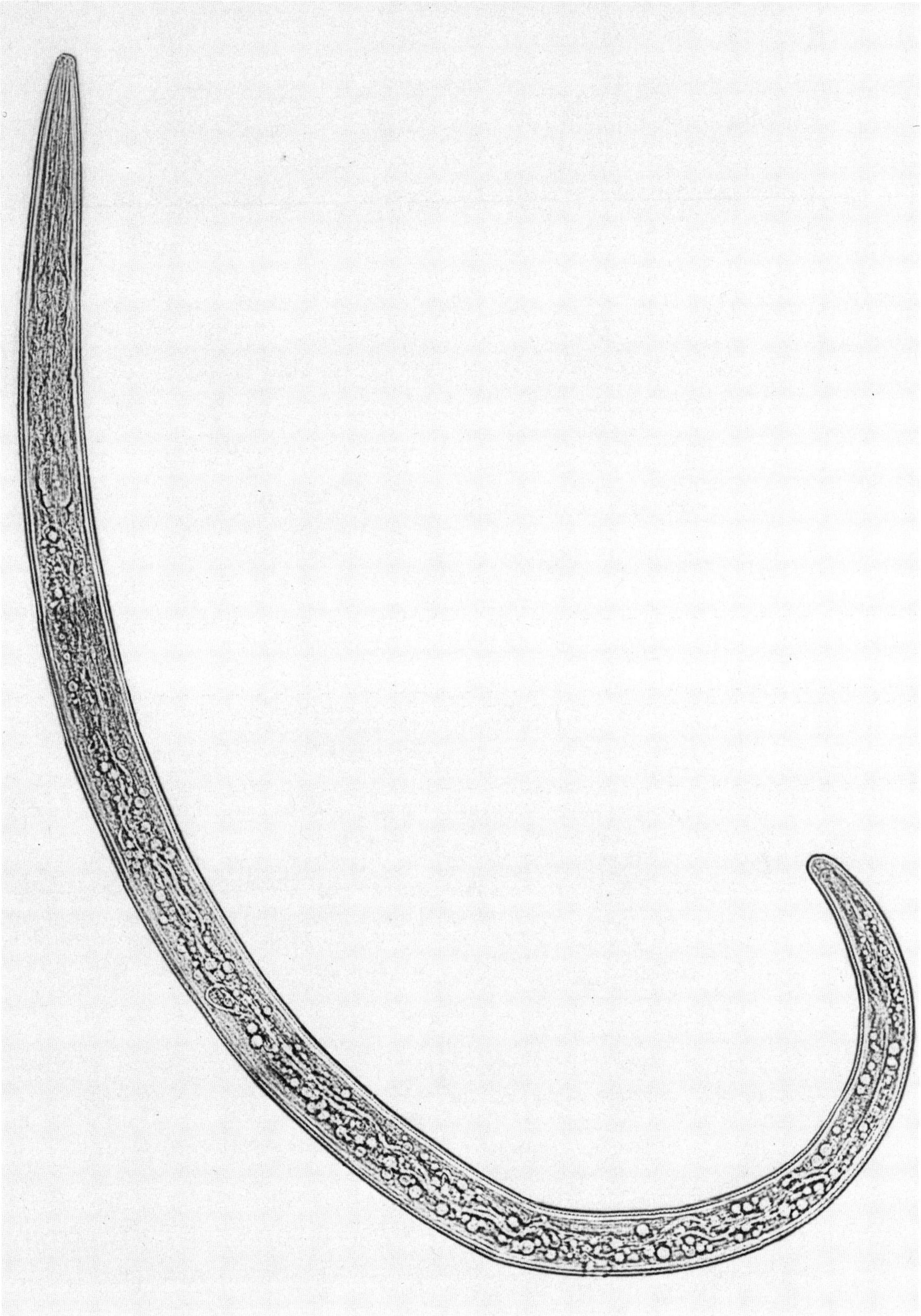

Fig. 55-3. Whole nematode, *Xiphenema americanum*.

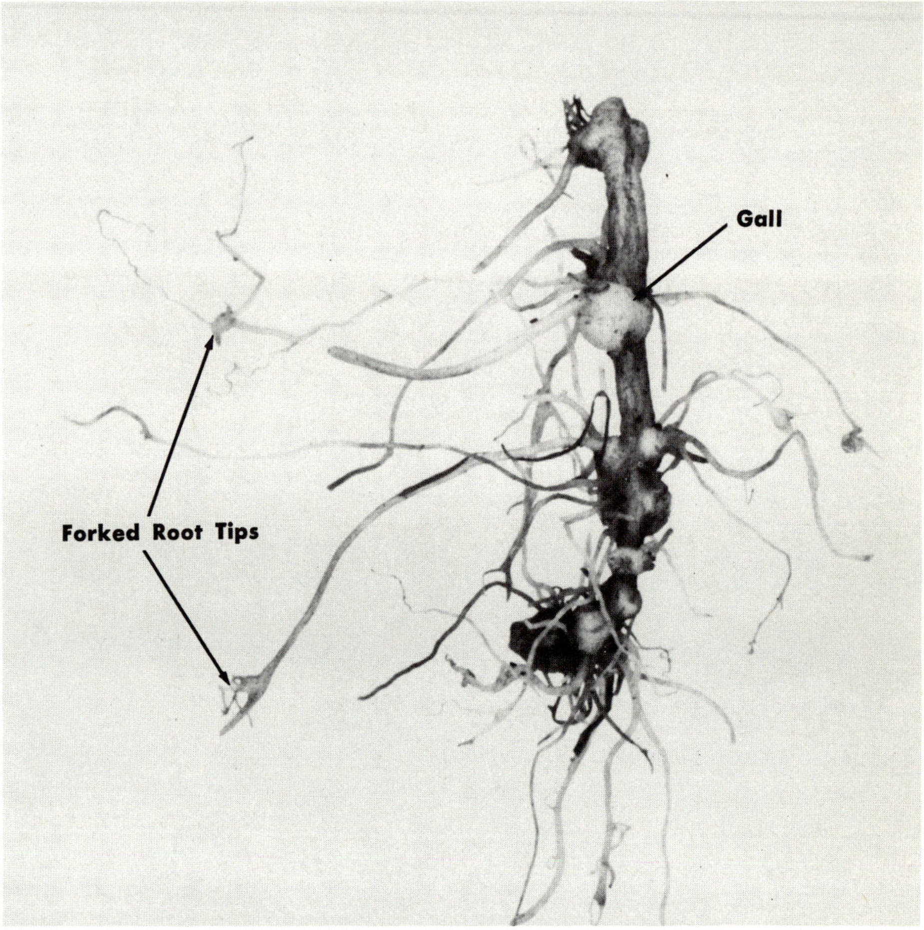

Fig. 55-4. Symptoms of root knot: galls and forked root tips.

A special kind of nematode problem is the so-called **replant problem** with orchard trees. Small trees which are planted in the prior sites of larger trees often are stunted. Large numbers of nematodes, built-up by the extensive root system of the larger trees, attack the limited root system of the newly transplanted trees and suppress their growth.

Control. The nematodes may be controlled by avoiding, eliminating, or reducing the number of nematodes. They can be avoided by purchase of inspected and certified plant materials and eliminated from others by treatment with hot water or chemicals. Those in the soil can be reduced by crop rotation and controlled by chemical fumigation, the cost of which is usually justified only with high value crops. The use of resistant varieties becomes more significant.

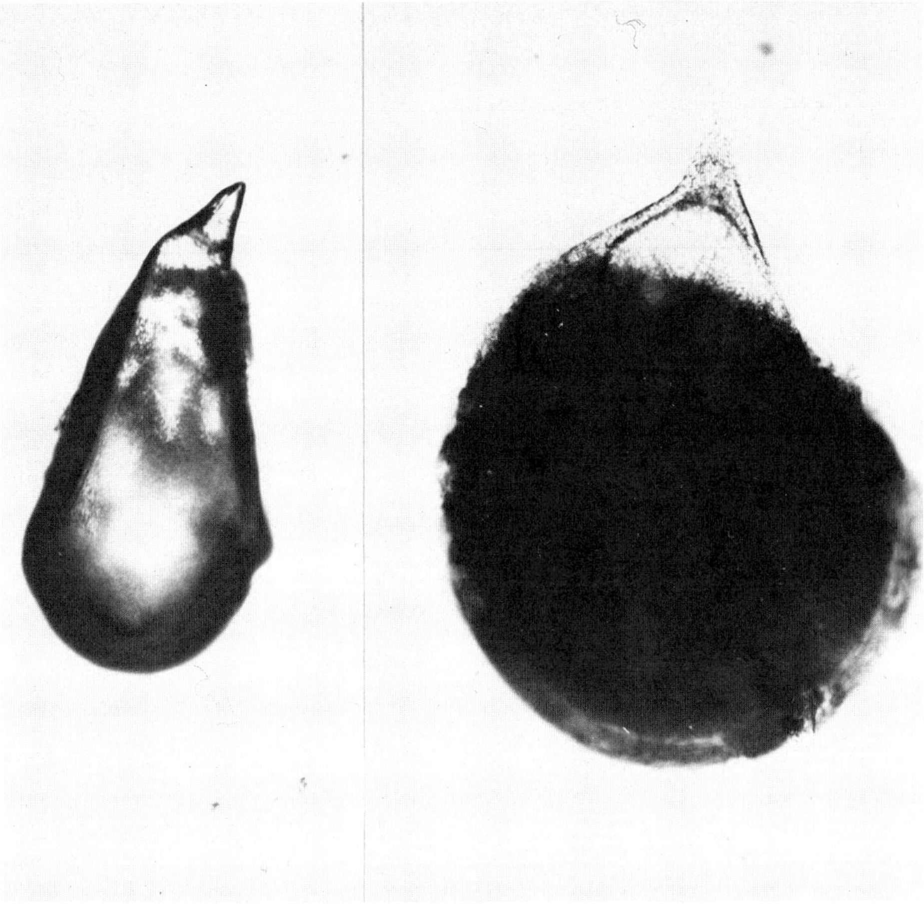

Fig. 55-5. Enlarged female root knot nematode: (A) partially dehydrated and (B) partially squashed.

GENERAL REFERENCES

Filipjer, I. N., *Nematodes Harmful and Beneficial to Agriculture.* Moscow and Leningrad, 1934.

Goodey, T. (rewritten by J. B. Goodey), *Soil and Fresh Water Nematodes.* New York, Wiley, 1963.

Mai, W. F., *Pictorial Key to Genera of Plant Parasitic Nematodes.* Ithaca, New York, Cornell University Mimeograph, 1964.

Sasser, J. N., and Jenkins, W. R., *Nematology.* Chapel Hill, University of North Carolina Press, 1960.

Thorne, G., *Principles of Nematology.* New York, McGraw-Hill, 1961.

Index